Springer Tracts in Modern Physics
Volume 152

Springer-Verlag Berlin Heidelberg GmbH

Springer Tracts in Modern Physics

Springer Tracts in Modern Physics provides comprehensive and critical reviews of topics of current interest in physics. The following fields are emphasized: elementary particle physics, solid-state physics, complex systems, and fundamental astrophysics.

Suitable reviews of other fields can also be accepted. The editors encourage prospective authors to correspond with them in advance of submitting an article. For reviews of topics belonging to the above mentioned fields, they should address the responsible editor, otherwise the managing editor.
See also http://www.springer.de/phys/books/stmp.html

Managing Editor

Gerhard Höhler

Institut für Theoretische Teilchenphysik
Universität Karlsruhe
Postfach 69 80
D-76128 Karlsruhe, Germany
Phone: +49 (7 21) 6 08 33 75
Fax: +49 (7 21) 37 07 26
Email: gerhard.hoehler@physik.uni-karlsruhe.de
http://www-ttp.physik.uni-karlsruhe.de/

Elementary Particle Physics, Editors

Johann H. Kühn

Institut für Theoretische Teilchenphysik
Universität Karlsruhe
Postfach 69 80
D-76128 Karlsruhe, Germany
Phone: +49 (7 21) 6 08 33 72
Fax: +49 (7 21) 37 07 26
Email: johann.kuehn@physik.uni-karlsruhe.de
http://www-ttp.physik.uni-karlsruhe.de/~jk

Thomas Müller

Institut für Experimentelle Kernphysik
Fakultät für Physik
Universität Karlsruhe
Postfach 69 80
D-76128 Karlsruhe, Germany
Phone: +49 (7 21) 6 08 35 24
Fax: +49 (7 21) 6 07 26 21
Email: thomas.muller@physik.uni-karlsruhe.de
http://www-ekp.physik.uni-karlsruhe.de

Roberto Peccei

Department of Physics
University of California, Los Angeles
405 Hilgard Avenue
Los Angeles, CA 90024-1547, USA
Phone: +1 310 825 1042
Fax: +1 310 825 9368
Email: peccei@physics.ucla.edu
http://www.physics.ucla.edu/faculty/ladder/
peccei.html

Solid-State Physics, Editor

Peter Wölfle

Institut für Theorie der Kondensierten Materie
Universität Karlsruhe
Postfach 69 80
D-76128 Karlsruhe, Germany
Phone: +49 (7 21) 6 08 35 90
Fax: +49 (7 21) 69 81 50
Email: woelfle@tkm.physik.uni-karlsruhe.de
http://www-tkm.physik.uni-karlsruhe.de

Complex Systems, Editor

Frank Steiner

Abteilung Theoretische Physik
Universität Ulm
Albert-Einstein-Allee 11
D-89069 Ulm, Germany
Phone: +49 (7 31) 5 02 29 10
Fax: +49 (7 31) 5 02 29 24
Email: steiner@physik.uni-ulm.de
http://www.physik.uni-ulm.de/theo/theophys.html

Fundamental Astrophysics, Editor

Joachim Trümper

Max-Planck-Institut für Extraterrestrische Physik
Postfach 16 03
D-85740 Garching, Germany
Phone: +49 (89) 32 99 35 59
Fax: +49 (89) 32 99 35 69
Email: jtrumper@mpe-garching.mpg.de
http://www.mpe-garching.mpg.de/index.html

Hendrik C. Kuhlmann

Thermocapillary Convection in Models of Crystal Growth

With 101 Figures

Springer

Dr. Hendrik C. Kuhlmann

ZARM-University of Bremen
Am Fallturm
D-28359 Bremen
Email: kuhl@zarm.uni-bremen.de

Physics and Astronomy Classification Scheme (PACS):
47.20.Dr, 47.20.Bp, 47.15.Gf, 81.10.Hs, 81.10.Fq, 81.10.Mx

ISSN 0081-3869
ISBN 978-3-662-14737-5

Library of Congress Cataloging-in-Publication Data.

Kuhlmann, Hendrik C. Thermocapillary convection in models of crystal growth/Hendrik C. Kuhlmann. p. cm. –
(Springer tracts in modern physics, 0081-3869; v. 152). Includes bibiographical references and index.
ISBN 978-3-662-14737-5   ISBN 978-3-540-49484-3 (eBook)
DOI 10.1007/978-3-540-49484-3

1. Crystal growth. 2. Heat–Convection. I. Title. II. Series: Springer tracts in modern
physics; 152.   QC1.S797   vol. 152   QD921   530 s–dc21   [548'.8]   98–44832

© Springer-Verlag Berlin Heidelberg 1999
Originally published by Springer-Verlag Berlin Heidelberg New York in 1999
Softcover reprint of the hardcover 1st edition 1999

Typesetting: Camera-ready copy by the author using a Springer LaTeX macro package
Cover design: *design & production* GmbH, Heidelberg

SPIN: 10646214   56/3144 - 5 4 3 2 1 0 – Printed on acid-free paper

For my parents

Preface

It has been known for a long time that fluid flow can be caused by a spatial variation of surface tension. A simple, everyday example of this is the motion of liquid from the hot center towards the cold periphery in a thin layer of heated oil.

Pioneering investigations of capillary flows in the last century were made by, among others, J. Thompson, R. Lüdtge, and C. Marangoni. Their discoveries, however, did not attract much attention in the following period. It was not until the early nineteen-sixties that the work of L. E. Scriven and C. V. Sternling revived the interest in interfacial instabilities and interfacial turbulence. Since a detailed examination of the hydrodynamic equations proved to be difficult, many aspects of surface-tension-driven flows still remained unexplained.

The situation changed in the seventies, when it was noticed that the surface flow induced by a variation of the surface temperature, also called *thermocapillary convection*, represents an important transport process in several technological applications involving liquid interfaces. The best-known example is probably crystal growth from the melt. Since then, many investigations have dealt with the oscillatory thermocapillary convection in the melt – a phenomenon that can have a profound influence on the homogeneity of the crystals grown. Another boost was created from the increasing power of computers. Today, numerical simulations have become one of the most important tools in fluid mechanics.

The aim of this book is to provide an introduction to thermocapillary flows, to present the most significant results, and to give a thorough discussion of the physical mechanisms of the observable phenomena. The emphasis is on the convection in cylindrical liquid bridges and the flow in rectangular open cavities, in which the flow is driven by tangential gradients of the surface tension. During the last two decades both systems have developed into paradigms of thermocapillary convection. The subject is analyzed with a view to providing a physical understanding of the flow and the pattern formation through hydrodynamic instabilities. The author's results, some presented in detail, are supplemented by those of other authors. Owing to the complexity of the topic, the large amount of literature, and to my personal preferences not all topics are covered exhaustively. However, many extensions to the fundamental models treated are outlined and references for further reading are provided. Results which do not contribute much to the general understanding of thermocapillary flows are not given much space. Furthermore, models of the technologically important Czochralski process have been excluded, as has pattern formation by the classical Marangoni instability which arises when the applied temperature gradient is aligned perpendicular to the free surface.

Following the introduction in Chap. 1 and a formulation of the basic equations in Chap. 2 the asymptotic behavior of thermocapillary flows in wedges and the scaling for several limiting cases of the governing parameters are presented in Chap. 3. Creeping flows in cylindrical liquid bridges are treated in detail in Chap. 4. As a preparation for the analyses to follow, Chap. 5 introduces some methods of classical stability theory. The convective thermocapillary flow instability in plane layers is presented and discussed in Chap. 6. A central part is Chap. 7, in which the instabilities and the nature of the three-dimensional flows in liquid bridges are analyzed. Thermocapillary flows in rectangular containers are reviewed in Chap. 8. Finally, several extensions to the basic thermocapillary models are presented in Chap. 9 and the possibility of flow control by external body forces is addressed.

A Feodor-Lynen scholarship of the Alexander-von-Humboldt foundation facilitated my first encounter with the problem of thermocapillary convection at Arizona State University in 1987–1988 as a postdoc of G. P. Neitzel. In 1991, at the Center of Applied Space Technology and Microgravity (ZARM) of the University of Bremen, I had the opportunity to take up these studies once more. This was made possible by the great freedom and support provided by H. J. Rath. I am very grateful to M. Wanschura and all members and guests of the *Hydrodynamic Stability* group at ZARM for their contributions to the work described in this book. It was a pleasure to work with them. Without the understanding and patience of my wife Kiki and my daughter Saskia this book would never have been possible.

Bremen,   H. C. Kuhlmann
August 1998

Contents

Notation

1 General Remarks

Owing to the wide variety of mathematical and physical variables no systematic attempt has been made to keep to a unified notation. Whenever possible the symbols commonly used in the literature are adopted. For dimensional and dimensionless quantities, the same symbols are used. The corresponding meaning and scaling are explained in the text. The most frequently employed symbols are listed below.

Matrices are generally indicated by sans serif symbols. Occasionally, the summation convention is used.

2 Latin Symbols

a	length of constant surface shear stress
A,B	matrices
$A,B,C,\ldots$	coefficients
A_i	amplitudes
$A_{ij}^{(n)},B_{ij}^{(n)}$	coefficients
b	negative temperature gradient, vibration amplitude
$\boldsymbol{B}$	magnetic induction
c	concentration (mass fraction), speed of sound
c_p	specific heat at constant pressure
C	surface concentration
d	length of liquid bridge, length scale
D	rate of dissipation, diffusion constant, radial derivative
D_T	rate of thermal diffusion
$D_* = D + 1/r$	
$\quad = \partial_r + 1/r$	radial derivative operators
$\boldsymbol{e}_z$	unit vector in z direction
E	energy
E_{kin}	kinetic energy
E_T	thermal energy
$\boldsymbol{E}$	electric field

F	function
F_{ij}	coefficient matrix
$\boldsymbol{g}, g_0$	acceleration due to gravity, acceleration
G	phase, function
h	heat transfer coefficient
$\tilde{H}$	space of kinematically admissible functions
i	imaginary unit
i	index
$\boldsymbol{i}$	diffusion current
I	unit matrix
$I_v, I_T, I_{\mathrm{Gr}}$	integral rates of change of energy
$\mathrm{I}_0, \mathrm{I}_1$	modified Bessel functions
j	heat current density
$\boldsymbol{j}$	electric current density
J	heat current
J_n, K_n, B_{nk}, C_{nk}	integrals
k	heat conductivity, wavenumber, segregation coefficient
$\boldsymbol{k}$	wave vector
k_{T}	thermal diffusion ratio
$\mathrm{K}_0, \mathrm{K}_1$	modified Bessel functions
l	length, characteristic length
L	length, linear operator, latent heat
$\boldsymbol{L}$	Lorentz force density
$\tilde{L}$	linear operator
$L^\dagger$	adjoint operator
L_x, L_y, L_z	lengths
m	azimuthal wavenumber
$N = (1 + \xi_z^2)^{1/2}$	normalizing denominator
$n, m, i, j, k, l \in \mathbb{N}$	indices
$\boldsymbol{n}$	surface-normal vector
p	pressure
p_{a}	ambient pressure
p_i	roots of $W(p) = 0$
$\boldsymbol{r} = (r, \varphi, z)$	cylindrical coordinates
r_{m}	minimum or maximum radius of a liquid bridge
R	radius
R_1, R_2	main radii of curvature
S	surface area
$\tilde{\mathsf{S}} = \eta\left[\nabla \boldsymbol{u} + (\nabla \boldsymbol{u})^{\mathrm{T}}\right]$	viscous stress tensor
$\mathsf{S} = -p\mathsf{I} + \tilde{\mathsf{S}}$	stress tensor
t	time
$\boldsymbol{t}$	tangent vector

T	temperature, eigenfunction
T	transposition, index for through-flow
T_a	ambient temperature
T_m	melting temperature
T_n	Chebyshev polynomial of order n
T_0	reference temperature
$\boldsymbol{u} = (u, v, w)$	velocity
U	characteristic velocity
V	volume
W	function
w_T	through-flow velocity
$\boldsymbol{x} = (x, y, z)$	Cartesian coordinates
$\boldsymbol{X} = (\boldsymbol{u}, p, \theta)$	variable of state
Y	eigenfunction

3 Greek Symbols

α	contact angle, angle of propagation, coefficient
β	thermal expansion coefficient, coefficient, through-flow paramenter
β_S	solutal expansion coefficient
γ	negative linear thermal coefficient of surface tension, coefficient, complex growth rate
γ_S	negative linear solutal coefficient of surface tension
$\tilde{\gamma}$	specific surface energy
Γ	aspect ratio
δ	characteristic length, boundary layer thickness, first variation, coefficient
δ_{ij}	Kronecker symbol
δ_kin	relative error in the kinetic-energy balance
δ_T	relative error in the thermal-energy balance
Δ	Laplace operator, boundary layer thickness
ΔT	temperature difference
ϵ	emissivity, normalized Reynolds number
ζ	axial position of the phase boundary, linear thermal Taylor coefficient of the kinematic viscosity
θ	normalized temperature, polar angle
ϑ	amplitude ratio
$\Theta = \theta_0 \pm z$	reduced temperature (deviation from the linear heat conduction profile), temperature perturbation
θ_a	normalized ambient temperature

$\kappa = k/\rho c_p$	thermal diffusivity
λ	eigenvalue, coupling parameter, wavelength
$\lambda_n, \lambda_1, \lambda_{1n}$	eigenvalues
$\mu = \rho\nu$	dynamic viscosity, magnetic susceptibility
ν, ν_0	kinematic viscosity, reference kinematic viscosity
ξ	radial position of the free surface
ϖ	vibrational frequency
ρ	density, radial coordinate, eigenvalue
ρ_a	density of the ambient fluid
σ	surface tension, real growth rate, electrical conductivity
σ_0	surface tension at reference temperature
τ	period
ϕ	eigenfunction
Φ	electrical potential
$\boldsymbol{\Phi} = (\phi_1, \phi_2)^{\mathrm{T}}, \ \boldsymbol{\Phi}^{(n)}$	biorthogonal vector functions
$\boldsymbol{\Psi} = (\psi_1, \psi_2), \ \boldsymbol{\Psi}^{(n)}$	adjoint biorthogonal vector functions
ψ	stream function
$\Omega, \ \Omega_i$	angular velocity of the rotating boundary
ω	vorticity
ω	circular frequency, vorticity

4 Other Symbols

c.c.	complex conjugate
$\mathbb{C}$	complex numbers
$\mathcal{F}$	functional
∂_t	partial derivative with respect to time
$\Im$	imaginary part
$\mathbf{N}$	positive integer numbers
$\mathcal{O}$	order of magnitude
$\mathbf{R}$	real numbers
$\Re$	real part
∇	nabla operator
$\nabla_\parallel = \mathsf{I} - \boldsymbol{nn}$	tangential nabla operator
$*$	complex conjugate

5 Similarity Groups

$$Bd = \frac{Gr}{Re} = \frac{\rho g \beta d^2}{\gamma} \qquad \text{dynamic Bond number}$$

$$Bi = \frac{hd}{k} \qquad \text{Biot number}$$

$$Bo = \frac{(\rho - \rho_a) g d^2}{\sigma_0} \qquad \text{static Bond number}$$

$$Ca = \frac{\gamma \Delta T}{\sigma_0} \qquad \text{capillary number}$$

$$Ek = \frac{\nu}{\Omega R^2} \qquad \text{Ekman number}$$

$$Ga = \frac{g d^3}{\nu^2} \qquad \text{Galilei number}$$

$$Gr = \frac{g \beta \Delta T d^3}{\nu^2} \qquad \text{Grashof number}$$

$$Ha = B_0 d \left(\frac{\sigma}{\rho \nu} \right)^{\frac{1}{2}} \qquad \text{Hartmann number}$$

$$Ma = PrRe = \frac{\gamma \Delta T d}{\rho \nu \kappa} \qquad \text{Marangoni number}$$

$$Nu = \frac{J_{total}}{J_{conductive}} \qquad \text{Nußelt number}$$

$$Pr = \frac{\nu}{\kappa} \qquad \text{Prandtl number}$$

$$R_\nu = \frac{\zeta \Delta T}{\nu_0} \qquad \text{viscosity group}$$

$$Ra = PrGr = \frac{g \beta \Delta T d^3}{\nu \kappa} \qquad \text{Rayleigh number}$$

$$Ra_v = \frac{(\beta \Delta T d \varpi b)^2}{\nu \kappa} \qquad \text{vibrational Rayleigh number}$$

$$Re = \frac{\gamma \Delta T d}{\rho \nu^2} \qquad \text{thermocapillary Reynolds number}$$

$$Re_m = \sigma \mu U d \qquad \text{magnetic Reynolds number}$$

$$Re_S = \frac{\gamma_S \Delta T d}{\rho \nu^2} \cdot \frac{k_T}{T_0} \qquad \text{solutocapillary Reynolds number}$$

$$\text{Re}_\text{v} = \frac{\varpi b^2}{\nu}$$ vibrational Reynolds number

$$\text{Sc} = \frac{\nu}{D}$$ Schmidt number

$$\text{We} = \frac{\rho \varpi^2 R^3}{\sigma_0}$$ Weber number

$$\psi_\text{S} = -\frac{\beta_\text{S}}{\beta} \cdot \frac{k_\text{T}}{T_0}$$ Soret separation ratio

1. Introduction

1.1 Marangoni Effect

Liquids are condensed fluids whose constituents are bound by intermolecular forces. The cohesion forces can be due to ionic, metallic, or hydrogen bonding, or van der Waals forces (Israelachvili (1992)). The forces on a molecule in the bulk of the liquid are isotropic on average and compensate each other. Molecules near the boundary region between a liquid and its vapor or between two immiscible liquids, however, are surrounded by fewer nearest neighbors of the same kind than the bulk molecules. Therefore, the molecules near the boundary are subject to a net force that is directed into the interior of the liquid. When the surface area is increased, work must be done against the attractive intermolecular forces to move molecules from the bulk to the surface. Thus the surface molecules have a higher potential energy than the molecules in the bulk. The work done per unit surface area created is the specific interfacial energy σ. In the case of two liquids in contact it is called the *interfacial tension*. When a liquid meets its vapor it is called the *surface tension*.

Far away from the critical point of the fluid, the interfacial region within which the energetic conditions vary has a thickness of only a few molecular diameters ($\approx 10\text{Å}$). Within a macroscopic continuum description this layer can thus be considered as a surface of discontinuity across which the thermophysical properties, e.g. the density and the heat conductivity, show a jump-like behavior. The interfacial tension generally depends on the temperature, the concentration of dissolved species, the electrical potential and, weakly, on the pressure (see e.g. Davies & Rideal (1963)).

In thermodynamic equilibrium the surface tension is constant and the free energy of the system is a minimum. For a given plane boundary of the interface, any curved interface has a higher energy than the plane interface. A curved interface thus experiences a force that can only be compensated by a pressure difference (Young (1805), Laplace (1806)) betweeen the two sides of the interface. The problem of the shape of stationary capillary surfaces as a function of the edge conditions was first addressed by Plateau (1873) and has developed into the theory of minimum surfaces, a branch of modern mathematics (Finn (1986), Fomenko (1989)).

In a nonequilibrium state the interfacial tension can vary along the phase boundary. Then a tangential gradient $\nabla_\parallel \sigma$ causes a force per unit surface area acting in the direction of highest surface tension; the system tends to minimize its surface energy. Under these conditions viscous liquids cannot remain at rest. The tangential shear stresses induce a flow on both sides of the interface (Fig. 1.1).

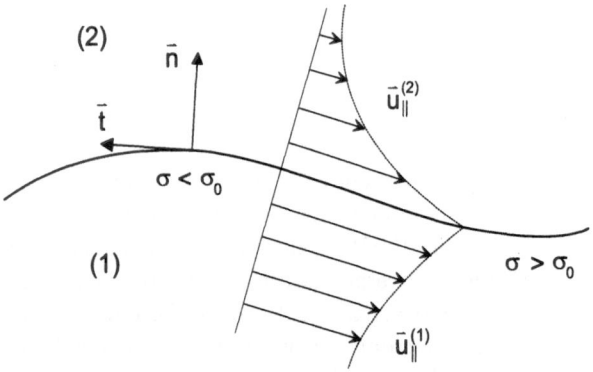

Fig. 1.1. Schematic illustration of the flow close to an interface with varying surface tension

Interfacial convection has probably been known for a long time. In his report on "certain curious motions observable at the surfaces of wine and other alcoholic liquors" Thomson (1855) was the first to give an explanation for the motion of droplets of wine (tears) on the wall of a wine glass. Owing to the volatility of alcohol the liquid in the meniscus adjacent to the wall becomes poorer in alcohol than the bulk liquid. The resulting variation of the alcohol concentration leads to surface tension gradients driving a surface flow towards and up the glass wall.[1] Later, Marangoni (1871), presumably not being aware of Thomson's (1855) work, investigated the spreading of oil droplets on the surface of clean water in a basin of the Tuileries of Paris which had a diameter of ≈ 70 m. He explained the fast spreading of the droplets (>2 m/s), as well as the different stages of the spreading, in terms of the smaller surface tension of oil and air compared to that of pure water and air. Today, all flows caused by interfacial-tension gradients are called *Marangoni convection*. Similar experiments were carried out by Lüdtge (1869). He mentioned capillary flows that appear when thin liquid layers of the same fluid but at different temperatures are brought into contact.

---

[1] Thomson (1855) also describes experiments in which a diverging flow emerges from a point on a liquid surface having a reduced surface tension. See also Bois-Reymond (1858), Weber (1855), and, for a theoretical treatment, Shtern & Hussain (1993).

Scriven & Sternling (1960) gave a historical overview of the Marangoni effects. They can occur in all multiphase liquid systems. Typical areas of application concern flows associated with the drying of thin liquid layers (Pearson (1958), VanHook et al. (1995)), the motion of droplets and bubbles (Subramanian (1992), Leppinen et al. (1996)), the spreading of droplets (Erhard & Davis (1991), Smith (1995)) and liquid films on solid substrates (Oron et al. (1997)), emulsions (Shinoda & Friberg (1986)), foams (Bikerman (1973), Hunter (1989)) and the stabilization of free thin films (Prevost & Gallez (1986)), the damping of surface waves by thin films (Levich (1962)), and interfacial turbulence (Sternling & Scriven (1959)). In biology the transport of bacteria and aerosols (Grotberg (1994)) have been discussed. Moreover, Marangoni effects are important in evaporation and boiling (McGillis & Carey (1996)), welding (Gieldt (1987), DebRoy & David (1995)) and in the spreading of flames over liquid fuels (Sirignano & Glassman (1970)). Electrokinetic effects are treated in the book by Levich (1962). Marangoni effects due to variations of the surface temperature or chemical composition and to adsorbed films have been reviewed by Kenning (1968). Levich & Krylov (1969) gave a general overview of flows driven by capillary forces.

Apart from the above areas, Marangoni convection also arises in a number of crystal growth techniques from the melt and contributes significantly to the transport of heat and mass. Owing to the technological importance of high-purity semiconductor single crystals, melt flows have recently received increased attention.

1.2 Thermocapillary Convection in Crystal Growth

The limits for the residual chemical-impurity concentrations in semiconductor single crystals are very tight. At present, the purest large-volume silicon single crystals are commercially grown using the floating zone method (Bohm et al. (1994)). During this process a melt zone is slowly moved through a cylindrical, vertically oriented crystal. The liquid zone between the polycrystalline feed rod and the grown single crystal is stabilized by surface tension (Fig. 1.2). The main advantage of this crucible-free method is the minimization of melt contamination by impurities. The surface tension is not reduced, for practical purposes, by the Gibbs adsorption of impurity atoms. The temperature distribution, however, varies along the liquid–gas interface, leading to surface-tension gradients which can drive a considerable thermal Marangoni flow. This flow is called *thermocapillary convection*. It is also present in other crystal-growth processes involving liquid–gas interfaces, such as the Czochralski and the open-boat methods (Hurle (1994)).

Another important transport process is buoyant convection induced by density variations. Both types of flow have a significant influence on the crystal growth. They modify the gradients of temperature and concentration of

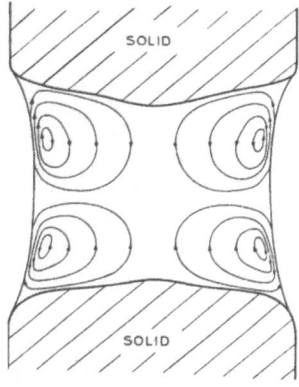

Fig. 1.2. Sketch of a melt zone after Schwabe et al. (1978)

dissolved species near the crystallization front and thus influence the local advancement of the liquid–solid interface as well as the large-scale distribution of dopants and impurities in the melt and, finally, in the crystal (*macro-segregation*). If, moreover, the flow is not stationary the growth conditions vary in time. Since the equilibrium concentration of minority components in the grown crystal depends on the growth rate, time-dependent flows lead to variations of the chemical composition of the crystal on small length scales (*micro-segregation*) (see e.g. Müller (1986)). These compositional variations are called *striations* and have a typical length scale of a few microns. An example is shown in Fig. 1.3. The striations cause inhomogeneities of the

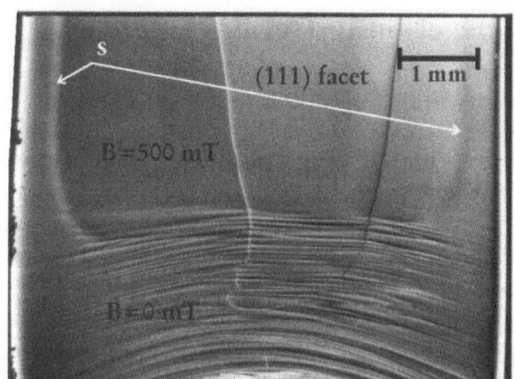

Fig. 1.3. Etched longitudinal section through a P-doped silicon single crystal. In the *lower part* of the figure, doping striations can be seen which are caused by a relative variation ($\approx 8\%$) of the phosphorus concentration. The crystal shown in the *upper part* of the figure has been grown in an axial magnetic field of 0.5 T. Concentration striations do not occur under this condition. In a boundary layer (s), however, a strong convective mixing has led to a concentration different from that one in the bulk. The width of the section shown is about 5 mm. After Cröll et al. (1994)

electronic properties of the single crystals and are responsible for mechanical stresses and dislocations (Völkl (1994)). In order to prevent or at least to suppress the striation formation a precise knowledge is required of the conditions under which oscillatory flows arise.

An experimental investigation of hydrodynamic processes in technically important semiconductor melts is very difficult owing to the high melting points and the opacity of the materials.[2] Therefore, the numerical modeling of crystal-growth processes has become important. One strategy is the numerical simulation of the whole system, including convection in the melt, heat conduction in the crystal, and heat transport in all components of the apparatus that influence the temperature distribution. Today, this approach is feasible in two dimensions (see e.g. Dupret & Bogaert (1994)) and is sufficient for many components of the crystal growth furnace.

The convection in the melt, however, is three-dimensional under typical growth conditions. For fundamental questions and three-dimensional problems a number of simple models have been developed. They enable, to a certain extent, parameter variations and separate investigations of thermocapillary convection, buoyant convection, surface waves, and other phenomena which may be nonlinearly coupled in real systems. Thermocapillary convection has been studied in infinitely extended liquid layers, rectangular open containers, and cylindrical liquid volumes with liquid–gas surfaces.

1.3 Previous Reviews

The number of publications related to convection problems in crystal growth has increased rapidly over the last two decades. Since a comprehensive overview would require too much space, only some of the main reviews will be mentioned in this section. Other relevant publications are discussed later in connection with the respective hydrodynamic problems.

The numerical treatment of thermocapillary flows in crystal growth was initiated by Chang & Wilcox (1975, 1976). At the beginning, fundamental questions of, for example, the strength and the basic structure of the induced flows were the focal point of the investigations. The magnitude of the characteristic flow velocity in thermocapillary systems is closely related to the boundary-layer thickness δ within which the velocity varies significantly. If the thickness of the liquid layer d is small compared to the length scale L on which the surface tension varies, convective motion penetrates the whole volume. Under such conditions Levich (1962) and Ostrach (1977) found the characteristic velocity

---

[2] Kakimoto et al. (1988) and Nakamura et al. (1996) employed X-rays to visualize the motion of small tracer particles. Baumgartl et al. (1993) suggested the use of magnetic-field measurements to determine the flow velocity in electrically conducting liquids.

$$U = \frac{\gamma \Delta T}{\rho \nu} \frac{d}{L}, \tag{1.1}$$

where ΔT is the typical temperature difference over the length L, ρ and ν denote the density and the kinematic viscosity, and $\gamma = -\partial\sigma/\partial T$. Equation (1.1) holds if $\mathrm{Re} \ll (L/d)^2$, where the *thermocapillary Reynolds number* is defined as

$$\mathrm{Re} = \frac{\gamma \Delta T d}{\rho \nu^2}. \tag{1.2}$$

For $\mathrm{Re} \gg (L/d)^2$ boundary layers arise (Ostrach (1977), Napolitano (1979)), with a characteristic velocity

$$U = \left(\frac{\gamma^2 \Delta T^2}{\rho^2 \nu L}\right)^{1/3}, \tag{1.3}$$

which varies within a free surface layer of thickness

$$\delta = \left(\frac{\nu L}{U d^2}\right)^{1/2} = \frac{L}{d} \left(\frac{\gamma \Delta T L}{\rho \nu^2}\right)^{-1/3}. \tag{1.4}$$

Further aspects of Marangoni boundary layers have been discussed by Batishchev et al. (1989), among others.

Apart from the scaling, end effects have also been investigated (Sen & Davis (1982), Strani et al. (1983)) and many two-dimensional numerical calculations have been carried out (see e.g. Shen (1989)). With an increase of the driving force, i.e. the surface temperature gradient, the thermocapillary convection does not generally become turbulent by a sudden transition. Typically, the transition is gradual and different regular spatio-temporal convection patterns arise.

Structure formation in infinitely extended liquid layers bounded from below by a rigid wall and from above by a liquid–gas interface has been discussed by Davis (1987). If buoyancy forces are absent and the layer is heated from the rigid bottom and cooled from the free surface, a critical threshold of the temperature difference across the layer exists above which Bénard–Marangoni convection sets in (Bénard (1900), Block (1956), Pearson (1958)). The convection patterns near the threshold have a hexagonal form in plan view in most cases (Riahi (1986), Dijkstra (1998)). When the layer is heated from the side a plane shear flow is generated. If, for given parameters, the tangential temperature gradient exceeds a critical value, three-dimensional *hydrothermal waves* appear (Smith & Davis (1983a)). These are plane convection waves which propagate at a certain angle with respect to the applied temperature gradient, depending on the Prandtl number. Smith & Davis (1983b) also treated surface waves induced by thermocapillary-driven flows.

Other overviews are more concerned with crystal-growth aspects. The major issues have been the mechanisms and the conditions for the onset of time-dependent flows. Langlois (1985) has reviewed natural convection in the Czochralski process. Pimputkar & Ostrach (1981) treated convective effects in crystal melts focusing on the Czochralski and the horizontal Bridgman method. They discussed the onset of oscillatory flows and other problems of crystal growth under weightlessness. This is also the main concern of the overview by Ostrach (1982) who, in addition to other topics, considered weak natural and thermocapillary convection and flows induced by vibrations. In a further article Ostrach (1983) gave an overview of natural and Marangoni convection in various crystal-growth configurations. Schwabe (1988) treated capillary flows driven by tangential temperature or concentration gradients in the most important crystal-growth setups (Czochralski, floating zone, open boat). In earlier contributions Schwabe (1981a,b) investigated Marangoni convection in a number of crystal-growth model systems using transparent liquids for flow visualization. For example, the stationary toroidal thermocapillary convection in liquid zones becomes time-dependent when a critical temperature difference is exceeded. For high-Prandtl-number fluids ($\mathrm{Pr} = \mathcal{O}(10)$), the empirical law $m = 2.2/\Gamma$ (see also Preisser et al. (1983)) was found, correlating the periodicity m of the wave in the azimuthal direction with the geometrical aspect ratio (Γ is the height/radius ratio). Müller (1989) was concerned with inhomogeneities in crystal growth, their creation, analysis, and prevention (see also Müller (1986)). To this end, experiments and numerical simulations were compared and the generation of inhomogeneities was discussed in the context of natural convection. An overview of the origin of melt convection, the measurement of convective structures, and a discussion of the results can be found in Müller & Ostrogorsky (1994).

Recently, the number of publications dealing with numerical calculations has increased more than proportionally. Today, the two-dimensional flow in complex geometries can be calculated to a high accuracy, as well as laminar three-dimensional flows in simple domains. The results obtained for finite geometries have been summarized by Kuhlmann (1994). Overviews of numerical methods for flows in systems with significant dynamic interface deformations have been given by Yeung (1982), Floryan & Rasmussen (1989), and Tsai & Yue (1996).

For an up-to-date review of all important topics of crystal growth, finally, the reader is referred to the comprehensive *Handbook of Crystal Growth* edited by Hurle (1994).

1.4 Scope

Considering thermal Marangoni effects, the angle made by the temperature gradient ∇T with respect to the surface-normal vector $\boldsymbol{n}$ depends on the boundary conditions and on several other factors. It is useful to distinguish two cases.

If the temperature gradient ($\nabla T \neq 0$) is parallel to the unit surface-normal vector (i.e. $\nabla T \times \boldsymbol{n} = 0$) the interface is isothermal and no a priori thermocapillary shear stresses arise along the interface. In the absence of other driving forces the trivial flow state $u = 0$ exists. Interfacial convection can arise in a viscous fluid only above a critical temperature gradient. Such a situation may occur, for example, during the drying of thin liquid layers involving heat and mass transfer through the interface. The first theoretical explanation of the onset of Marangoni convection in a liquid layer on a heated plane wall and surrounded by a cold gas was given by Pearson (1958). This configuration will not be a subject here.[3]

Here we are concerned with the case where the temperature gradient has major components tangential to the interface ($\nabla T \times n \neq 0$). Except for special cases[4] even a minute tangential surface-tension gradient causes a flow parallel to the interface. We refer to this as *thermocapillary flow*.[5] The objective is an introduction to and analysis of the fundamental qualitative and quantitative properties of laminar thermocapillary flows. The discussion will primarily be based on numerical investigations of convective processes. Surface waves are excluded from the analysis (but see Sect. 9.1.2). Thermocapillary flows in plane layers, cylinders, and rectangular containers will be investigated.[6] The two latter geometries are idealizations of the floating-zone and open-boat methods and are therefore directly related to crystal growth from the melt.

When the flow is weak the symmetries of the equations of motion and the boundary conditions are reflected by the convection structures. For higher driving forces, symmetry breaking usually occurs through hydrodynamic instabilities. In crystal-growth problems the breaking of time invariance is of particular importance. The investigations focus on three subjects: stationary patterns of creeping flows, hydrodynamic instabilities under strong driving forces, and the influence of side walls on the otherwise two-dimensional stationary flow.

The basic equations are provided in Chap. 2. For the laminar flows under consideration it is sufficient in most cases to take into account only the temperature dependence of the surface tension and buoyancy forces and to assume constant material properties otherwise (Boussinesq-type approximation). Possible improvements are briefly addressed in Sect. 9.1.5.

In Chap. 3 some general properties of thermocapillary flows are considered in several limiting cases. First, a similarity solution for the viscous flow in the vicinity of the contact line is considered in the corner between a rigid plane isothermal wall and a plane thermocapillary surface. It will be shown

---

[3] The reader is referred to Davis (1987) and the references cited therein. Some more recent publications are due to Thess & Orszag (1995), Colinet et al. (1995), Nepomnyashchy & Simanovskii (1995), and Dijkstra (1998).

[4] The tangential variation of the surface tension due to thermal gradients may be compensated, e.g. by suitable concentration gradients (cf. Sect. 9.1.6).

[5] Here we use the term *thermocapillary flow* when the surface temperature gradient is non-zero and *Marangoni convection* when the externally applied temperature gradient is perpendicular to the interface.

[6] Thermocapillary convection in droplets and bubbles is not treated. The reader is referred to, e.g., Subramanian (1992).

that the Stokes approximation to the flow exhibits a singular behavior in the corner. Moreover, the normal stress condition cannot be satisfied by a *plane* thermocapillary surface and the approximation must break down in a small vicinity of the contact line. The discontinuity of the boundary conditions at the contact line may cause problems in a numerical solution of the Navier–Stokes equations. These considerations are followed by scaling analyses of the thermocapillary boundary layer flow in the vicinity of cold and hot corners.

Chapter 4 is concerned with the structure of two-dimensional flows in circular cylinders driven by thermocapillary forces due to a linear axial temperature variation on the mantle surface. By means of the method of biorthogonal functions, the creeping thermocapillary flow is calculated and the influence of the aspect ratio on the vortex structures is discussed for several limiting cases. By an asymptotic expansion, convection-induced dynamic surface deformations of the cylindrical free surface can be computed to first order. Within an Oseen-like approximation, the method is extended to include the case where a homogeneous axial through flow is present in addition to the weak thermocapillary flow.

In Chap. 5 methods for stability analyses are introduced. Linear stability analyses will be used to solve several problems of pattern formation in cylindrical volumes. The equations governing the energy transport are derived and applied to the energy exchange between symmetric basic flow states and pattern-forming perturbation modes, elucidating the physical instability mechanisms. Energy stability theory will also be addressed.

As a paradigm for the convective instability of thermocapillary systems, the flow in an infinitely extended liquid layer driven by a constant thermocapillary surface stress is treated in Chap. 6 for the case of zero lateral through flow. The typical convective instability in this system occurs in the form of hydrothermal waves.

A central section is Chap. 7. Following a brief discussion of axisymmetric basic flows in cylindrical liquid bridges, their stability with respect to three-dimensional convective modes is investigated. In the absence of gravity, pure thermocapillary flow instabilities are analyzed for two representative Prandtl numbers, $Pr = 0.02$ (liquid metals) and $Pr = 4$ (transparent fluids). In general, buoyancy forces must be included. Before mixed convection is analyzed, the limiting case of pure buoyant convection in liquid bridges without thermocapillarity is considered. Particular attention is paid to the transition from a two-dimensional axisymmetric state to a three-dimensional flow state. Thereafter, two-dimensional mixed convection flows and their linear stability are treated. A comparison between available experimental and numerical results for the critical data is followed by a discussion of the nonlinear dynamics of supercritical three-dimensional flows. The section closes with a few remarks on the results of energy stability calculations.

Chapter 8 is concerned with convection in rectangular open cavities heated from the side. Two-dimensional stationary and oscillatory flows, driven by thermocapillary and buoyancy forces, are discussed. Three-dimensional flows caused by near side walls are analyzed in detail for the thermocapillary flow in a cubic container, and three-dimensional flows arising through bulk flow instabilities are considered.

Extensions to the common models are introduced in Chap. 9. One aspect is the generalization and improvement of the models themselves, using more realistic boundary conditions that take into account static (Sect. 9.1.1) and dynamic (Sect. 9.1.2) deviations from the ideal plane or cylindrical interface shape. As another improvement, dynamic melting and solidifying crystallization fronts (Sect. 9.1.4) may be included. In addition, the manipulation of thermocapillary and buoyant convection by the use of other body and surface forces is discussed. The most promising means of flow control are rotation and, for metallic melts, the application of magnetic fields. The topics addressed in Chap. 9 suggest possible extensions and further developments in the field of thermocapillary flows.

2. Basic Equations

2.1 Volume Equations

The flow velocities U in buoyancy- or surface-tension-driven flows close to the first pattern-forming instability are typically small compared to the speed of sound c. As a good approximation the fluid flow can then be considered incompressible with a constant density ρ. Corrections are of the order of magnitude $\mathcal{O}(M^2)$, where $M = U/c$ denotes the Mach number (see e.g. Schlichting (1968)). The momentum, mass, and energy conservation equations for an incompressible fluid with kinematic viscosity ν, thermal diffusivity κ, and thermal expansion coefficient $\beta = -1/\rho \, (\partial\rho/\partial T)_p$ in a homogeneous gravity field $\boldsymbol{g} = -g\boldsymbol{e}_z$ are (Landau & Lifshitz (1959))

$$\partial_t \boldsymbol{u} + \boldsymbol{u} \cdot \nabla \boldsymbol{u} = -\frac{1}{\rho}\nabla p + \nu \Delta \boldsymbol{u} + g\beta T \boldsymbol{e}_z \,, \tag{2.1}$$

$$\nabla \cdot \boldsymbol{u} = 0 \,, \tag{2.2}$$

$$\partial_t T + \boldsymbol{u} \cdot \nabla T = \kappa \, \Delta T \,, \tag{2.3}$$

where (2.2) is the continuity equation. The velocity, pressure, and temperature fields are denoted by $\boldsymbol{u}$, p, and T, respectively. Equations (2.1–2.3) are called the Oberbeck–Boussinesq equations. Within this approximation temperature-induced density variations are retained only in the buoyancy term $g\beta T \boldsymbol{e}_z$. The approximation requires the temperature variations to be sufficiently small.[1] The Oberbeck–Boussinesq equations are the basis for the following analyses of buoyant and thermocapillary convection.

2.2 Boundary Conditions

2.2.1 Velocity Field

The fluid volumes under consideration are bounded by rigid stationary walls as well as by free surfaces. Since a fluid in contact with a rigid wall may not slip, the velocity components parallel and perpendicular to the wall must vanish:

---

[1] More accurate criteria have been given by Gray & Giorgini (1976).

$$\boldsymbol{u} = 0 \,. \tag{2.4}$$

Along the free surface between two immiscible fluids (1) and (2) the forces on adjacent surface elements of (1) and (2) must be the same. If the surface is plane and the surface tension is constant, this balance reads

$$\mathsf{S}^{(1)} \cdot \boldsymbol{n} = \mathsf{S}^{(2)} \cdot \boldsymbol{n} \,, \tag{2.5}$$

where the pressure and the viscous forces per unit surface area enter the stress tensor

$$\mathsf{S} = S_{ij} = -p\delta_{ij} + \eta \left(\frac{\partial u_i}{\partial x_j} + \frac{\partial u_j}{\partial x_i} \right) = -p\mathsf{I} + \eta \left[\nabla \boldsymbol{u} + (\nabla \boldsymbol{u})^{\mathsf{T}} \right] \,. \tag{2.6}$$

Here, $\eta = \rho \nu$ is the dynamic viscosity, I is the identity matrix, and $\boldsymbol{n}$ is the unit normal vector directed out of liquid (1) into the ambient fluid (2), which is assumed to be gaseous. Generally, the free surface is not plane, and the surface tension varies along the interface. In this case the appropriate boundary condition is

$$\mathsf{S}^{(1)} \cdot \boldsymbol{n} + \sigma(\nabla \cdot \boldsymbol{n}) \boldsymbol{n} - (\mathsf{I} - \boldsymbol{nn}) \cdot \nabla \sigma = \mathsf{S}^{(2)} \cdot \boldsymbol{n} \,. \tag{2.7}$$

The term $\sigma (\nabla \cdot \boldsymbol{n})$ is the Laplace pressure. The mean curvature of the interface,

$$\nabla \cdot \boldsymbol{n} = \frac{1}{R_1} + \frac{1}{R_2} \,, \tag{2.8}$$

can be expressed as the sum of the inverse principle radii of curvature R_1 and R_2. We take a radius of curvature to be positive if the origin of the corresponding circle lies on the side of fluid (1) and the body of fluid (1) is barrel-shaped; otherwise, it is negative. The second additional term in (2.7) describes a surface force acting tangentially to the interface. It is proportional to the negative (tangential) gradient of the surface tension σ (σ is the energy per unit area). The operator $\mathsf{I} - \boldsymbol{nn}$ represents the orthogonal projection of a vector onto the tangent plane defined by $\boldsymbol{n}$.

In addition to the above force balance, a kinematic boundary condition is required for dynamic surface deformations. If, for example, the position of the interface is given by $x_1 = \xi(x_2, x_3)$,

$$u_1 = \frac{d\xi}{dt} \tag{2.9}$$

guarantees that a material surface element on the interface will remain on the interface for later times.

2.2.2 Temperature Field

On the boundary of the fluid volume (1) the temperature must be continuous. The same applies to the heat current in the absence of heat sources.[2] Since

---

[2] Heat sources on the interface may be due to latent heat during phase changes or due to surface chemical reactions.

the surrounding medium (2) generally has a non-zero heat conductivity, the temperature of the ambient atmosphere must be known in order to determine the surface temperature of the fluid. In many cases the ambient temperature distribution close to the interface is not known exactly. Then Newton's law of heat transfer

$$k\boldsymbol{n} \cdot \nabla T = -h(T - T_{\mathrm{a}}) \tag{2.10}$$

is frequently employed[3] (see, for instance, Gröbner et al. (1963)). It relates the heat current through the interface to the surface temperature and some known ambient temperature $T_{\mathrm{a}}(\boldsymbol{x})$ at some distance from the interface. Here, k denotes the heat conductivity of the fluid (1) and the phenomenological parameter h is called the heat transfer coefficient. In the limit $h \to \infty$ the exterior temperature T_{a} is imposed on the surface of fluid (1),

$$T = T_{\mathrm{a}}. \tag{2.11}$$

This is a good approximation if the heat diffusivity of the exterior medium (2) is large compared to that of fluid (1). For this reason, copper is often used as a wall material to realize nearly constant temperatures on rigid boundaries. The other limit, $h \to 0$ in (2.10), describes a thermally insulating boundary, for instance, a liquid with low vapor pressure surrounded by its own vapor.

In crystal growth processes the temperature of the melt is usually so high that radiative heat transfer cannot be neglected.[4] For an approximate treatment of the heat transfer due to radiation, the Stefan–Boltzmann law

$$k\boldsymbol{n} \cdot \nabla T = -\epsilon s_0(T^4 - T_{\mathrm{a}}^4) \tag{2.12}$$

can be used. Here, ϵ is the emissivity of the liquid surface, s_0 the Stefan–Boltzmann constant, and T_{a} the surface temperature of the body in radiative exchange with the liquid surface. Depending on the thermal conditions, (2.10) and (2.12) may be combined.

2.3 Dimensionless Equations

To be more specific, we consider a fluid volume V bounded axially by two rigid, parallel disks of equal radii $r = R$ at $z = \pm d/2$, a distance d apart. The radial boundary, given by a free surface of mean surface tension σ_0, supports the liquid. This configuration is called a *liquid bridge*. Its geometry is characterized by the aspect ratio $\Gamma = d/R$. For the particular case $V = \pi R^2 d$ and $\boldsymbol{g} = 0$, sketched in Fig. 2.1, the fluid volume takes an upright cylindrical shape. The disks are kept at constant but different temperatures $T(\pm d/2) =$

---

[3] An extension of this model for free surfaces, including the effects of changes of surface tension due to surface dilatation and variation of the temperature, has been discussed by Chen & Tsamopoulos (1993).
[4] The melting point of silicon is $T_{\mathrm{m}} = 1410°\mathrm{C}$.

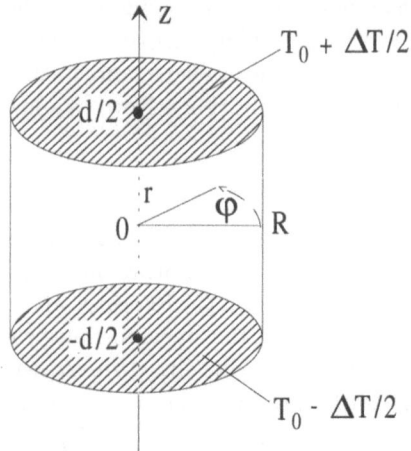

Fig. 2.1. Geometry and coordinate system of a heated upright cylindrical liquid bridge

$T_0 \pm \Delta T/2$. In the case where gravity is resent we assume $\boldsymbol{g} \parallel \boldsymbol{e}_z$. If the z axis is oriented antiparallel to the acceleration due to gravity, this configuration corresponds to heating from above.

Besides the temperature dependence of the density, which is part of the Oberbeck–Boussinesq approximation, the temperature dependence of the surface tension is also taken into account here. For small temperature variations it suffices to expand the surface tension up to first order in $T-T_0$,

$$\sigma(T) = \sigma_0(T_0) - \gamma(T - T_0) + \mathcal{O}\left[(T - T_0)^2\right] , \qquad (2.13)$$

where T_0 is the reference temperature (mean temperature of the liquid bridge), $\sigma_0(T_0)$ the reference surface tension, and γ the negative linear Taylor coefficient of the surface tension. For most liquid–gas interfaces γ is positive.

When the field quantities $(\boldsymbol{u} = (u, v, w), p, T)$ are written in cylindrical coordinates r, φ, z and the scales for length, time, velocity, pressure, and temperature given in Table 2.1 are used, the volume equations are obtained in the form

$$\partial_t \boldsymbol{u} + \text{Re}\,\boldsymbol{u} \cdot \nabla \boldsymbol{u} = -\nabla p + \Delta \boldsymbol{u} + \text{Bd}\,\theta \boldsymbol{e}_z , \qquad (2.14)$$

$$\nabla \cdot \boldsymbol{u} = 0, \qquad (2.15)$$

$$\partial_t \theta + \text{Re}\,\boldsymbol{u} \cdot \nabla \theta = \frac{1}{\text{Pr}}\,\Delta \theta , \qquad (2.16)$$

Table 2.1. Scales used to nondimensionalize the Oberbeck–Boussinesq equations

Variable	$r,\ z$	t	$\boldsymbol{u} = (u, v, w)$	p	T
Scale	d	d^2/ν	$\gamma\Delta T/\rho\nu$	$\gamma\Delta T/d$	ΔT

where the normalized temperature $\theta = (T-T_0)/\Delta T$ has been introduced. The dimensionless groups (characteristic numbers) are

$$\text{Re} = \frac{\gamma \Delta T d}{\rho \nu^2} \qquad \text{(Reynolds number)}, \qquad (2.17)$$

$$\text{Gr} = \frac{g\beta \Delta T d^3}{\nu^2} \qquad \text{(Grashof number)}, \qquad (2.18)$$

$$\text{Pr} = \frac{\nu}{\kappa} \qquad \text{(Prandtl number)}, \qquad (2.19)$$

$$\text{Bd} = \frac{\text{Gr}}{\text{Re}} = \frac{\rho g \beta d^2}{\gamma} \qquad \text{(dynamic Bond number)}. \qquad (2.20)$$

The Reynolds and Grashof numbers, Re and Gr, measure the relative strength of the thermocapillary surface forces ($\gamma \Delta T/d$) and buoyancy forces ($\rho g \beta \Delta T d$), respectively, compared to the viscous forces per unit surface area ($\rho \nu^2/d^2$). The Prandtl number is the ratio of viscous diffusion (ν) to heat diffusion (κ) and the dynamic Bond number measures the strength of buoyancy forces relative to that of thermocapillary forces. Instead of Re and Gr, the Marangoni number Ma=RePr and the Rayleigh number Ra=−GrPr can be used.[5]

The boundary conditions on the rigid isothermal walls at $z = \pm 1/2$ read

$$\boldsymbol{u} = 0, \qquad \theta = \pm \frac{1}{2}, \qquad \text{on} \qquad z = \pm \frac{1}{2}. \qquad (2.21)$$

With the position of the free surface $r = \xi(\varphi, z)$, the dimensionless force balance on the free surface taking the thermocapillary effect (2.13) into account is obtained as

$$- p\boldsymbol{n} + \left[\nabla \boldsymbol{u} + (\nabla \boldsymbol{u})^{\mathrm{T}} \right] \cdot \boldsymbol{n} + \left(\frac{1}{\text{Ca}} - \theta \right)(\nabla \cdot \boldsymbol{n})\boldsymbol{n} + (\mathsf{I} - \boldsymbol{nn}) \cdot \nabla \theta$$

$$= \frac{\text{Bo}}{\text{Ca}} z\boldsymbol{n}, \qquad \text{on} \quad r = \xi(\varphi, z), \qquad (2.22)$$

where, up to the constant ambient value $p_a(z=0)$, p is the pressure in the liquid bridge. Viscous shear forces in the gas phase (2) have been disregarded owing to the small dynamic viscosity of gases. The term on the right-hand side of (2.22) is caused by the hydrostatic pressure difference. The Bond number Bo and the capillary number Ca are defined as[6]

$$\text{Bo} = \frac{(\rho - \rho_a)g d^2}{\sigma_0}, \qquad (2.23)$$

$$\text{Ca} = \frac{\gamma \Delta T}{\sigma_0}. \qquad (2.24)$$

---

[5] The minus sign has been introduced in order that the definition of the Rayleigh number conforms to the usual convention for thermal convection according to which a positive Rayleigh number corresponds to an unstable density stratification. See also Chandrasekhar (1961).

[6] Sometimes the ratio of viscous force per unit area to the typical capillary pressure, Ca/Re=$(\rho \nu^2)/(\sigma_0 d)$, is denoted as the capillary number.

These numbers measure the relative importance of hydrostatic and hydrody-namic pressure differences to the characteristic capillary pressure σ_0/d. The kinematic boundary condition (2.9) in dimensionless form is

$$u = \frac{1}{\mathrm{Re}}\partial_t\xi + \frac{v}{r}\partial_\varphi\xi + w\partial_z\xi\,. \tag{2.25}$$

Neglecting radiative heat transport henceforth, the dimensionless thermal boundary condition is

$$\boldsymbol{n}\cdot\nabla\theta = -\mathrm{Bi}\left[\theta - \theta_\mathrm{a}(z)\right]\,, \tag{2.26}$$

where $\mathrm{Bi} = hd/k$ is the Biot number.[7] It is a measure of the heat transport through the interface. In an analogous manner, a Biot number for radiative heat transfer can be defined.

A note is in order on the use of the Oberbeck–Boussinesq approximation for flows involving free surfaces. The main assumption in the derivation of the Boussinesq equations (2.14–2.16) is that relative density changes due to thermal expansion are small, $\epsilon = \beta\Delta T \ll 1$, while the Grashof number $\mathrm{Gr} = \epsilon\mathrm{Ga}$ remains nonzero and finite. This implies that the Galilei number $\mathrm{Ga} = gd^3/\nu^2 = \mathcal{O}(\epsilon^{-1}) \gg 1$ becomes very large. With $\mathrm{Bo}/\mathrm{Ca} = (\mathrm{Ga}/\mathrm{Re})\times(\rho - \rho_\mathrm{a})/\rho$ the normal component of the stress equation (2.22) may be written as

$$-p + \boldsymbol{n}\cdot\left[\nabla\boldsymbol{u} + (\nabla\boldsymbol{u})^\mathrm{T}\right]\cdot\boldsymbol{n} + \left(\frac{1}{\mathrm{Ca}} - \theta\right)(\nabla\cdot\boldsymbol{n})$$

$$= \frac{\mathrm{Ga}}{\mathrm{Re}}\frac{\rho - \rho_\mathrm{a}}{\rho}z\,, \quad \text{on} \quad r = \xi(\varphi, z)\,. \tag{2.27}$$

In the limit $\epsilon \to 0$, with finite Reynolds number and $(\rho - \rho_\mathrm{a})/\rho = \mathcal{O}(1)$, the hydrostatic pressure diverges. To prevent mechanical instability the surface tension must then be asymptotically large, i.e. $\mathrm{Ca} \sim \mathrm{Ga}^{-1} = \mathcal{O}(\epsilon) \to 0$. The leading-order terms can only be balanced by the pressure, yielding the Young–Laplace equation

$$p = \frac{\nabla\cdot\boldsymbol{n}}{\mathrm{Ca}} - \frac{\mathrm{Bo}}{\mathrm{Ca}}z\,, \quad \text{on} \quad r = \xi(\varphi, z)\,, \tag{2.28}$$

for the shape $\xi(r, \varphi)$ of the free surface. This is independent of the velocity field. If flow-induced surface deformations are to be taken into account for $\mathrm{Gr} \neq 0$, higher corrections to the Boussinesq equations must be included for consistency (see e.g. Simanovskii & Nepomnyashchy (1993)). In the framework of the Oberbeck–Boussinesq equations we shall, therefore, consider only static surface shapes whenever buoyancy is present.

The tangential stress balance

$$\boldsymbol{t}\cdot\left[\nabla\boldsymbol{u} + (\nabla\boldsymbol{u})^\mathrm{T}\right]\cdot\boldsymbol{n} + \boldsymbol{t}\cdot\nabla\theta = 0\,, \quad \text{on} \quad r = \xi(\varphi, z)\,, \tag{2.29}$$

---

[7] Occasionally the dimensionless number characterizing the heat transfer is called the Nußelt number. The Nußelt number is used here to describe the normalized heat current through the rigid walls (cf. Sect. 7.1).

where t stands for both of the linearly independent orthogonal tangent vectors, is not affected by this restriction.

The above limitation does not apply for $Gr = 0$. Under zero-gravity conditions $Bo = 0$ and the flow field can induce large surface deformations if the capillary number is large. A major simplification for the analysis of the flow, however, is obtained in the limit $Ca \to 0$. In this case the term $Ca^{-1} \nabla \cdot n$ in the normal component of the stress equation (2.22) can only be balanced by the pressure p in the liquid bridge and we obtain the Young–Laplace equation for the case of weightlessness,

$$p = \frac{\nabla \cdot n}{Ca} . \tag{2.30}$$

Thus the pressure in the liquid bridge is given by the capillary pressure. For $\nabla \cdot n \neq 0$ it is asymptotically large compared to the hydrodynamic pressure. Using the unit surface-normal vector

$$n = \frac{1}{N} \left(e_r - \frac{\xi_\varphi}{\xi} e_\varphi - \xi_z e_z \right) \tag{2.31}$$

with the normalizing denominator $N = [1 + \xi_z^2 + (\xi_\varphi/\xi)^2]^{1/2}$, the surface curvature can be expressed as

$$\nabla \cdot n = \frac{-1}{\xi^3 N^3} \left[\xi \xi_{zz} \left(\xi^2 + \xi_\varphi^2 \right) + 2\xi_z \xi_\varphi \left(\xi_z \xi_\varphi - \xi \xi_{z\varphi} \right) \right.$$
$$\left. - \left(1 + \xi_z^2 \right) \left(\xi^2 + 2\xi_\varphi^2 - \xi \xi_{\varphi\varphi} \right) \right] . \tag{2.32}$$

Equation (2.30) is second order in z and φ. The solutions $\xi(\varphi, z)$ depend on the volume of the liquid bridge as well as on the boundary conditions at $z = \pm 1/2$.[8] Here we assume that all contact points where liquid, gas, and solid meet are pinned and form a fixed contact line[9]

$$\xi \left(\varphi, z = \pm \frac{1}{2} \right) = \frac{1}{\Gamma} . \tag{2.33}$$

In the limit $Ca \to 0$ the surface takes a cylindrical shape, if the volume is $V = \pi R^2 d$. Then the boundary conditions read

$$\left. \begin{array}{r} \partial_r w + \partial_z \theta = 0 \\[4pt] r \partial_r \left(\frac{v}{r} \right) + \frac{1}{r} \partial_\varphi \theta = 0 \\[4pt] u = 0 \\[4pt] \partial_r \theta = -Bi \left(\theta - z \right) \end{array} \right\} \quad \text{on} \quad r = \frac{1}{\Gamma} , \tag{2.34}$$

where a linear temperature profile $\theta_a = z$ in the gas phase has been assumed. Equations (2.14–2.16), (2.21), and (2.34) are the basis of the analyses to

---

[8] See Myshkis et al. (1987) for isothermal equilibrium free-surface shapes and their stability.

[9] Experimentally, a fixed contact line can be realized by a solid boundary with a sharp edge and/or an antiwetting coating.

follow. Typical material parameters of metallic melts and model liquids are given in Sect. A.1.

Finally, it may be noted that for small capillary numbers, higher corrections to the lowest-order field quantities $(\boldsymbol{u}, p, \theta, \xi)_0$ may be obtained by a perturbation expansion (see Sect. 4.2.4),

$$(\boldsymbol{u}, p, \theta, \xi) = (\boldsymbol{u}, p, \theta, \xi)_0 + \sum_{n=1}^{\infty} \mathrm{Ca}^n \, (\boldsymbol{u}, p, \theta, \xi)_n \, . \tag{2.35}$$

3. Asymptotic Behavior

Before calculating thermocapillary flows in a fully numerical way, it is useful to consider possible singularities and the scaling behavior of the flow. A fundamental problem is the fluid flow near the contact line, i.e. the flow in a wedge made by an isothermal wall and a plane thermocapillary surface. Since the velocity field must vanish on the rigid wall ($u=0$) one is tempted to assume that the velocity in a small vicinity of the corner remains small. If the derivatives of the velocity field also remain small the flow should be well approximated by the Stokes equation. This can be solved locally, close to the corner. One major difficulty is posed, however, by the discontinuity of the boundary conditions at the contact line.

Another aspect is the appearance of different boundary layers when the Reynolds and Marangoni numbers are large. When convective heat transport becomes important, for instance, the free-surface isotherms are convectively compressed towards the cold contact line. With the help of scaling analyses it is possible to calculate the orders of magnitude of the boundary layer thicknesses and velocity fields.

3.1 Flow Near the Contact Line

Consider, for the moment, the local flow in the vicinity of a cold corner with contact angle α.[1] In the limit $r \to R$, $z \to -d/2$, a Cartesian coordinate system can be used. Corrections are of vanishing order $\mathcal{O}[(R-r)/R]$. The fixed contact point $(r, z) = (R, -d/2)$ in the (r, z) plane is taken as the origin of a local system of cylindrical polar coordinates[2] (r, θ) shown in Fig. 3.1. The cold corner is made by an isothermal plane wall at $\theta = \alpha$ on which $u = 0$, and by a plane interface at $\theta = 0$ along which the temperature, to lowest order, varies linearly from the origin and thus induces a constant tangential surface force.

Since the velocity vanishes at $r = 0$, it must be small in the direct vicinity of the contact line. If we use the extension $a \ll d$ of the region over which

---

[1] The behavior near the hot corner is analogous.

[2] In this section θ denotes the polar angle rather than the temperature and r is the distance from the contact line rather than the distance from the axis of a liquid bridge.

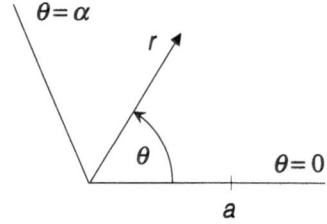

Fig. 3.1. Geometry and local coordinates in the cold corner

the temperature varies linearly as the length scale, the total temperature difference being finite, the Reynolds number goes to zero linearly with a and inertia terms in the Navier–Stokes equation can be neglected. Moreover, thermal convection is insignificant, since $Bd = Gr/Re$ vanishes like a^2. Similarly, the convective temperature transport vanishes. The problem is then reduced to solving the Stokes equation and the heat conduction equation. Although the velocity field is driven by the surface temperature gradient, the temperature field decouples from the velocity field and it depends on the thermal conditions far away from the corner when the Biot number is finite ($Bi < \infty$).

Since creeping flow is unique (Leray (1933)), it must reflect the symmetry of the boundary conditions and therefore be axisymmetric in a liquid bridge. Assuming stationary flow ($\partial_t = 0$) and taking the curl of (2.1), we are left with the biharmonic equation

$$\nabla^4 \psi = \left(\frac{1}{r} \partial_r r \partial_r + \frac{1}{r^2} \partial_\theta^2 \right)^2 \psi(r, \theta) = 0, \tag{3.1}$$

where the stream function ψ has been introduced such that

$$u = \frac{1}{r} \partial_\theta \psi, \tag{3.2}$$

$$v = -\partial_r \psi. \tag{3.3}$$

The linear temperature variation along the free surface leads to a constant shear stress $\eta \omega_0$ and the boundary conditions for the stream function can be written as

$$\psi = \frac{1}{r^2} \partial_\theta^2 \psi - \left\{ \begin{matrix} \omega_0 \,, \text{ if } r \leq a \\ 0 \,, \text{ if } r > a \end{matrix} \right\} = 0 \qquad \text{on} \quad \theta = 0, \tag{3.4}$$

$$\psi = \partial_\theta \psi = 0 \qquad \text{on} \quad \theta = \alpha. \tag{3.5}$$

It is well known that (3.1) admits similarity solutions for $a \to \infty$ of the form

$$\psi = r^\lambda f_\lambda(\theta), \quad \lambda \in \mathbb{C}. \tag{3.6}$$

For homogeneous boundary conditions λ is complex and determined by conditions far away from the corner (Moffatt (1964a)).[3] For the present inhomo-

---

[3] Hasimoto & Sano (1980) have given an overview of vortices in creeping flow. For a recent treatment of the flow in a corner between a rigid wall and a stress-

geneous problems λ is positive and must be compatible with the boundary conditions, i.e. $\lambda=2$. Therefore, the similarity solution of (3.1) takes the form

$$\psi = \omega_0 r^2 f(\theta),$$ (3.7)

where f must satisfy the boundary conditions

$$f(0) = f''(0) - 1 = f(\alpha) = f'(\alpha) = 0.$$ (3.8)

The differential equation for f,

$$(\partial_\theta^2 + 4)\, \partial_\theta^2 f(\theta) = 0,$$ (3.9)

leads to the solution

$$f(\theta) = A \cos 2\theta + B \sin 2\theta + C\theta + D,$$ (3.10)

with the four integration constants

$$A = -\frac{1}{4}, \qquad D = \frac{1}{4},$$ (3.11)

$$B = \frac{1}{4} \frac{\cos 2\alpha + 2\alpha \sin 2\alpha - 1}{\sin 2\alpha - 2\alpha \cos 2\alpha},$$ (3.12)

$$C = \frac{1}{2} \frac{\cos 2\alpha - 1}{\sin 2\alpha - 2\alpha \cos 2\alpha},$$ (3.13)

determined by the boundary conditions (3.8). This solution has also been given by Canright (1994). For the important case $\alpha=\pi/2$ the solution is

$$\psi(r, \theta, \alpha=\pi/2) = \frac{\omega_0}{4} r^2 \left[1 - \cos 2\theta - \frac{2}{\pi} (\sin 2\theta + 2\theta)\right].$$ (3.14)

Typical streamline patterns for $\alpha=\pi/4$ and $\pi/2$ are shown in Fig. 3.2. The vorticity is

$$\omega = \nabla \times \boldsymbol{u} = -\nabla^2 \psi \, \boldsymbol{e} = -\omega_0 \, (4f + f'') \, \boldsymbol{e} = -4\omega_0 \, (C\theta + D) \, \boldsymbol{e},$$ (3.15)

where $\boldsymbol{e}$ is the unit vector parallel to the contact line. The surface on which the vorticity vanishes is given by the angle

$$\theta_0 = \theta(\omega=0) = -\frac{D}{C} = \frac{1}{2} \frac{\sin 2\alpha - 2\alpha \cos 2\alpha}{1 - \cos 2\alpha}.$$ (3.16)

This surface is shown as dashed lines in Fig. 3.2.

Since the vorticity depends on θ only, it takes different values if the origin is approached along different paths, e.g. along $r \to 0$ with θ fixed. Therefore, the vorticity is always singular at the origin $r=0$. The same applies to other derivatives of the velocity field. As can be seen from (3.10–3.13), the solution becomes singular everywhere if α is a root of the equation

---

free surface and of the flow of two viscous fluids in a rigid wedge, see Anderson & Davis (1993). Slow flow near tri-junctions with two rigid surfaces and one stress-free surface, with application to crystal growth, including heat transfer and solidification, has been treated by Anderson & Davis (1994).

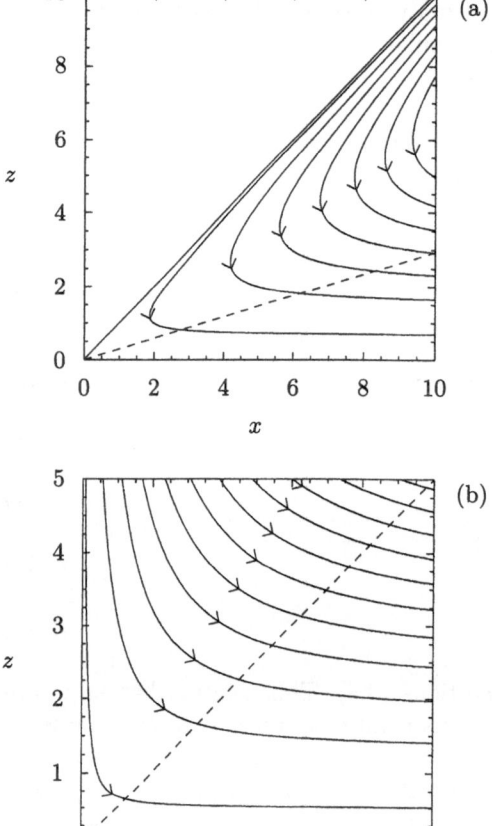

Fig. 3.2. Streamlines in a cold corner for $\alpha = \pi/4$ (**a**) and $\pi/2$ (**b**). In this figure the cold rigid wall coincides with the x axis. On the *dashed lines*, $\omega = 0$

$$\sin 2\alpha - 2\alpha \cos 2\alpha = 0. \tag{3.17}$$

The only nontrivial root in the interval $[0, \pi]$ is $\alpha_c = 0.715\pi = 2.2467 = 128.7°$. At this contact angle the surface $\theta = \theta_0$ (3.16) on which the vorticity vanishes coincides with the free surface. The boundary conditions, however, impose a constant nonzero vorticity $\omega = -\omega_0$ on the free surface. Therefore, the vorticity field must exhibit strong gradients all along the free surface when $\alpha \approx \alpha_c$. As a result high velocity gradients arise and the streamlines become asymptotically dense for $\alpha \to \alpha_c$. For $\alpha < \alpha_c$ and $\omega_0 > 0$ the flow on the free surface is directed towards the cold corner ($u(\theta = 0) < 0$), while the similarity solution predicts a flow away from it ($u(\theta = 0) > 0$) for $\alpha > \alpha_c$.[4]

---

[4] This counterintuitive prediction of the Stokes flow solution motivated Shevtsova et al. (1996) to numerically investigate the full nonlinear Navier–Stokes flow near the corner in a noncylindrical liquid bridge in the limit Ca $\to 0$. By successive

For a resolution of the problem associated with the critical angle, note that the similarity solution (3.7) can be valid only locally for any finite-length-scale problem, such as the flow in a liquid bridge. It breaks down for $\alpha \geq \alpha_c$ and we must search for a local solution of more general form. This can be done using a Mellin transformation for a modified corner problem in which the driving shear stress is constant and nonzero within a restricted distance a from the origin. By an analysis in the manner of Tranter (1948), Moffatt (1964b), and Moffatt & Duffy (1980), it can be shown (see Sect. A.2) that the leading-order contribution to the flow field for $\alpha > \alpha_c$ is of a nonsimilarity type and that inertia terms cannot be neglected in any neighborhood of $r = 0$ when $a \to \infty$. The asymptotic form of the leading-order contribution to the stream function for $r \to 0$ is

$$\psi(r,\theta) \sim \begin{cases} \omega_0 r^2 f(\theta), & \alpha < \alpha_c, \\ \dfrac{\omega_0}{2} \dfrac{\partial_p F(\theta,0)}{W''(0)} r^2 \ln \dfrac{r}{a}, & \alpha = \alpha_c, \\ \dfrac{\omega_0}{4} r^2 \left(\dfrac{r}{a}\right)^{p_1} \dfrac{F(\theta,p_1)}{W'(p_1)(p_1+1)p_1}, & \alpha > \alpha_c, \end{cases} \qquad (3.18)$$

where $0 > p_1 \geq -1/2$ for $\alpha_c < \alpha \leq \pi$ and $p_1, F, \partial_p F, W', W''$ are given in Sect. A.2.

The local similarity solution (3.7) for $\alpha < \alpha_c$ is scale-independent. In order that the Stokes flow solution is a valid approximation to the solution of the Navier–Stokes equation, inertia terms must be small compared to viscous ones. With the flow velocities being of order $\mathcal{O}(\omega_0 r)$ this requires

$$|\boldsymbol{u} \cdot \nabla \boldsymbol{u}| = \mathcal{O}(\omega_0^2 r) \ll \mathcal{O}\left(\nu \frac{\omega_0}{r}\right) = \nu|\nabla^2 \boldsymbol{u}|, \qquad (3.19)$$

from which follows

$$r^2 \ll \frac{\nu}{\omega_0}. \qquad (3.20)$$

For any given finite, nonzero ν and ω_0 there always exists a radial range for which this inequality is satisfied. Equation (3.20) may also be expressed as

$$r \ll \mathrm{Re}^{-\frac{1}{2}}, \qquad (3.21)$$

where we have introduced the Reynolds number $\mathrm{Re} = \omega_0 a^2 / \nu$ and the length scale a. The radial range within which inertia effects are negligible grows as the Reynolds number tends to zero. On the other hand, the spatial range is limited by the requirement of a locally Cartesian geometry ($a \ll d$) and a linear variation of the surface temperature. Since the stream function diverges like $\psi = \mathcal{O}(|\alpha - \alpha_c|^{-1}) \to \infty$, the radius of validity of (3.7) shrinks to zero as $\alpha \to \alpha_c$.

For $\alpha > \alpha_c$ the velocities scale as $a^{-p_1} r^{1+p_1}$, i.e. they increase with infinite slope from the contact point, and we get the balance

---

subdivisions of the integration area for $\Gamma = 1$ and $\alpha = 135° > \alpha_c$, they did not find any flow reversal in the cold or hot corner on a scale down to $10^{-4}d$.

$$|u \cdot \nabla u| = \mathcal{O}(a^{-2p_1} r^{1+2p_1}) \ll \mathcal{O}\left(\nu a^{-p_1} r^{-1+p_1}\right) = \nu |\nabla^2 u|, \qquad (3.22)$$

or

$$r \ll \nu^{1/(2+p_1)} a^{p_1/(2+p_1)}. \qquad (3.23)$$

Since $-1/2 \leq p_1 < 0$, the exponent in $a^{p_1/(2+p_1)}$ is negative and the radial range of validity of the creeping-flow solution shrinks to zero as the range a over which the free-surface shear stress is nonzero tends to infinity. Thus the corner flow for $\alpha > \alpha_c$ with $a \to \infty$ cannot be approximated by the Stokes equation. For $\alpha = \alpha_c$ the requirement is $r^2 \ln(r/a) \ll \nu$ and the Stokes approximation likewise fails for any nonzero r when $a \to \infty$.

It should be noted that the radial distance within which the Stokes flow solution (3.7) is a valid approximation is also restricted for $r \to 0$. Integrating the Stokes equations, it is seen that the pressure does not depend on θ and diverges logarithmically as the corner is approached:

$$p(r) = \eta \omega_0 \left(4f' + f'''\right) \ln\left(\frac{r}{a}\right) + p(a) = 4\eta \omega_0 C \ln\left(\frac{r}{a}\right) + p(a). \qquad (3.24)$$

This local solution holds in the limit $Ca \to 0$. For any real fluid, however, the capillary number is finite. The flow-induced large pressure close to the contact line must then be balanced by the Laplace pressure. This requires the curvature of the interface to become very large for $r \to 0$. Hence, the assumption of a statically determined meniscus close to the contact line loses its validity. For a related problem, the extrusion flow from nozzles, Salamon et al. (1995) found a highly curved meniscus close to the contact line (see also Salamon et al. (1997)). Their finding supports the hypothesis of Schultz & Gervasio (1990) that the diverging normal stress in (2.22) is balanced by the term $Ca^{-1} \nabla \cdot n$, indicating a divergence of the curvature of the free surface.[5]

The restrictions of the range of validity of the Stokes flow solution indicate the type of problem that may arise in a numerical treatment in the framework of the Navier–Stokes equations. For a contact angle $\alpha = \pi/2$ the singularity is restricted to a single point. Ignoring the problem of the diverging normal stress, several numerical calculations have shown that the errors in the velocity fields associated with this kind of singularity are restricted to the very vicinity of the corners.[6] Therefore, the flow in the bulk can be reliably calculated. As can be seen from (2.30) and (3.24), the range of large surface curvature is exponentially small for asymptotically small capillary numbers. This is an important empirical result, on which most of the numerical calculations are based, particularly those for flows in liquid bridges

---

[5] The problem of a moving contact line is singular too (see Dussan V. (1979), Dussan V. et al. (1991)), Chen et al. (1995b), and Shikhmurzaev (1997)).

[6] See also Koplik & Banavar (1995) for the singularity in the lid-driven cavity problem.

and rectangular containers with $\alpha = \pi/2$. The real structure of the thermo-capillary flow close to the singular corners in the presence of a dynamically deformable surface has not yet been investigated.

3.2 Boundary-Layer Scalings

Qualitative flow properties can often be obtained by order-of-magnitude estimates leading to scaling laws for the field quantities and the boundary layers. Here we consider the plane steady flow near rectangular corners ($\alpha = \pi/2$) of semi-infinite domains, cavities, or liquid bridges for $Ca = Gr = Bi = 0$. To carry out an order-of-magnitude analysis the velocity and temperature distributions must be known on the boundaries. On the free surface, however, they are not known beforehand but, rather are part of the solution. For this reason we must resort to some a priori assumptions about the magnitude of the boundary values. These have to be justified later, e.g. by numerical simulations.

Since the flow is driven by temperature-induced surface tension gradients, the regions of high surface temperature variation are of prime importance for the flow. Numerical calculations (see Sects. 7.1 and 8.1, Canright (1994), and Zebib et al. (1985)) have revealed the following behavior.

For low-Prandtl-number creeping flow, the temperature gradient is $\mathcal{O}(1)$ everywhere on the free surface. As the flow becomes stronger and the Prandtl number remains sufficiently small, viscous boundary layers develop on the free surface and the rigid walls. When convective effects become important the isotherms on the free surface become compressed towards the cold boundary and the driving forces become more localized there.

For high-Prandtl-number fluids thermal boundary layers develop on the hot wall owing to the cold internal return flow. The resulting large temperature variations extend to the free surface and provide the major driving force. As a result fluid is sucked from the bulk and accelerated along the free surface away from the hot corner. Owing to the suction effect the thermal-wall boundary layer also remains thin close to the free surface. Except for both corner regions, where there are strong thermal gradients, the temperature attains a nearly constant intermediate value over the remainder of the free surface.

Scaling laws can be derived by considering asymptotic limits of the governing parameters. Only two of the control parameters Re, Pr, and Ma are independent. If we use Re and Ma, four different limits exist depending on whether the velocity field is viscous ($Re \to 0$) or inertial ($Re \to \infty$) and whether the temperature field is conductive ($Ma \to 0$) or convective ($Ma \to \infty$). Depending on the particular limit, one of the two corner regions will dominate in driving the flow, while the contribution of the other corner is asymptotically small. It is then sufficient to consider only a single corner with suitable assumptions about the bulk flow.

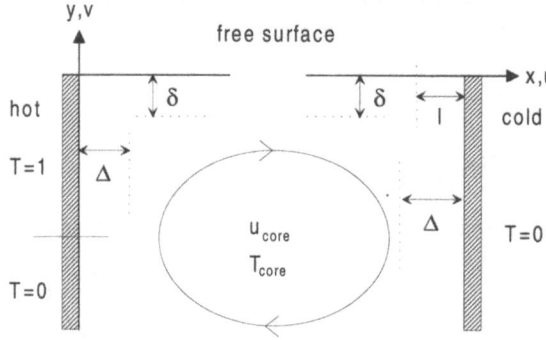

Fig. 3.3. Structure and notation used for the discussion of the boundary layer scalings

Consider the thermocapillary flow in rectangular corners depicted in Fig. 3.3. Using the nondimensionalization of Table 2.1, the governing steady-state equations for zero gravity read

$$\text{Re } \boldsymbol{u} \cdot \nabla \boldsymbol{u} = -\nabla p + \nabla^2 \boldsymbol{u}, \tag{3.25}$$

$$\text{Ma } \boldsymbol{u} \cdot \nabla \theta = \nabla^2 \theta, \tag{3.26}$$

$$\nabla \cdot \boldsymbol{u} = 0. \tag{3.27}$$

The stress boundary condition at $y=0$ is

$$\partial_y u + \partial_x \theta = 0. \tag{3.28}$$

Since the inverses of the Reynolds and Marangoni numbers appear in front of the highest derivatives (∇^2), the limit of large parameters is singular and a boundary-layer character is expected. The diffusive terms are thus important only on decreasingly small scales close to the boundaries. Within an order-of-magnitude analysis the derivatives in (3.25–3.28) are replaced by quotients of the respective field quantity and the length scale on which it varies.[7] When both Re and Ma are small, boundary layers are absent.

3.2.1 Low-Prandtl-Number Inertial Flow

Free Surface Layer. In the limit Ma $\to 0$ thermal boundary layers are absent and the temperature on the free surface varies on an $\mathcal{O}(1)$ length scale. If we denote the free-surface shear-layer thickness by δ, the thermocapillary stress condition (3.28) yields $\partial_y u = \mathcal{O}(1)$, which implies $u \sim \delta$. From the continuity equation we obtain (see Schlichting (1968))

$$\partial_x u + \partial_y v \sim \frac{u}{1} + \frac{v}{\delta} \sim 0 \quad \Rightarrow \quad v \sim \delta u. \tag{3.29}$$

The primary momentum balance parallel to the interface (x direction) is determined by

---

[7] For an order-of-magnitude analysis we can disregard the pressure term, since it is equivalent to a quadratic nonlinearity $(\nabla \cdot (3.25))$. Alternatively, it may be eliminated by taking the curl of (3.25).

$$\text{Re}(\underbrace{u\partial_x u + v\partial_y u}_{\mathcal{O}(\delta^2)}) \sim \frac{1}{\delta^2}u\,. \tag{3.30}$$

With $u \sim \delta$ we obtain for the thickness of the free-surface shear layer

$$\delta \sim \text{Re}^{-1/3}\,, \tag{3.31}$$

and the velocity scale

$$u \sim \text{Re}^{-1/3}\,. \tag{3.32}$$

This is the most important scaling law (compare with (1.3) and (1.4)). It has been derived by Ostrach (1977) and Napolitano (1979).[8]

Rigid-Wall Layers in Thermocapillary Scaling. On the rigid hot and cold walls we expect conventional viscous boundary layers. In the present scaling the free-stream velocity is $v \sim \text{Re}^{-1/3}$. With the thickness of the viscous boundary layer at the wall denoted by Δ, the continuity equation (3.27) requires

$$\frac{u}{\Delta} + \frac{\text{Re}^{-1/3}}{1} \sim 0\,, \tag{3.33}$$

and hence $u \sim \Delta\text{Re}^{-1/3}$. The streamwise momentum balance (y direction) yields

$$\text{Re}(\underbrace{u\frac{v}{\Delta} + v\partial_y v}_{\sim \text{Re}^{-2/3}}) \sim \frac{1}{\Delta^2}v\,, \tag{3.34}$$

and we get the viscous boundary-layer thickness

$$\Delta \sim \text{Re}^{-1/3}\,, \tag{3.35}$$

together with the velocity scale (Table 2.1)

$$u \sim \text{Re}^{-2/3}\,. \tag{3.36}$$

Thus the viscous boundary-layer thickness on the rigid walls scales like the free-surface boundary-layer thickness $\sim \text{Re}^{-1/3}$. The relative cross-stream velocity $u/v \sim \text{Re}^{-1/3}$ scales like the boundary-layer thickness, as must be the case for viscous boundary layers at a wall.

3.2.2 Boundary Layers Near the Cold Corner

Now we investigate more general flows near the cold corner. There is numerical evidence that the thermal gradient in the immediate vicinity of the cold corner is characterized by a single length scale l and the temperature change over this length is of the order $\mathcal{O}(1)$ in units of the applied temperature difference ΔT. We use the length ld ($l < 1$ is dimensionless) and the corresponding

---

[8] See also Batishchev et al. (1989) and Zebib et al. (1985).

temperature drop to define the Marangoni number $\mathrm{Ma} = (\gamma \Delta T l d)/(\rho \kappa \nu)$ in this subsection. For an insulating free surface the temperature field varies only weakly in the y direction perpendicular to the free surface and we set $\partial \theta / \partial y = \mathcal{O}(1)$. Under these assumptions Canright (1994) derived several scaling limits for $\{\mathrm{Re}, \mathrm{Ma}\} \to \{0, \infty\}$.

Since all driving is located within a distance l from the corner, the shear stress condition (3.28) on the free surface yields (cf. Fig. 3.3)

$$0 = \partial_y u + \partial_x \theta \sim \frac{u}{\delta} + \frac{(\theta \sim 1)}{l} \quad \Rightarrow \quad u \sim \frac{\delta}{l}. \tag{3.37}$$

From the continuity equation (3.27) we get

$$0 = \partial_x u + \partial_y v \sim \frac{u}{l} + \frac{v}{\delta} \quad \Rightarrow \quad v \sim u\frac{\delta}{l}. \tag{3.38}$$

With the substitutions $u \sim \delta/l$, $v \sim \delta^2/l^2$, $\partial_x \sim 1/l$, $\partial_y u \sim u/\delta$, and $\partial_y \theta \sim 1$, the balances (3.25) and (3.26) take the form

$$\mathrm{Re}\,\frac{\delta}{l^2} \sim \frac{1}{\delta^2} + \frac{1}{l^2}, \tag{3.39}$$

$$\mathrm{Ma}\,\frac{\delta}{l^2}\,(1 + \delta) \sim 1 + \frac{1}{l^2}. \tag{3.40}$$

These equations are now considered in the four limits of Re and Ma.

Ma $\to$ 0, Re $\to$ 0: Conductive–Viscous Limit. In this case the left-hand sides of (3.39) and (3.40) vanish. One obtains $l \sim \delta = \mathcal{O}(1)$. From (3.37) it follows that $u = \mathcal{O}(1)$. Therefore, the relevant scales for the conductive–viscous limit are just given by the scales of Table 2.1.

Ma $\to$ 0, Re $\to \infty$: Conductive–Inertial Limit. This limit has been considered in Sect. 3.2.1, with the result $\delta \sim u \sim \mathrm{Re}^{-1/3}$. If the δ scale is inserted into (3.40) one obtains ($l = \mathcal{O}(1)$)

$$\mathrm{Ma}\,\mathrm{Re}^{-1/3}\left(1 + \mathrm{Re}^{-1/3}\right) \ll 1. \tag{3.41}$$

Since $\mathrm{Re}^{-1/3} \to 0$, a less restrictive condition on Ma follows. Instead of $\mathrm{Ma} \to 0$ it must only be guaranteed that $\mathrm{Ma} \ll \mathrm{Re}^{1/3}$ in order to obtain the conductive–inertial limit.

Ma $\to \infty$, Re $\to$ 0: Convective–Viscous Limit. For $\mathrm{Re} \to 0$ and from (3.39) we obtain $\delta \sim l$. Equation (3.37) yields $u = \mathcal{O}(1)$. From (3.40) and with $\mathrm{Ma} \to \infty$ the length scale in the x direction must be small, $l \ll 1$. This leads to $\delta \sim l \sim \mathrm{Ma}^{-1}$. As before, the limit for Re may be relaxed. This is seen by inserting the scales into (3.39):

$$\mathrm{Re}\,\frac{1}{l} \ll \frac{1}{\delta^2} + \frac{1}{l^2} \sim \frac{1}{l^2}, \tag{3.42}$$

which gives $\mathrm{Re} \ll \mathrm{Ma}$.

Ma → ∞, Re → ∞: Convective–Inertial Limit. Since $Ma \to \infty$, the boundary-layer thickness l will be small, $l \ll 1$. Equation (3.40) together with (3.37) yields

$$Ma\,\delta\,(1+\delta) \sim 1. \tag{3.43}$$

Therefore, $\delta \sim Ma^{-1}$. Using this result and (3.39) we obtain ($Pr = Ma/Re$)

$$\frac{1}{l^2 Pr} \sim Ma^2 + \frac{1}{l^2} \quad \Rightarrow \quad l^2 \sim \frac{Pr^{-1} - 1}{Ma^2}. \tag{3.44}$$

Now one has to distinguish between large and small Prandtl numbers. For $Pr \ll 1$ we obtain $l \sim Ma^{-1}Pr^{-1/2}$. In order that the condition $l \ll 1$ remains valid, the Marangoni number must satisfy $Ma \gg Pr^{-1/2}$. If, on the other hand, $Pr \gg 1$, one obtains from (3.44) $l \sim Ma^{-1}$. This scale is identical with that for the convective–viscous case above. All resulting scales are summarized in Table 3.1.

Table 3.1. Scaling of the boundary layers in the vicinity of a thermocapillary cold corner with contact angle $\alpha = \pi/2$; after Canright (1994)

Type	Limit	l	δ	u	r_0
Conductive–viscous	$Ma \to 0$ $Re \to 0$	1	1	1	1
Conductive–inertial	$Ma \ll Re^{1/3}$ $Re \to \infty$	1	$Re^{-1/3}$	$Re^{-1/3}$	$Re^{-1/2}$
Convective–viscous	$Ma \to \infty$ $Re \ll Ma$	Ma^{-1}	Ma^{-1}	1	Ma^{-1}
Convective–inertial	$Ma^3 \gg Re$ $Re \to \infty$	$Ma^{-1}Pr^{-1/2}$	Ma^{-1}	$Pr^{1/2}$	$Ma^{-1}Pr^{1/4}$

Independently of the limits considered, the velocities tend to zero as the cold corner is approached. For small distances r from the cold corner the creeping-flow regime is entered. The radial distance r_0 within which the flow is creeping is determined by conditions far away from the vertex. For the conductive–viscous case the usual restriction applies, i.e. $r \ll 1$. For convection-dominated flows the Stokes flow range is reduced to $r \ll l$, since the surface temperature varies linearly only within a distance l from the corner. For inertial flows the local inertia terms must be small compared to the viscous ones. Since the creeping flow (3.14) scales with r/l (the driving shear forces are $\sim l^{-1}$) we obtain the condition (compare with (3.21))

$$Re\,|\boldsymbol{u} \cdot \nabla \boldsymbol{u}| \sim \frac{rRe}{l^2} \ll \frac{1}{rl} \sim |\nabla^2 \boldsymbol{u}|, \tag{3.45}$$

from which we get

$$r^2 \ll \frac{l}{\text{Re}}. \tag{3.46}$$

The extension of the Stokes domain shrinks like $\sim l^{1/2}$ when l becomes small.

Canright (1994) validated these scales by numerical calculations of the flow for a semi-infinite model problem with a cold corner. Numerical calculations of thermocapillary flows in *finite-size* containers (Sect. 8.1, Zebib et al. (1985)) have shown that the temperature gradient on the free surface varies considerably within a small distance from the cold and the hot walls. Nevertheless, the temperature gradient does not completely vanish over the rest of the free surface. According to Canright (1994) this behavior could be related to a feedback of the flow over the finite volume, and this may be the reason why the conductive–inertial scaling is observed in finite volumes instead of the expected convective–inertial scaling, even for very high Prandtl numbers. In fact, for $\text{Re} = 10^4$ and $\text{Ma} = 10^5$, Zebib et al. (1985) found $u \sim \text{Re}^{-1/3}$. It is still an open question whether the convective–inertial scaling can be reached in finite-size systems.

3.2.3 High-Prandtl-Number Thermocapillary Flow in a Hot Corner (Viscous–Convective Limit)

Next we consider the thermocapillary flow in a rectangular hot-corner region. The analysis is due to Cowley & Davis (1983). In their model problem the hot wall is kept at a constant high temperature within a finite distance from the free surface, and this distance determines the length scale. The thermal scale is given by the temperature difference between the heated wall section and the remainder of the wall, which is at the fluid's core temperature (see Fig. 3.3). The boundary conditions on the hot wall $x=0$ are

$$\boldsymbol{u} = 0, \qquad T = \begin{cases} 1, & y > -1, \\ 0, & y < -1. \end{cases} \tag{3.47}$$

For a finite cavity or a liquid bridge the length scale is fixed by the geometry, and the temperature far away from the hot wall (core temperature) may be taken as the cold-wall temperature. We assume that the Reynolds number is $\mathcal{O}(1)$, while the Marangoni number is large, $\text{Ma} \to \infty$. Then only thermal boundary layers are present. The thicknesses of the free-surface and hot-wall boundary layers are denoted by δ and Δ, respectively. For scaling purposes the exponents b, f, and d are defined according to

$$|\boldsymbol{u}_{\text{core}}| = \mathcal{O}(\text{Ma}^{-b}), \quad \Delta = \mathcal{O}(\text{Ma}^{-f}), \quad \delta = \mathcal{O}(\text{Ma}^{-d}). \tag{3.48}$$

Thermal Boundary Layer on the Rigid Wall. Since $v_{\text{core}} \sim \text{Ma}^{-b}$ and there are no viscous boundary layers, the velocity v parallel to the hot wall increases linearly from it:

$$v \sim \Delta \text{Ma}^{-b} = \text{Ma}^{-b-f}. \tag{3.49}$$

The scaling of u in the thermal boundary layer is obtained from the continuity equation:

$$0 = \partial_x u + \partial_y v \sim \frac{u - 0}{\Delta} + \frac{\Delta \mathrm{Ma}^{-b}}{1} \quad \Rightarrow \quad u \sim \mathrm{Ma}^{-b-2f} . \tag{3.50}$$

Free-Surface Layer. Owing to the absence of a no-slip boundary condition on the free surface we have $u \sim u_{\mathrm{core}} \sim \mathrm{Ma}^{-b}$ far away from the hot wall. Moreover, there is no viscous boundary layer, and $\partial_y u \sim \mathrm{Ma}^{-b}$. From continuity we get

$$0 = \partial_x u + \partial_y v \sim \frac{\mathrm{Ma}^{-b}}{1} + \frac{0 - v}{\delta} \quad \Rightarrow \quad v \sim \mathrm{Ma}^{-b-d} . \tag{3.51}$$

Heat Transport From the Hot Wall. The balance (3.26) between convective and diffusive heat transport in the hot-wall boundary layer is given by

$$\mathrm{Ma}\, u\, \partial_x \sim \partial_x^2 \quad \Rightarrow \quad \mathrm{Ma}^{1-b-f} \sim \mathrm{Ma}^{2f} , \tag{3.52}$$

yielding the first equation,

$$1 - b - 3f = 0 . \tag{3.53}$$

Heat Transport Through the Free-Surface Layer. The thermal balance in the free-surface layer, with (3.51), leads to

$$\mathrm{Ma}\, v\, \partial_y \sim \partial_y^2 \quad \Rightarrow \quad \mathrm{Ma}^{1-b} \sim \mathrm{Ma}^{2d} , \tag{3.54}$$

and provides the second condition,

$$1 - b - 2d = 0 . \tag{3.55}$$

Thermocapillary Boundary Condition. Using the core velocity $u \sim \mathrm{Ma}^{-b}$ and assuming that the temperature of the free surface far away from the hot wall scales like $T \sim \mathrm{Ma}^{-a}$, we obtain from the free-surface boundary condition

$$0 = \partial_y u + \partial_x T \sim \mathrm{Ma}^{-b} + \frac{Ma^{-a}}{1} \quad \Rightarrow \quad a = b . \tag{3.56}$$

Global Heat Conservation. The total heat transport in the x direction is given by

$$\mathrm{Nu}(x) = \int_{-\infty}^{0} j_x \, dy = \int_{-\infty}^{0} (\mathrm{Ma}\, u\, T - \partial_x T) \, dy , \tag{3.57}$$

where j_x denotes the heat current density in x direction. On the hot wall ($u = 0$) we only have diffusive transport,

$$\mathrm{Nu}(0) = - \int_{-\infty}^{0} \partial_x T \, dy \sim \frac{1}{\Delta} . \tag{3.58}$$

Since the free surface is insulating, the Nußelt number Nu is independent of x and Nu(0) must equal the lateral heat flow far away from the wall,

where all heat is transported convectively in the thermal boundary layer and
streamwise diffusion is negligible:

$$\mathrm{Nu}(0) = \mathrm{Nu}(x \to \infty) = \int_{-\delta}^{0} \mathrm{Ma}\, u\, T \, dy \sim \mathrm{Ma}^1 \mathrm{Ma}^{-b} \mathrm{Ma}^{-a} \delta\,. \qquad (3.59)$$

From the equality of the two fluxes we obtain the third condition,

$$\mathrm{Nu} \sim \frac{1}{\Delta} \sim \mathrm{Ma}^{1-a-b-d} \quad \Rightarrow \quad 1 - a - b - d - f = 0\,. \qquad (3.60)$$

Exponents. Solving the three equation (3.53), (3.55), and (3.60) yields the
solution

$$a = b = \frac{1}{7}, \quad f = \frac{2}{7}, \quad d = \frac{3}{7}, \qquad (3.61)$$

and we obtain the viscous convective hot-corner scaling

$$u_{\mathrm{core}} \sim \mathrm{Ma}^{-1/7}, \qquad T_{\mathrm{surf}} \sim \mathrm{Ma}^{-1/7},$$
$$\Delta \sim \mathrm{Ma}^{-2/7}, \qquad\qquad \delta \sim \mathrm{Ma}^{-3/7}, \qquad \mathrm{Nu} \sim \mathrm{Ma}^{2/7}\,. \qquad (3.62)$$

The scaling $u_{\mathrm{core}} \sim \mathrm{Ma}^{-1/7}$ has been confirmed numerically by Kamotani
& Ostrach (1998) for several high-Prandtl-number flows in finite-size liquid
bridges even though the flow was not entirely viscous (Fig. 3.4). This behavior suggests that for high Prandtl numbers and Reynolds numbers not
too large, the hot corner determines the scaling of the surface temperature,
bulk velocity, and Nußelt number. The reason probably is that the driving

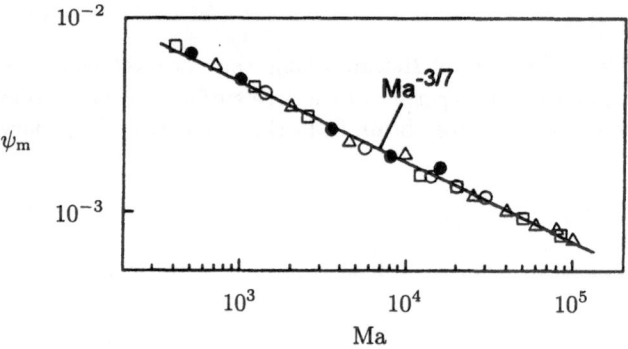

Fig. 3.4. Scaling $\psi_{\mathrm{m}} \sim u_{\mathrm{core}} \Delta \sim \mathrm{Ma}^{-3/7}$ of the maximum absolute value of the
stream function $\psi_{\mathrm{m}} = \max |\psi(r, z)|$ of the steady flow in a thermocapillary liquid
bridge with $\Gamma = 1.4$ (cf. Chap. 7). (●) $\mathrm{Pr} = 10$; (○) $\mathrm{Pr} = 20$; (□) $\mathrm{Pr} = 50$; (△)
$\mathrm{Pr} = 100$; after Kamotani & Ostrach (1998)

near the cold corner arises within a region $\mathcal{O}(\mathrm{Ma}^{-1})$, which tends to zero as Ma increases. Since the fluid is accelerated towards the wall, the locally high velocity decays rapidly away from the corner. On the other hand, the fluid near the hot corner is accelerated away from the rigid wall within a region of $\mathcal{O}(\mathrm{Ma}^{-2/7})$, which is much larger.[9]

---

[9] This argument requires that the extended region of constant temperature between the heated walls remains at a temperature sufficiently different from the two wall temperatures.

4. Creeping Flow
in Thermocapillary Liquid Bridges

The full nonlinear problem (2.14–2.16) for finite-size systems must be solved numerically, even for zero gravity (Bo = 0) and for large surface tension (Ca → 0). If, however, thermocapillary forces are sufficiently weak and the flow is slow, nonlinear effects may be neglected throughout the whole volume V. For stationary conditions the resulting creeping flow is then obtained as the solution of a biharmonic equation with inhomogeneous boundary conditions. It can be given analytically in form of an infinite series.

The biharmonic equation can be solved using the method of biorthogonal functions. The technique is based on the work of Smith (1952), who considered the bending of a semi-infinite strip clamped at its ends. He solved the biharmonic equation by means of an infinite series. Joseph (1977) and Joseph & Sturges (1978) generalized the method and proved the convergence of the series solution for a whole class of boundary conditions. They demonstrated that the method of biorthogonal functions can be used very efficiently to calculate creeping flows in finite-size containers that are driven by moving walls (Joseph & Sturges (1978)) or by buoyancy forces (Joseph & Sturges (1975)). In the following this method is applied to calculate the creeping flow in a liquid bridge driven by thermocapillary surface forces.

A closed-form analytical solution for the flow in a thermocapillary liquid bridge in the limit Re → 0 is not known. In this limit the momentum balance (2.14) is viscous-dominated and the convective terms can be neglected. A further simplification is obtained in the limit of zero Prandtl number Pr → 0. Then the temperature field (2.16) is purely diffusive and decouples from the velocity field. The latter assumption is not too restrictive for most applications in crystal growth, since the Prandtl numbers of liquid metals are usually small $(Pr = \mathcal{O}(10^{-2}))$. To avoid difficulties associated with a deformation of the free surface we assume, in addition, Bo=0 and the limit Ca → 0. Higher corrections to the temperature field (Pr $\neq$ 0) and to the shape of the interface (Ca $\neq$ 0) are given in Sects. 4.2.3 and 4.2.4.

In the next section the method of biorthogonal functions is formally developed. It is then applied to thermocapillary convection in liquid bridges in Sect. 4.2 (Kuhlmann (1989)).

4.1 Method of Biorthogonal Functions

In the limit $(\mathrm{Re}, \mathrm{Pr}, \mathrm{Ca}) \to 0$ the temperature field is given by the conducting profile $\theta = z$. The stationary flow is described by the two-dimensional biharmonic equation for the stream function. In cylindrical coordinates, this equation reads

$$\left(DD_* + \partial_z^2\right)^2 \psi(r, z) = 0,\tag{4.1}$$

where the stream function[1] ψ is defined by

$$u = \partial_z \psi,\tag{4.2}$$
$$w = -D_* \psi,\tag{4.3}$$

with $D_* = D + 1/r = \partial_r + 1/r$. In the above limit and for a given volume $V = \pi R^2 d$ the fluid occupies a cylindrical domain. The symmetry conditions at $r=0$ for axisymmetric flows and the boundary conditions (2.21) and (2.34) read

$$\begin{aligned}
\psi = DD_*\psi = 0, &\qquad \text{on } r = 0, \\
\psi = DD_*\psi - 1 = 0, &\qquad \text{on } r = 1/\Gamma, \\
\psi = \partial_z \psi = 0, &\qquad \text{on } z = \pm 1/2.
\end{aligned}\tag{4.4}$$

Seeking a separable solution of (4.1) of the form

$$\psi(r, z) = T(r)\, Y(z),\tag{4.5}$$

and inserting this ansatz, we get

$$\left(\frac{1}{T}DD_*T\right)\left(\frac{1}{T}DD_*T\right) + 2\left(\frac{1}{T}DD_*T\right)\left(\frac{1}{Y}\partial_z^2 Y\right)$$
$$+ \left(\frac{1}{Y}\partial_z^2 Y\right)\left(\frac{1}{Y}\partial_z^2 Y\right) = 0.\tag{4.6}$$

The factors appearing depend either on r or on z and can thus be written as

$$\frac{1}{T}DD_*T = \alpha f(r) + \beta,\tag{4.7}$$

$$\frac{1}{Y}\partial_z^2 Y = \gamma g(z) + \delta.\tag{4.8}$$

Since the mixed term in (4.6) cannot depend on both r and z, we may write

$$[\alpha f(r) + \beta]\, [\gamma g(z) + \delta] = \begin{cases} F(r) \\ \text{or} \\ G(z). \end{cases}\tag{4.9}$$

---

[1] The stream function ψ used here differs from the Stokes stream function. It is related to the Stokes stream function, for which the contour lines coincide with the streamlines, by $\psi_{\mathrm{Stokes}} = r\psi$.

Several possibilities exist to satisfy these equations. The choice depends on whether one selects $\alpha=0$ with $\beta=0$, $\beta>0$, or $\beta<0$, or whether one takes $\gamma=0$ with corresponding signs of δ (see also Yoo & Joseph (1978)). Each of these choices leads to a set of functions satisfying (4.1). We are free to select those functions which are best suited to satisfy the boundary conditions (4.4). Here we use $\alpha=0$ and $\beta=4\lambda^2$. Then the solutions T of (4.7) are the modified Bessel functions of first order $I_1(2\lambda r)$ and $K_1(2\lambda r)$. Since $K_1(2\lambda r)$ is singular at $r=0$ these latter functions are discarded. The main advantage of the functions $I_1(2\lambda r)$ is that they automatically satisfy the boundary conditions at $r=0$. We thus have

$$\psi \propto I_1(2\lambda r)\,\phi(z,\lambda)\,. \tag{4.10}$$

Inserting this form of ψ into (4.1), the differential equation

$$\left(\partial_z^2 + 4\lambda^2\right)^2 \phi = \left[\partial_z^4 + 2(4\lambda^2)\partial_z^2 + (4\lambda^2)^2\right]\phi = 0 \tag{4.11}$$

for ϕ is obtained. The solution is a superposition of harmonics and products of harmonics with z,

$$\psi = I_1(2\lambda r)\,[\,A\cos(2\lambda z) + B\sin(2\lambda z)$$
$$+ 2Cz\cos(2\lambda z) + 2Dz\sin(2\lambda z)]\,. \tag{4.12}$$

The four integration constants A, B, C, and D are determined by the boundary conditions. Since $\phi(1/2)=\phi(-1/2)=\phi'(1/2)=\phi'(-1/2)=0$, the integration constants must satisfy

$$\begin{pmatrix} \cos\lambda & \sin\lambda & \cos\lambda & \sin\lambda \\ \cos\lambda & -\sin\lambda & -\cos\lambda & \sin\lambda \\ -\lambda\sin\lambda & \lambda\cos\lambda & \cos\lambda-\lambda\sin\lambda & \sin\lambda+\lambda\cos\lambda \\ \lambda\sin\lambda & \lambda\cos\lambda & \cos\lambda-\lambda\sin\lambda & -\sin\lambda-\lambda\cos\lambda \end{pmatrix} \begin{pmatrix} A \\ B \\ C \\ D \end{pmatrix} = 0\,. \tag{4.13}$$

The solvability condition for this system of equations, $\det(...)=0$, yields the characteristic roots $\{\lambda_n\}$ as solutions of the transcendental equation[2]

$$(\lambda - \sin\lambda\cos\lambda)(\lambda + \sin\lambda\cos\lambda) = 0\,. \tag{4.14}$$

Owing to the symmetry of the differential equation, the solutions of (4.11) separate into even and odd functions of z. This can be easily confirmed by setting either $B=C=0$ or $A=D=0$ in (4.13). Since the boundary conditions (4.4) for ψ are symmetrical in z, only the even functions are required. The odd functions can be obtained in an analogous way (Joseph (1977)). The characteristic roots belonging to the even functions are solutions of

$$\lambda + \sin\lambda\cos\lambda = 0\,. \tag{4.15}$$

---

[2] Hillman & Salzer (1943) and Robbins & Smith (1948) were the first to calculate the roots.

They must be determined numerically. The asymptotic formula (Joseph (1977))

$$2\lambda_n \sim \left(2n - \frac{1}{2}\right)\pi + i\ln\left[(4n-1)\pi\right] \qquad (4.16)$$

provides good initial values to calculate the roots. If λ is a root then $-\lambda$ and λ^* are also roots of (4.15). We define $\lambda_{-n} = \lambda_n^*$ and order the index n according to the real part of λ_n. From (4.13) the even functions are obtained as

$$\phi^{(n)}(z) = \phi(\lambda_n, z) = \lambda_n\left[\sin\lambda_n \cos(2\lambda_n z) - 2z\cos\lambda_n \sin(2\lambda_n z)\right], \quad (4.17)$$

and the symmetric solution can be written as a superposition of all modes,

$$\psi(r,z) = \sum_{n=-\infty}^{\infty} A_n I_1(2\lambda_n r)\,\phi^{(n)}(z). \qquad (4.18)$$

The boundary conditions (4.4) on the free surface are now used to determine the unknown coefficients A_n. Since two boundary conditions must be satisfied simultaneously, it is of advantage to write the fourth-order equation (4.11) as two second-order equations. The coefficients A_n will then be obtained by projecting the equations onto suitably constructed orthonormal modes. In this formulation the modes consist of two-component vector functions. Before the projection is carried out, orthogonality must first be proved. To that end we define

$$\phi_1 = \phi, \qquad (4.19)$$

$$\phi_2 = \frac{1}{4\lambda^2}\phi_1''. \qquad (4.20)$$

The differential equation (4.11) for ϕ may then be written as

$$L\boldsymbol{\Phi} = 0, \qquad (4.21)$$

where $\boldsymbol{\Phi} = (\phi_1, \phi_2)^{\mathrm{T}}$, $L = \partial_z^2 + 4\lambda^2 \mathsf{A}$, and

$$\mathsf{A} = \begin{pmatrix} 0 & -1 \\ 1 & 2 \end{pmatrix}. \qquad (4.22)$$

Next, a scalar product and the functions adjoint to $\boldsymbol{\Phi}$ are required. For this purpose (4.21) is multiplied with a function $\boldsymbol{\Psi}$ and integrated twice by parts. One obtains

$$0 = \int_{-1/2}^{1/2} \boldsymbol{\Psi}\cdot L\boldsymbol{\Phi}\,\mathrm{d}z = \int_{1/2}^{1/2}\left(\boldsymbol{\Psi}\cdot\partial_z^2\boldsymbol{\Phi} + 4\lambda^2\boldsymbol{\Psi}\cdot\mathsf{A}\cdot\boldsymbol{\Phi}\right)\mathrm{d}z$$

$$= \int_{-1/2}^{1/2}\left(\partial_z^2 + 4\lambda^2\mathsf{A}^{\mathrm{T}}\right)\boldsymbol{\Psi}\cdot\boldsymbol{\Phi}\,\mathrm{d}z = \int_{-1/2}^{1/2}\left(L^\dagger\boldsymbol{\Psi}\right)\cdot\boldsymbol{\Phi}\,\mathrm{d}z. \qquad (4.23)$$

Here we have defined $L^\dagger = \left(\partial_z^2 + 4\lambda^2\mathsf{A}^{\mathrm{T}}\right)$. From (4.23) the adjoint problem is

$$L^\dagger \boldsymbol{\Psi} = 0 . \tag{4.24}$$

To carry out the integrations by parts in (4.23) the following boundary conditions for $\boldsymbol{\Phi}$ have been used:

$$\phi_1 = \partial_z \phi_1 = 0 , \qquad \text{for} \quad z = \pm 1/2 . \tag{4.25}$$

In addition, we have required

$$\psi_2 = \partial_z \psi_2 = 0 , \qquad \text{for} \quad z = \pm 1/2 . \tag{4.26}$$

Equations (4.26) represent the proper boundary conditions for the adjoint functions $\boldsymbol{\Psi}$. If (4.24) is written as a fourth-order equation for ψ_2, it can be seen that ψ_2 satisfies the same differential equation as ϕ_1. With this result and by integrating the first component of the adjoint problem (4.24) twice, all functions can be given explicitly:

$$\phi_1^{(n)}(z) = \lambda_n \sin \lambda_n \cos(2\lambda_n z) - 2z\lambda_n \cos \lambda_n \sin(2\lambda_n z) ,$$
$$\phi_2^{(n)}(z) = - (\lambda_n \sin \lambda_n + 2 \cos \lambda_n) \cos(2\lambda_n z) + 2z\lambda_n \cos \lambda_n \sin(2\lambda_n z) ,$$
$$\psi_1^{(n)}(z) = (\lambda_n \sin \lambda_n - 2 \cos \lambda_n) \cos(2\lambda_n z) - 2z\lambda_n \cos \lambda_n \sin(2\lambda_n z) ,$$
$$\psi_2^{(n)}(z) = \phi_1^{(n)}(z) . \tag{4.27}$$

These are called the Papkovich–Fadle functions after Papkovich (1940) and Fadle (1941). To obtain the orthogonality relation, consider the expression

$$0 = \int_{-1/2}^{1/2} \left(\boldsymbol{\Psi}^{(m)} \cdot L\boldsymbol{\Phi}^{(n)} - \boldsymbol{\Phi}^{(n)} \cdot L^\dagger \boldsymbol{\Psi}^{(m)} \right) \mathrm{d}z . \tag{4.28}$$

Integrating twice by parts, the terms containing second derivatives cancel and one is left with

$$0 = \int_{-1/2}^{1/2} \left(4\lambda_n^2 \boldsymbol{\Psi}^{(m)} \cdot \mathbf{A} \cdot \boldsymbol{\Phi}^{(n)} - 4\lambda_m^2 \boldsymbol{\Phi}^{(n)} \cdot \mathbf{A}^T \cdot \boldsymbol{\Psi}^{(m)} \right) \mathrm{d}z$$
$$= 4(\lambda_n^2 - \lambda_m^2) \int_{-1/2}^{1/2} \boldsymbol{\Psi}^{(m)} \cdot \mathbf{A} \cdot \boldsymbol{\Phi}^{(n)} \, \mathrm{d}z . \tag{4.29}$$

It follows that

$$\int_{-1/2}^{1/2} \boldsymbol{\Psi}^{(m)} \cdot \mathbf{A} \cdot \boldsymbol{\Phi}^{(n)} \, \mathrm{d}z = K_m \delta_{nm} . \tag{4.30}$$

The normalizing contants are given by

$$K_n = \int_{-1/2}^{1/2} \boldsymbol{\Psi}^{(n)} \cdot \mathbf{A} \cdot \boldsymbol{\Phi}^{(n)} \, \mathrm{d}z = -2 \cos^4 \lambda_n . \tag{4.31}$$

Equation (4.30) is the required *biorthogonality relation* for the two-component vector functions $\boldsymbol{\Phi}^{(n)}$. The unknown amplitudes A_n can now be determined by projection of the two boundary conditions (4.4) for $r = 1/\Gamma$. Using this technique, the stream function will be calculated in the next section to the lowest order of an expansion in small Reynolds, Prandtl, and capillary numbers.

4.2 Thermocapillary Flow and Small Surface Deformations

4.2.1 Formulation

The analysis starts with the two-dimensional Navier–Stokes equations (2.14–2.16) for zero gravity (Gr = Bo = 0). It is appropriate for creeping flows to use the viscous scales ν/d and $\rho\nu^2/d^2$ for the velocity and the pressure instead of the scales in Table 2.1, since the viscous scales are independent of Re. For the rescaling of the equations, $(\boldsymbol{u}, \psi, p)$ are simply replaced by $(\boldsymbol{u}, \psi, p)/\text{Re}$. With help of the stream function (4.2–4.3) one obtains the equations for ψ and Θ:

$$\tilde{L}^2\psi = -\left[\frac{2}{r}(\partial_z\psi) + (D_*\psi)\partial_z - (\partial_z\psi)D_*\right]\tilde{L}\psi\,, \tag{4.32}$$

$$\Delta\Theta = -\text{Pr}\left\{\left[(D_*\psi)\partial_z - (\partial_z\psi)D\right]\Theta + D_*\psi\right\}\,. \tag{4.33}$$

Here $\Theta = \theta - z$ is the deviation of the temperature from the conductive profile and $\tilde{L} = DD_* + \partial_z^2$ denotes a linear differential operator. The two-dimensional Laplace operator is given by $\Delta = D_*D + \partial_z^2$. From (2.22), with (2.31) and (2.32), we obtain for the tangential (axial) and the normal stress balance on the free surface $r = \xi(z)$

$$\left\{\partial_z^2 - DD_* + 2\xi_z D_*\partial_z + \xi_z\left[2D\partial_z - \xi_z\left(\partial_z^2 - DD_*\right)\right]\right\}\psi$$
$$+ N\text{Re}\left(\xi_z D\Theta + \partial_z\Theta + 1\right) = 0\,, \tag{4.34}$$

and

$$\text{Ca}\left[D\partial_z - \xi_z^2 D_*\partial_z - \xi_z\left(\partial_z^2 - DD_*\right)\right]\psi - \frac{N^2}{2}\text{Ca}\,p$$
$$+ \frac{N}{2}\text{Re}\left[1 - \text{Ca}(\Theta + z)\right]\left(\frac{1}{\xi} - \frac{\xi_{zz}}{N^2}\right) = 0\,. \tag{4.35}$$

Here, $N = (1 + \xi_z^2)^{1/2}$. The kinematic condition (2.25) reads

$$(\partial_z + \xi_z D_*)\,\psi = 0\,, \qquad \text{on} \quad r = \xi(z)\,, \tag{4.36}$$

and the thermal boundary condition (2.26) is

$$(D - \xi_z\partial_z + \text{Bi}\,N)\,\Theta = 0\,, \qquad \text{on} \quad r = \xi(z)\,, \tag{4.37}$$

where we have chosen $\theta_a = z$. To obtain the leading-order contributions to ψ, Θ, p, and ξ, all field quantities are formally expanded in terms of the small parameters Re, Pr, and Ca:

$$\psi(r,z) = \sum_{\substack{n=1\\m=k=0}}^{\infty} \text{Re}^n\text{Pr}^m\text{Ca}^k\psi_{nmk}(r,z)\,, \tag{4.38}$$

$$\Theta(r,z) = \sum_{\substack{n=1\\m=k=0}}^{\infty} \text{Re}^n\text{Pr}^m\text{Ca}^k\Theta_{nmk}(r,z)\,, \tag{4.39}$$

$$p(r,z) = \frac{\mathrm{Re}}{\mathrm{Ca}} p_{\mathrm{s}}(z) + \sum_{\substack{n=1 \\ m=k=0}}^{\infty} \mathrm{Re}^n \mathrm{Pr}^m \mathrm{Ca}^k\, p_{nmk}(r,z) \,, \tag{4.40}$$

$$\xi(z) = \xi_{\mathrm{s}}(z) + \sum_{\substack{n=m=0 \\ k=1}}^{\infty} \mathrm{Re}^n \mathrm{Pr}^m \mathrm{Ca}^k \xi_{nmk}(z) \,. \tag{4.41}$$

Inserting these expansions into the normal stress balance (4.35) one obtains, at lowest order $\mathcal{O}(\mathrm{Re}^1\mathrm{Pr}^0\mathrm{Ca}^0)$, the static meniscus problem (index "s"; compare (2.30))

$$p_{\mathrm{s}} = \frac{1}{N_{\mathrm{s}}} \left(\frac{1}{\xi_{\mathrm{s}}} - \frac{\partial_z^2 \xi_{\mathrm{s}}}{N_{\mathrm{s}}^2} \right) \,. \tag{4.42}$$

The order of magnitude of the pressure jump $\mathrm{Re}/\mathrm{Ca} = \sigma_0 d/\rho\nu^2$ is independent of ΔT. In the following we consider a cylindrical liquid bridge with a dimensionless volume $V = \pi/\Gamma^2$. Then $\xi_{\mathrm{s}} = 1/\Gamma$, and the pressure jump $p_{\mathrm{s}} = \Gamma$ is independent of z.

4.2.2 Flow Field

If the expansions (4.38–4.39) are inserted into the equations for the stream function and temperature field, (4.32) and (4.33), we obtain, at lowest order $\mathcal{O}(\mathrm{Re}^1\mathrm{Pr}^0\mathrm{Ca}^0)$,

$$\tilde{L}^2\psi_{100} = 0 \,, \tag{4.43}$$

$$\Delta\Theta_{100} = 0 \,, \tag{4.44}$$

with boundary conditions (4.4) for ψ_{100} and

$$\Theta_{100}(z=\pm 1/2) = D\Theta_{100}(r=0) = (D+\mathrm{Bi})\Theta_{100}(r=1/\Gamma) = 0 \tag{4.45}$$

for the temperature field. It follows that $\Theta_{100} = 0$ vanishes identically, i.e. the temperature field is purely conductive: $\theta = z$. For this reason the lowest-order velocity field is independent of the Biot number.

According to the solution developed in the preceding section, the stream function ψ_{100} is represented as an infinite series (4.18) of symmetric Papkovich–Fadle functions. The boundary conditions (4.4) at $r = 1/\Gamma$ can be written as

$$\begin{pmatrix} 1 \\ 0 \end{pmatrix} = \begin{pmatrix} DD_*\psi_{100} \\ \partial_z^2\psi_{100} \end{pmatrix}_{r=1/\Gamma} = \sum_{n=-\infty}^{\infty} 4\lambda_n^2 A_n \mathrm{I}_1(2\lambda_n/\Gamma) \begin{pmatrix} \phi_1^{(n)} \\ \phi_2^{(n)} \end{pmatrix} \,. \tag{4.46}$$

Multiplying this equation with $\boldsymbol{\Psi}^{(m)} \cdot \mathbf{A}$, integrating over $[-\frac{1}{2}, \frac{1}{2}]$, and making use of the biorthogonality condition (4.30), one arrives at

$$\sum_{n=-\infty}^{\infty} 4\lambda_n^2 A_n \mathrm{I}_1(2\lambda_n/\Gamma) K_n \delta_{nm} = \int_{-1/2}^{1/2} \boldsymbol{\Psi}^{(m)} \cdot \mathbf{A} \cdot \begin{pmatrix} 1 \\ 0 \end{pmatrix} \mathrm{d}z$$

$$= \int_{-1/2}^{1/2} \psi_2^{(m)} \, \mathrm{d}z = 2 \,, \tag{4.47}$$

from which the amplitudes

$$A_n = \frac{-1}{4\lambda_n^2 \cos^4 \lambda_n I_1(2\lambda_n/\Gamma)} \qquad (4.48)$$

follow.

For an evaluation of the infinite sum over n, it must be truncated at some finite value N to give the stream function approximation

$$\psi_{100}(r,z) \approx \sum_{n=-N}^{N} \frac{-I_1(2\lambda_n r)}{4\lambda_n^2 \cos^4 \lambda_n I_1(2\lambda_n/\Gamma)} \phi_1^{(n)}(z). \qquad (4.49)$$

It turns out that the sequence of finite series converges rapidly for increasing $N \to \infty$. The corresponding flow structures will be explained in Sects. 4.2.5 and 4.2.6.

4.2.3 Temperature Field

The first nonvanishing correction to the temperature field is of order $\mathcal{O}(\mathrm{Re}^1 \mathrm{Pr}^1 \mathrm{Ca}^0)$. The equations for ψ_{110} and Θ_{110} read

$$\tilde{L}^2 \psi_{110} = 0, \qquad (4.50)$$
$$\Delta\Theta_{110} = -D_* \psi_{100}, \qquad (4.51)$$

with the homogeneous boundary conditions (4.45), as in the $\mathcal{O}(\mathrm{Re}^1 \mathrm{Pr}^0 \mathrm{Ca}^0)$ field. The inhomogeneous condition $DD_* \psi_{100} = 1$ at $r = 1/\Gamma$ in (4.4) is merely replaced by $DD_* \psi_{110} = 0$. Thus, $\psi_{110} = 0$. Combining a particular solution with the solution of the homogeneous problem, one obtains for the temperature field

$$\Theta_{110}(r,z) = \sum_{n=-\infty}^{\infty} 2a_n \lambda_n \Big[I_0(2\lambda_n r)\Big(z^2 \cos(2\lambda_n z) + b_n z \sin(2\lambda_n z)$$
$$+ c_n \cos(2\lambda_n z)\Big) + \sum_{k=0}^{\infty} B_{nk} I_0(\kappa_k r) \cos(\kappa_k z)\Big], \qquad (4.52)$$

where

$$a_n = \big[16\lambda_n^2 \cos^3 \lambda_n I_1(2\lambda_n/\Gamma)\big]^{-1}, \qquad (4.53)$$

$$b_n = -\frac{1 + 2\sin^2 \lambda_n}{2\lambda_n}, \qquad (4.54)$$

$$c_n = -\frac{1}{4}\left(1 + 2b_n \tan \lambda_n\right), \qquad (4.55)$$

$$\kappa_k = (2k+1)\pi, \qquad (4.56)$$

and

$$B_{nk} = -2 \frac{2\lambda_n I_1(2\lambda_n/\Gamma) + \mathrm{Bi}\, I_0(2\lambda_n/\Gamma)}{\kappa_k I_1(\kappa_k/\Gamma) + \mathrm{Bi}\, I_0(\kappa_k/\Gamma)} \tag{4.57}$$

$$\times \int_{-1/2}^{1/2} \cos(\kappa_k z) \left[z^2 \cos(2\lambda_n z) + b_n z \sin(2\lambda_n z) + c_n \cos(2\lambda_n z) \right]\, dz\,.$$

The solution in the limit $\mathrm{Ca} \to 0$ has the following asymptotic form:

$$\psi(r,z) = \mathrm{Re}\,\psi_{100}(r,z) + \mathcal{O}(\mathrm{Re}^2)\,, \tag{4.58}$$
$$\Theta(r,z) = \mathrm{RePr}\,\Theta_{110}(r,z) + \mathcal{O}(\mathrm{Re}^2)\,. \tag{4.59}$$

4.2.4 Dynamic Surface Deformations

When the capillary number Ca is small the asymptotic form of the thermo-capillary flow-induced surface deformation ξ_{001} can be calculated. To this end, we consider the normal stress balance (4.35) at $\mathcal{O}(\mathrm{Re}^1\mathrm{Pr}^0\mathrm{Ca}^1)$,

$$\left(\partial_z^2 + \Gamma^2\right)\xi_{001} = -\Gamma z + (2\partial_z D\psi_{100} - p_{100})_{r=1/\Gamma}\,. \tag{4.60}$$

To derive (4.60) for ξ_{001}, we have had to expand ξ^{-1}. This expansion is valid only if the surface deflections are small compared to the radius of the static meniscus of the liquid bridge, i.e. if $\mathrm{Ca}|\xi_{001}| \ll \xi_s = 1/\Gamma$. The stationary shape of the interface depends on the structure of the flow-induced normal stresses at $\mathcal{O}(\mathrm{Re}^1\mathrm{Pr}^0\mathrm{Ca}^0)$. The pressure p_{100} is readily obtained from ψ_{100} by integrating the Navier–Stokes equation to yield, at $\mathcal{O}(\mathrm{Re}^1\mathrm{Pr}^0\mathrm{Ca}^0)$, $Dp_{100} = \partial_z \tilde{L}\psi_{100}$. This can be used to solve (4.60). After some calculations one arrives at

$$\xi_{001}(z) = \frac{1}{\Gamma}\left(\frac{1}{2}\frac{\sin(\Gamma z)}{\sin(\Gamma/2)} - z\right) \tag{4.61}$$

$$+ \sum_{n=-\infty}^{\infty} \left[C_n\left(z\cos(2\lambda_n z) - \frac{1}{2}\frac{\sin(\Gamma z)}{\sin(\Gamma/2)}\right) \right.$$

$$\left. + D_n\left(\sin(2\lambda_n z) - \sin\lambda_n \frac{\sin(\Gamma z)}{\sin(\Gamma/2)}\right)\right]\,, \quad \text{for } \Gamma/2 \neq \begin{cases} n\pi\,, \\ \tan(\Gamma/2) \end{cases}.$$

The constants are given by

$$C_n = \frac{4\lambda_n}{(4\lambda_n^2 - \Gamma^2)\cos^3\lambda_n}\left(\frac{\Gamma}{2\lambda_n} - \frac{I_0(2\lambda_n/\Gamma)}{I_1(2\lambda_n/\Gamma)}\right) \tag{4.62}$$

and

$$D_n = \frac{2}{(4\lambda_n^2 - \Gamma^2)\cos^4\lambda_n}\left[\left(\lambda_n\sin\lambda_n + \cos\lambda_n - \frac{8\lambda_n^2\cos\lambda_n}{4\lambda_n^2 - \Gamma^2}\right)\right.$$

$$\left. \times \left(\frac{\Gamma}{2\lambda_n} - \frac{I_0(2\lambda_n/\Gamma)}{I_1(2\lambda_n/\Gamma)}\right) - \cos\lambda_n \frac{I_0(2\lambda_n/\Gamma)}{I_1(2\lambda_n/\Gamma)}\right]\,. \tag{4.63}$$

The surface deflection (4.61) is volume-preserving at $\mathcal{O}(\mathrm{Ca}^1)$. The singular points, $\Gamma/2 = n\pi$ and $\Gamma/2 = \tan(\Gamma/2)$, have been identified also as bifurcation points by Rybicki & Floryan (1987b), albeit by a different method, at which the static stability of an isothermal liquid bridge changes under axisymmetric perturbing deformations. When the aspect ratio Γ approaches the Rayleigh limit ($\Gamma \to 2\pi$) the solution (4.61) diverges. The differential equation (4.60) loses its validity in any case, however, for $\mathrm{Ca} \gg \mathcal{O}(2\pi - \Gamma)$, since we have assumed $\mathrm{Ca}|\xi_{001}| \ll 1/\Gamma$. Rybicki & Floryan (1987b) have shown, and it has been confirmed by numerical calculations by Chen et al. (1990), that the maximum stable length of a nonisothermal liquid bridge is essentially restricted by the Rayleigh limit.[3]

4.2.5 Flow Patterns

The stream function, temperature, and surface deflection are calculated by truncating the infinite sums over n and k at finite values N and K. For the moderate aspect ratio $\Gamma = 1$, a truncation order $N = 20$ is sufficiently accurate, and the difference between the maximum velocities obtained for N and $N+1$ is less than 1%. If $K = N$ the solutions are well converged in the bulk, while the convergence near $r \approx 1/\Gamma$ is slightly slower. Since the relative length of the end zones near $z \approx \pm 1/2$, within which the flow is forced to return, decreases with $1/\Gamma$, high gradients in the z direction must be resolved if the aspect ratio Γ is large, and an increasing number of modes N must be retained. More general considerations on the convergence of biorthogonal series can be found in Joseph & Sturges (1978).

As an example, the streamlines (contour lines of $\psi_{\mathrm{Stokes}} = r\psi_{100}$) are shown in Fig. 4.1. The flow pattern consists of a single toroidal vortex with a stream function minimum $\psi_{\mathrm{Stokes,min}} = -0.0108$ at $r = 0.80$.[4] In Fig. 4.2 the associated temperature field Θ_{110} is shown. As expected for small Prandtl numbers, the regions of elevated and lowered temperatures are caused by the axial flow components perpendicular to the conduction isotherms.

The form of the surface deflection ξ_{001}, its magnitude, and its dependence on Γ agree with the results of Rybicki & Floryan (1987b), who used a finite-difference method. Characteristic profiles of ξ_{001} are given in Fig. 4.3 for different values of Γ. The constriction near the hot wall of the liquid bridge ($z = 0.5$) and the bulging near the cold wall ($z = -0.5$) are typical for all aspect ratios.

To estimate the range of parameters for which the creeping-flow solution is a valid approximation the results have been compared with the numerical calculations of Shen (1989) for the full Navier–Stokes equations in the limit

---

[3] It is possible, however, that the surface deflections of a nonisothermal liquid bridge become time-dependent even for $\Gamma < 2\pi$, if the bridge is heated antisymmetrically (Rybicki & Floryan (1987b)).

[4] The minimum $\psi_{100,\mathrm{min}} = -0.0140$ is located at $r = 0.75$

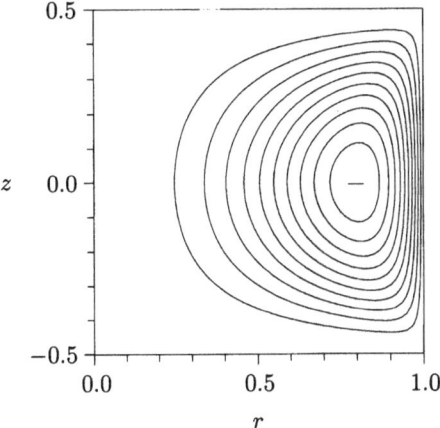

Fig. 4.1. Streamlines of the Stokes flow ($\psi_{\text{Stokes}} = r\psi_{100}$) in a thermocapillary liquid bridge for $\Gamma = 1$. Contour lines at $-0.001 \times n$ with $n \in [1, 10]$; $N = 20$

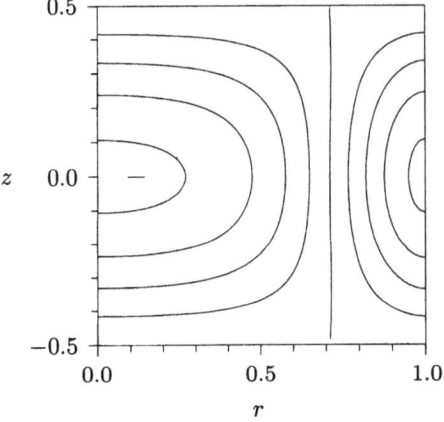

Fig. 4.2. Isotherms of Θ_{110} for creeping flow and small Prandtl number; $\Gamma = 1$, $N = 20$, and $K = 40$

$Ca \to 0$ (no surface deformations). Figures 4.4 and 4.5 show the axial velocity w and the temperature Θ at the midplane ($z = 0$) for $\Gamma = 1$, $Pr = 0.1$, and $Bi = 0$. Both results agree well up to a Reynolds number $Re \approx 100$. For higher values of the Reynolds number ($Re = 500$) the approximate solution loses its validity. The axial dependence of w and Θ is displayed in Figs. 4.6 and 4.7 for the same parameters. Larger deviations are visible. Even at $Re = 100$ the numerical data are slightly asymmetric with respect to $z = 0$. The asymmetry becomes even more pronounced for $Re = 500$ because of a combined inertia effect of order $\mathcal{O}(Re^2)$ and a thermocapillary coupling at the free surface of order $\mathcal{O}(Re^2 Pr)$. The latter effect diminishes for lower Prandtl numbers. For larger aspect ratios the asymmetry of the fields is weaker. This can be

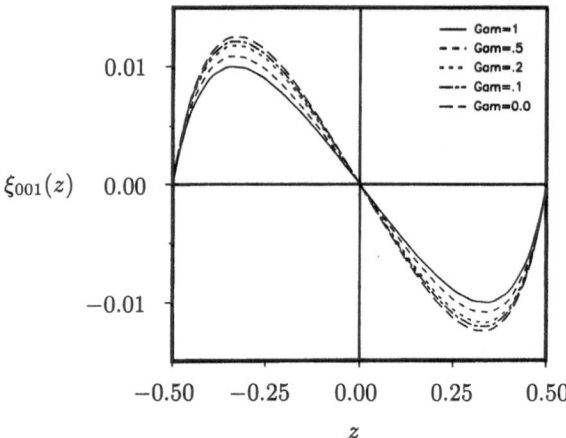

Fig. 4.3. Surface deflection ξ_{001} for $\Gamma=0$, 0.1, 0.2, 0.5, and 1.0; $N=20$

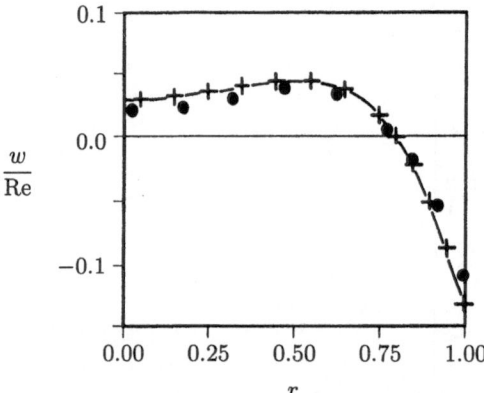

Fig. 4.4. Radial dependence ($z=0$) of the axial velocity w of the Stokes flow solution according to (4.49) in comparison with numerical results (Y. Shen (private communication, 1988)) for the full Navier–Stokes equations for $\Gamma=1$, Pr$=0.1$, Bi$=0$. (——) Stokes flow ($N=K=20$); (+) numerical results (Re$=100$); ($\bullet$) numerical results (Re$=500$)

seen from Fig. 4.8, where the axial velocity on the free surface is plotted for $\Gamma=5$, Pr$=0.1$, and Bi$=0$. In this case the deviations in the region $z<0$ are quite small even for Re$=500$. Apart from the turning zones near $z \approx \pm 1/2$ the axial velocity at the free surface has already reached the asymptotic value $w/$Re$=-1/(4\Gamma)=-0.05$ for Poiseuille flow.[5]

The influence of a nonlinear ambient temperature distribution ($\theta_a \neq z$) can easily be investigated. In order that such a modified temperature distribution θ_a influences the flow the Biot number must be nonzero, though. If θ_a has

---

[5] For $\Gamma \to \infty$ the temperature field vanishes on the free surface ($\Theta(r=1/\Gamma) \to 0$).

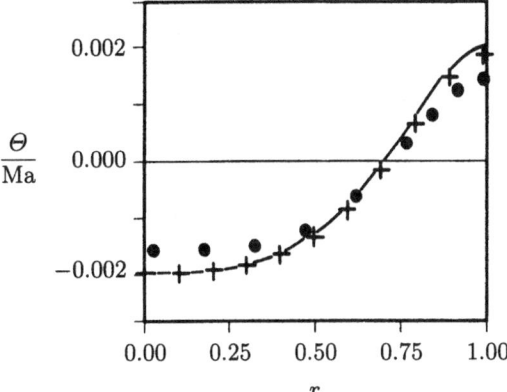

Fig. 4.5. Radial dependence of the temperature Θ of the Stokes flow solution according to (4.52). Symbols as in Fig. 4.4

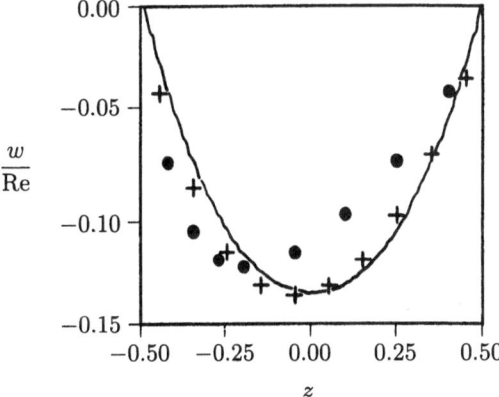

Fig. 4.6. Axial dependence of the axial velocity w on the free surface $r = 1/\Gamma = 1$ in comparison with numerical results. All parameters and symbols as in Fig. 4.4

contributions symmetric with respect to z, the antisymmetric basis functions have to be taken into account in (4.18). The influence of the Biot number on the temperature field Θ_{100}, essentially that the amplitude of Θ_{100} is reduced by increasing Bi, is in agreement with the behavior observed in the numerical simulations (Y. Shen (private communication, 1988)) as long as $\mathrm{Re} < 200$.

The first calculation of the creeping flow in a thermocapillary cylindrical domain in the limit $\mathrm{Pr} \to 0$ was due to Bauer (1982). He employed periodic boundary conditions in the z direction, a parabolic ambient temperature, and stress-free boundary conditions at $z = \pm 1/2$. The influence of different ambient-temperature profiles has been investigated by Rybicki & Floryan (1987a) using finite differences, and by Davis (1989) using analytical techniques. Immediately below the free surface the convection structures consist

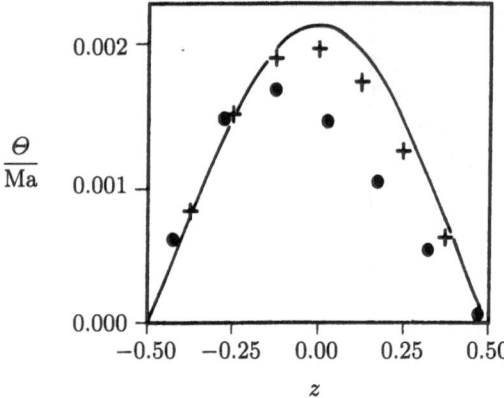

Fig. 4.7. Axial dependence of the surface temperature Θ in comparison with numerical results. All parameters and symbols as in Fig. 4.4

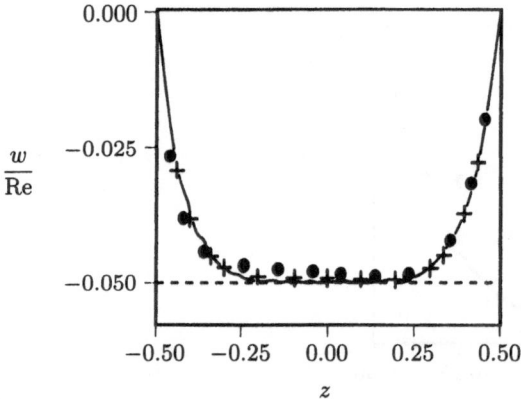

Fig. 4.8. Axial velocity w on the free surface $r = 1/\Gamma$ for $\Gamma = 5$, $\mathrm{Pr} = 0.1$, and $\mathrm{Bi} = 0$. Symbols as in Fig. 4.4. The truncation parameter for w_{100} is $N = 80$. (- - -) asymptotic value for $\Gamma \to \infty$

of $n + 1$ vortices stacked in the z direction, where n is the number of interior zeros of $\partial_z \theta_a$. The stacked vortices vanish on a radial length scale $\mathcal{O}(1)$ in the interior of the liquid bridge, giving way to a sequence of Moffatt eddies to be discussed in Sect. 4.2.6. Using the method of biorthogonal functions Chen & Roux (1991) calculated the creeping flow in a symmetrically heated liquid bridge for which the temperature distribution θ_a reaches its maximum between the ends, in analogy to the floating-zone configuration. Viviani & Cioffi (1992) employed the biorthogonal series method too, and extended the calculations to a liquid bridge consisting of two different immiscible liquids stratified radially and subject to thermocapillary forces along both the interior interface and the outer surface.[6]

4.2.6 Aspect Ratio Limits

$\Gamma \to 0$. In the limit of a shallow liquid bridge ($\Gamma \to 0$) the expression for the stream function can be simplified far away from the lateral boundaries $r = 0$ and $r = 1/\Gamma$. For large arguments ($1 \ll r$), the Bessel functions in (4.49) can be expanded to yield an exponential dependence on r. Owing to the rapid decay of these functions from the free surface, only the mode $n = 1$ contributes substantially to the flow for $r \ll 1/\Gamma$, resulting in the approximation

$$\psi(1 \ll r \ll 1/\Gamma, z) \cong \frac{-1}{4\lambda_1^2 \cos^4 \lambda_1} \frac{\phi_1^{(1)}(z)}{\sqrt{\Gamma r}} e^{2\lambda_1(r-1/\Gamma)} + \text{c.c.} \qquad (4.64)$$

The flow consists of a sequence of radially nested vortices having a self-similar shape, and the amplitude decays nearly exponentially for $r \to 0$. The radial diameter of the vortices is $\pi/[2\Im(\lambda_1)] = 1.396$. As an example, the streamlines for the exact stream function $\psi_{\text{Stokes}} = r\psi_{100}$ according to (4.49) are shown for $\Gamma = 0.25$ in Fig. 4.9. The nested vortices have the same diameter

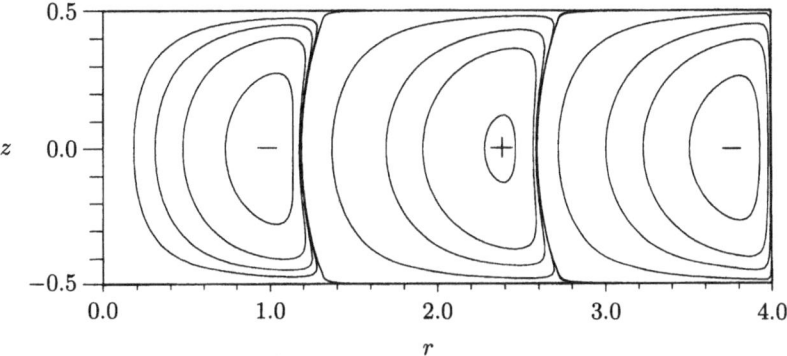

Fig. 4.9. Creeping thermocapillary flow in a liquid bridge with $\Gamma = 0.25$; $N = 20$. Streamlines ψ_{Stokes} are drawn for $(-0.03, -0.01, -3 \times 10^{-3}, -3 \times 10^{-4})$, $(10^{-4}, 3 \times 10^{-5}, 10^{-5}, 10^{-6})$ and $(-10^{-7}, -3 \times 10^{-8}, -10^{-8}, -3 \times 10^{-9})$

as the rectilinear viscous Moffatt eddies between two semi-infinite, parallel, rigid planes (Moffatt (1964a), Pan & Acrivos (1967)). While the rectilinear vortices decay exponentially with a spatial decay rate $2\Re(\lambda_1) = 4.212$, the axisymmetric vortices decay somewhat slower, owing to the additional factor $r^{-1/2}$ in (4.64).

---

[6] This problem is related to the so-called encapsulated-crystal-growth technique, during which a second liquid is used to cover the melt zone in order to protect the interior liquid from oxidation and contamination (Monberg (1994), Hurle (1994)). See also Sect. 9.1.7.

In the limit $\Gamma \to 0$ the surface deflection takes the form

$$\xi_{001}(z, \Gamma \to 0)$$
$$= \sum_{n=-\infty}^{\infty} \left(\frac{\sin(2\lambda_n z) - 2z \sin \lambda_n}{2 \cos^5 \lambda_n} - \frac{z \cos(2\lambda_n z) - z \cos \lambda_n}{\lambda_n \cos^3 \lambda_n} \right). \quad (4.65)$$

The deformation has a sinusoidal shape with extrema $|\xi_{001}(z = \pm 0.33)| = 0.0125$. This is 6% less than the value extrapolated by Rybicki & Floryan (1987b). The curve is nearly identical to the one for $\Gamma = 0.1$ in Fig. 4.3.

$\Gamma \to \infty$. An isothermal liquid bridge breaks because of the Rayleigh instability if $\Gamma > \Gamma_c = 2\pi$ (Rayleigh (1879), Drazin & Reid (1981)). The critical aspect ratio Γ_c merely changes slightly in the presence of weak thermocapillary flow (Sect. 4.2.4). However, strong thermocapillary shear forces may have a stabilizing effect (Xu & Davis (1985)). The nonlinear dynamics of the decay of isothermal and thermocapillary jets has been investigated by Ashgriz & Mashayek (1995) and Mashayek & Ashgriz (1995). Even though long liquid bridges are unstable, the limit of an infinite thermocapillary liquid bridge is a useful mathematical model, because the analysis of convective instabilities is substantially simplified in this case (see Sect. 7.4.3 and Xu & Davis (1984)).

For large aspect ratios the flow far away from the rigid boundaries ($z = \pm 1/2$) is independent of z and nearly axial ($u \approx 0$). In the limit $\Gamma \to \infty$, $\mathrm{Ca} \to 0$, the Navier–Stokes equations reduce to

$$(DD_*)^2 \psi = 0, \quad (4.66)$$
$$D_* D\Theta = -\mathrm{Pr}\, D_* \psi. \quad (4.67)$$

Because these equations are identical to the perturbation expansion up to order $\mathcal{O}(\mathrm{Re}^1 \mathrm{Pr}^1 \mathrm{Ca}^0)$ ((4.43–4.44), (4.50–4.51)), the perturbation expansion is exact at $\mathcal{O}(\mathrm{Re}^1 \mathrm{Pr}^1 \mathrm{Ca}^0)$ in the limit $\Gamma \to \infty$, $\mathrm{Ca} \to 0$. The axial velocity w_{100} at $z = 0$, for example, asymptotically approaches the exact axial velocity (Fig. 4.8). Considering the stream function (4.49) at $z = 0$ and expanding the Bessel functions into ascending series for small $1/\Gamma$ (Abramowitz & Stegun (1972)) one obtains, as expected, the well-known Hagen–Poiseuille profile

$$w_{100}(r, \Gamma \to \infty) = \frac{\Gamma}{2} \left(\frac{1}{2\Gamma^2} - r^2 \right) + \mathcal{O}\left(\Gamma^{-3} \right), \quad (4.68)$$

where we have used the relations[7]

$$\sum_{n=-\infty}^{\infty} \cos^{-5} \lambda_n = 0 \quad \text{and} \quad \sum_{n=-\infty}^{\infty} \sin^2 \lambda_n \cos^{-3} \lambda_n = 1. \quad (4.69)$$

It is not possible to calculate the asymptotic surface deformations by taking the formal limit $\xi_{001}(\Gamma \to \infty)$, since the resulting expressions diverge periodically (cf. (4.61)). Between the poles of ξ_{001}, however, there exist ranges of Γ

---

[7] The relations can be derived from the boundary conditions at $r = 1/\Gamma$ for $z = 0$.

within which $Ca|\xi_{001}| \ll 1/\Gamma$, such that ξ_{001} is a valid approximation to the exact surface shape, regardless of its stability.

4.3 Axial Through Flow

The method of biorthogonal functions can be extended to thermocapillary flows in liquid bridges subject to axial through flow. The interest in this problem is motivated by propagating melt and solidification fronts in crystal growth. Some aspects of weak through flow have been treated fully numerically by Alexander et al. (1989) for the Bridgman–Stockbarger configuration (Monberg (1994)).

In the following we shall make use of the biorthogonal functions for the problem

$$\Delta (\Delta - \beta \partial_z) \psi = 0, \qquad \beta \in \mathbb{R}, \tag{4.70}$$

to calculate the convection in cylindrical volumes in the presence of axial through flow and weak thermocapillary surface forces. The associated biorthogonal functions are derived in Sect. A.3 in analogy to Sect. 4.1. More details can be found in Kuhlmann & Adabala (1993).

For a mathematical formulation of the problem the stream function (4.2–4.3) is employed and the equations are considered on viscous scales as in Sect. 4.2. It is further assumed that the axial through flow is homogeneous at the boundaries $z = \pm 1/2$. The boundary conditions (4.4) need merely to be modified to

$$\left.\begin{array}{l} \partial_z \psi = 0 \\ D_* \psi = -4\beta \\ \Theta = 0 \end{array}\right\} \quad \text{on} \quad z = \pm 1/2, \tag{4.71}$$

corresponding to a homogeneous axial flow with velocity $w_T = 4\beta$ through the isothermal walls. The volume equations (4.32–4.33) remain unaltered.

It has been shown in Sect. 4.2.1 that the thermocapillary flow field decouples from convection-induced surface deflections at the lowest order in Ca. This no longer holds in the presence of through flow ($\beta \neq 0$). Therefore, we take the strict limit $Ca \to 0$ a priori, and the stream function must satisfy the boundary conditions (4.4).

For weak thermocapillary driving ($Re \ll 1$), the flow field decouples from the temperature field. If, moreover, the through flow is weak ($\beta \ll 1$), i.e. if the through flow is small compared to the diffusive velocity ν/d, then the total flow is a superposition of the axial through flow $\boldsymbol{u}_T = 4\beta \boldsymbol{e}_z$ and the thermocapillary convection $\boldsymbol{u}_{100} = (\partial_z \psi_{100}, -D_* \psi_{100})$; cf. Sect. 4.2.2.

In their silicon crystal-growth experiments, Cröll et al. (1991) have used axial-through-flow rates up to $\beta \approx 0.8$. For germanium, even higher values of the through-flow parameter are possible, because the kinematic viscosity is lower than that of silicon by a factor of four (see Sect. A.1). Hence, a superposition of the two flow fields is no longer allowed and nonlinear terms must be taken into account. These will be treated here in the lowest order of a perturbation expansion.

Expanding the stream function and the temperature field for small Re and Pr,

$$\psi = \psi_T + \sum_{n=1}^{\infty}\sum_{m=0}^{\infty} \text{Re}^n \text{Pr}^m \psi_{nm}\,, \tag{4.72}$$

$$\Theta = \Theta_T + \sum_{n=1}^{\infty}\sum_{m=0}^{\infty} \text{Re}^n \text{Pr}^m \Theta_{nm}\,, \tag{4.73}$$

where the subscript T denotes the solution for pure axial through flow (Re = 0), and inserting the expansions into (4.32) and (4.33), one obtains at lowest order (Re = 0)

$$\psi_T = -2\beta r\,, \tag{4.74}$$

$$\Delta\Theta_T = 4\beta\text{Pr}\,(\partial_z\Theta_T + 1)\,. \tag{4.75}$$

The solution of (4.75) with the boundary conditions (4.45) reads, at the lowest nonvanishing order of the Prandtl number,

$$\Theta_T = \text{Pr}\left[2\beta\left(z^2 - \frac{1}{4}\right) + \text{Bi}\sum_{k=0}^{\infty} a_k I_0(2\mu_k r)\cos(2\mu_k z)\right] + \mathcal{O}(\text{Pr}^2)\,, \tag{4.76}$$

where

$$a_k = \frac{2\beta\,(-1)^k}{\mu_k^3\,[2\mu_k I_1(2\mu_k/\Gamma) + \text{Bi}\,I_0(2\mu_k/\Gamma)]}\,, \quad \mu_k = \left(k + \frac{1}{2}\right)\pi\,. \tag{4.77}$$

The sum in (4.76) describes the effect of heat transfer to the ambient atmosphere, for which we have assumed a linear temperature profile $\theta_a = z$.

To calculate the influence of the basic flow ψ_T and Θ_T on the thermocapillary flow, we consider (4.32) and (4.33) at order $\mathcal{O}(\text{Re}^1\text{Pr}^0)$ to obtain

$$\tilde{L}\left(\tilde{L} - 4\beta\partial_z\right)\psi_{10} = 0\,, \tag{4.78}$$

$$\Delta\Theta_{10} = 0\,, \tag{4.79}$$

subject to the boundary conditions (4.4) and (4.45). The solution for the temperature field is $\Theta_{10} = 0$. The thermocapillary convection at $\mathcal{O}(\text{Re}^1\text{Pr}^0)$ has no influence on the temperature field, even in the presence of axial through flow. Buoyancy forces could have been taken into account in (4.78) for Ca = 0. At lowest order, however, they do not influence the flow field, since $\Theta_{10} = 0$ and $\Theta_T = \mathcal{O}(\text{Pr})$.

Equation (4.78) resembles the problem of Oseen flow (Oseen (1910)). Here, the solution ψ_{10} is written as a sum of biorthogonal functions $\phi_1^{(n)}$ derived in Sect. A.3,[8]

$$\psi_{10}(r,z) = \sum_{n=-\infty}^{\infty} c_n I_1(2\lambda_n r) \frac{\phi_1^{(n)}(z)}{\lambda_n^2}. \tag{4.80}$$

As in Sect. 4.2.2 the unknown coefficients c_n must be determined with the help of the two boundary conditions (4.4) for $r = 1/\Gamma$:

$$\begin{pmatrix} 1 \\ 0 \end{pmatrix} = \begin{pmatrix} DD_* \\ \partial_z^2 \end{pmatrix} \psi_{10}\left(\Gamma^{-1}, z\right) = \sum_{n=-\infty}^{\infty} 4 c_n I_1\left(\frac{2\lambda_n}{\Gamma}\right) \boldsymbol{\Phi}^{(n)}(z). \tag{4.81}$$

Applying the operator

$$\int_{-1/2}^{1/2} \left[\psi_1^{(m)}, \psi_2^{(m)}\right] \cdot \begin{pmatrix} 0 & -1 \\ 1 & 2 \end{pmatrix} dz \tag{4.82}$$

to (4.81) and making use of the biorthogonality condition (A.58), one obtains

$$J_m = \int_{-1/2}^{1/2} \psi_2^{(m)} \, dz = 4 c_m I_1\left(\frac{2\lambda_m}{\Gamma}\right) K_m. \tag{4.83}$$

The normalizing constants K_m and the integrals J_m are given in the Appendix ((A.59) and (A.61)), and the solution for ψ_{10} can be written as

$$\psi_{10}(r,z) = \lim_{N\to\infty} \sum_{n=-N}^{N} \frac{J_n}{4\lambda_n^2 K_n} \frac{I_1(2\lambda_n r)}{I_1(2\lambda_n/\Gamma)} \phi_1^{(n)}(z). \tag{4.84}$$

In the limit $\beta \to 0$ one obtains $J_n \to 2$, $K_n \to -2\cos^4\lambda_n$, and the stream function ψ_{100} of Sect. 4.2.2 is recovered consistently.

Having calculated ψ_{10} it is easy to specify the lowest-order contributions to the temperature field. As for $\beta = 0$, the equation for the temperature field at order $\mathcal{O}(\mathrm{Re}^1\mathrm{Pr}^1)$ is given by (4.51) together with the boundary conditions (4.45). After some calculations one obtains

$$\Theta_{11}(r,z) = \sum_{n=-\infty}^{\infty} \frac{J_n}{2\lambda_n K_n I_1(2\lambda_n/\Gamma)}$$

$$\times \left\{ I_0(2\lambda_n r)\left[-\bar{\phi}_1^{(n)}(z) + A_*^{(n)}\cos(2\lambda_n z) + B_*^{(n)}\sin(2\lambda_n z)\right] \right.$$

$$\left. + \sum_{k=0}^{\infty} B_{nk} I_0(2\mu_k r)\cos(2\mu_k z) + \sum_{k=1}^{\infty} C_{nk} I_0(2\nu_k r)\sin(2\nu_k z) \right\},$$

$$\tag{4.85}$$

---

[8] We use the same symbols for the biorthogonal functions, their adjoint functions, and the characteristic roots as in Sects. 4.1 and 4.2.

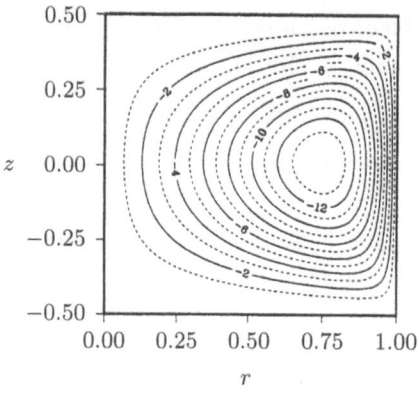

Fig. 4.10. Stream function $\psi_{10} \times 10^3$ for $\Gamma = 1$, $\beta = 0.2$, and $N = 40$

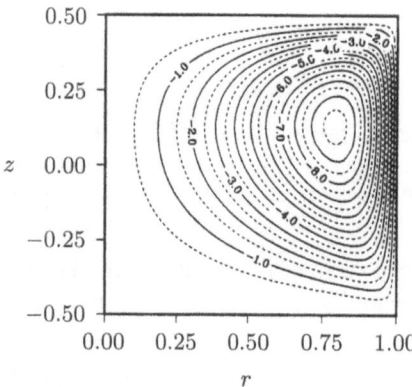

Fig. 4.11. Stream function $\psi_{10} \times 10^3$ for $\Gamma = 1$, $\beta = 4$, and $N = 40$

where $\nu_k = k\pi$ and $\mu_k = (k + \frac{1}{2})\pi$. The constants $A_*^{(n)}$, $B_*^{(n)}$, B_{nk}, C_{nk}, and the functions $\bar{\phi}_1^{(n)}$ are listed in Sect. A.3. Summarizing, we can write

$$\psi = \psi_T + \mathrm{Re}\,\psi_{10} + \mathcal{O}(\mathrm{Re}^2, \mathrm{RePr})\,, \tag{4.86}$$

$$\Theta = \mathrm{Pr}\,\Theta_T + \mathrm{RePr}\,\Theta_{11} + \mathcal{O}(\mathrm{Re}^2, \mathrm{Pr}^2)\,. \tag{4.87}$$

Examples of the thermocapillary stream function ψ_{10} are shown in Figs. 4.10–4.12 for a mode truncation $N = 40$. With increasing through flow ($\beta > 0$, flow in the positive z direction) the center of the toroidal vortex is shifted towards the hot corner ($z = 0.5$, $r = 1/\Gamma$), and the vortex becomes weaker. This behavior is typical for all parameter combinations. Since the temperature field Θ_{11} is driven by ψ_{10}, it depends on Γ and β in a similar way (Fig. 4.13). Owing to the asymmetry of the streamlines with respect to $z = 0$, ever more modes must be kept in the series for increasing β. For long liquid bridges ($\Gamma \gg 1$) the flow has a Poiseuille profile with finite mean w_T. Only close to $z = \pm 1/2$ are asymmetric modifications of the Poiseuille flow present, similar to those in Figs. 4.10–4.12.

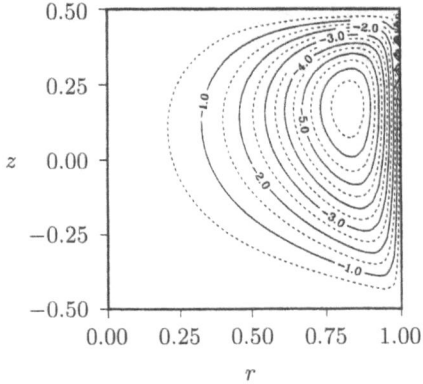

Fig. 4.12. Stream function $\psi_{10} \times 10^3$ for $\Gamma = 1$, $\beta = 8$, and $N = 40$

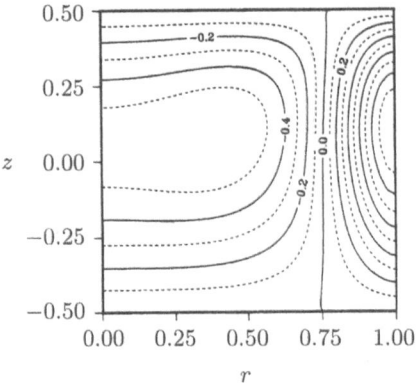

Fig. 4.13. Temperature field $\Theta_{11} \times 10^3$ for $\Gamma = 1$, $\beta = 8$, and $N = K = 40$

If $\Gamma \ll 1$, a sequence of vortices is obtained as a generalization of (4.64),

$$\psi_{10}(1 \ll r \ll 1/\Gamma, z) \cong \frac{J_1}{4\lambda_1^2 K_1} \frac{\phi_1^{(1)}(z)}{\sqrt{\Gamma r}} e^{2\lambda_1(r - 1/\Gamma)} + \text{c.c.} \qquad (4.88)$$

The vortices have a common shape and, for large values of β, their centers are shifted in the positive z direction because of the asymmetry of $\phi_1^{(1)}(z)$. Moreover, the separating streamlines between neighboring vortices with opposite sense of rotation are very asymmetric. Typical flow structures are shown in Fig. 4.14. Figure 4.14a shows the contour lines of the basic mode ($N = 1$ in (4.84)) for $\beta = 1$. The shape of the streamlines is similar to that for $\beta = 0$ (cf. Fig. 4.9 or Rybicki & Floryan (1987a)). For $\beta = 8$, the flow pattern has become quite asymmetric and the radial vortex diameters have grown. The strength of circulation for $\beta = 8$ is reduced by a factor of about 40 compared to $\beta = 1$. From (4.88) and Table A.3 it can be seen that the exponential factor $e^{2\Re(\lambda_1)r}$ determining the radial decay rate is responsible for the flow suppression ($2\Re[\lambda_1(\beta = 0)] = 4.212$ compared to $2\Re[\lambda_1(\beta = 8)] = 5.968$). Similarly, the increased radial diameter of the vortices can be explained

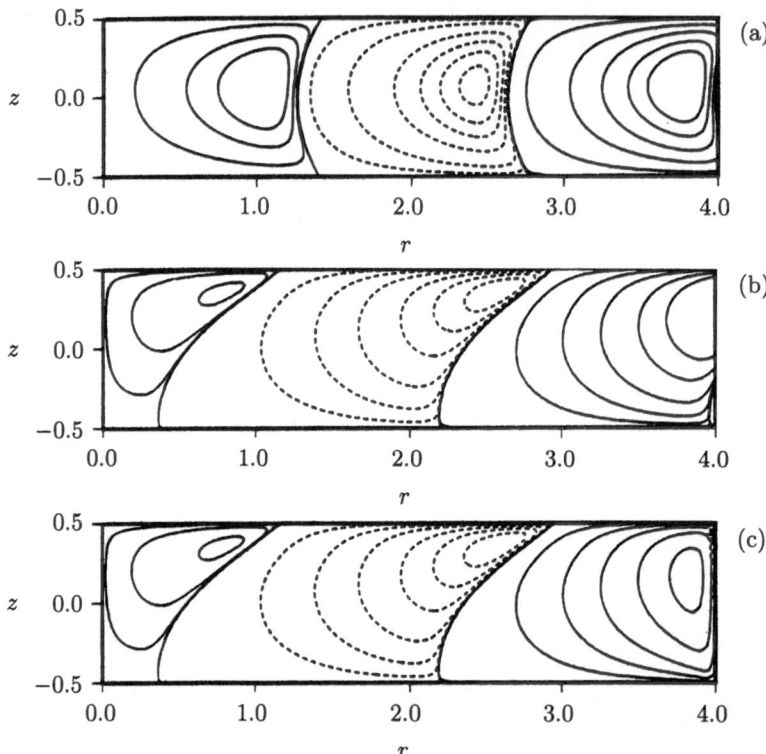

Fig. 4.14. Stream function $\psi_{10}\times10^3$ for $\Gamma=0.25$ and $\beta=1$, $N=1$ (**a**), $\beta=8$, $N=1$ (**b**), and $\beta=8$, $N=40$ (**c**). (——) Negative and (- - -) positive values of the stream function. Contour lines are drawn at 0, $(-9\times10^{-3}, -6\times10^{-3}, -3\times10^{-3}, -10^{-3}, -10^{-4})$, $(4\times10^{-5}, 3\times10^{-5}, 2\times10^{-5}, 10^{-5}, 2\times10^{-6}, 2\times10^{-7})$ and at $(-10^{-7}, -5\times10^{-8}, -10^{-8})$ for (**a**). For (**b**) and (**c**) contour lines are shown at 0, $(-5\times10^{-3}, -2\times10^{-3}, -5\times10^{-4}, -10^{-4}, -10^{-5})$, $(10^{-6}, 5\times10^{-7}, 2\times10^{-7}, 5\times10^{-8}, 10^{-8}, 10^{-9})$, and $(-5\times10^{-11}, -10^{-11}, -10^{-12})$

$(\pi/2\Im[\lambda_1(\beta=0)] = 1.396$ compared to $\pi/2\Im[\lambda_1(\beta=8)] = 1.773)$. The intersection of the contour lines with the free-surface boundary ($r=4$) in Fig. 4.14b indicates the poor convergence rate for increasing β. Therefore, only the middle vortices in Figs. 4.14a,b exhibit the typical small-aspect-ratio form ($\Gamma\ll1$). While increasing the truncation order up to $N=40$ improves the convergence near the free surface $1/\Gamma=1$, the vortices in the interior are hardly affected by the value of N.

In conclusion, the through-flow-induced asymmetry of the fields leads to high gradients, particularly close to the downstream corner (the hot corner at $z=0.5$, $r=1/\Gamma$ for $\beta>0$). This effect slows down the convergence for high values of β. The relative strength of Re and β determines whether the flow is dominated by toroidal vortices or by axial flow. For $\beta=8$, this transition occurs near $|\text{Re}|\approx400$. As an example, the total flow $\psi_T+\text{Re}\,\psi_{10}$ is illustrated in Fig. 4.15 for Re$=\pm400$.

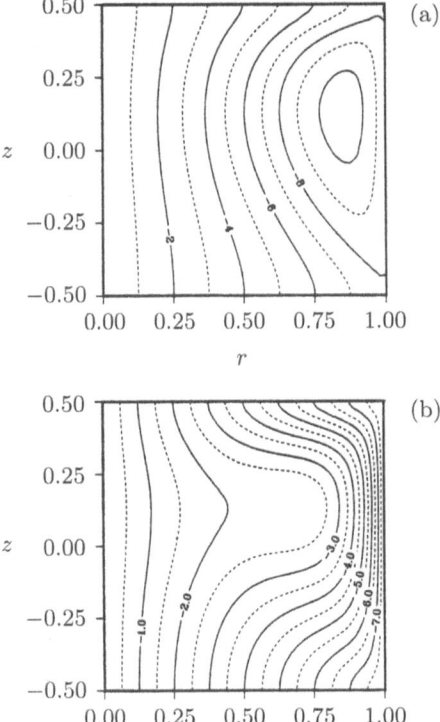

Fig. 4.15. Stream function $\psi_T + \mathrm{Re}\psi_{10}$ for $\Gamma = 1$, $\beta = 8$, $N = 40$, for $\mathrm{Re} = 400$ (**a**) and $\mathrm{Re} = -400$ (**b**)

5. Concepts of Stability Analysis

In the limit of small Reynolds and Marangoni numbers the thermocapillary flow is laminar and unique. However, in many applications, such as crystal growth (Schwabe (1988)), combustion (Sirignano & Glassman (1970)), welding (DebRoy & David (1995)), and bubble migration (Subramanian (1992)), the driving of the flow is strong and the characteristic nondimensional numbers are large. Nonlinear terms in the Navier–Stokes equations become important then, and different flow states may exist for $t \to \infty$. Not all of these solutions must necessarily be observable in experiments. Which flow state is realized in a given experiment depends on the initial conditions and on the stability of the respective solution. The types of solutions and their stability properties are determined by the external parameters. Experiments on thermocapillary liquid bridges carried out by Velten et al. (1991) (see Fig. 7.47), for instance, demonstrated that many different types of flow, ranging from regular to chaotic and turbulent states, may be observed depending on the geometry and the Reynolds number.

To understand the laminar pattern formation the solutions of the Navier–Stokes equations must be calculated, their stability investigated, and possible instability mechanisms identified. Regarding stability, the reaction of the system to external perturbations is crucial. In experiments, perturbations may be induced by vibrations, fluctuations, or deviations from the ideal geometry. The general stability problem is not easy to solve. For closed flow systems, it is of advantage to consider the dynamical evolution of the system when it is perturbed at some time, say $t = 0$, i.e. when the inital conditions deviate from the solution of the Navier–Stokes equations in question. To investigate the stability of a basic state (u_0, p_0, θ_0) in this manner two complementary methods have been developed: *linear stability analysis* and *energy stability analysis*. The basics of both methods will be explained in the following. For further studies the reader is referred to Chandrasekhar (1961), Drazin & Reid (1981), and Joseph (1976).

5.1 Linear Stability

Linear stability analysis deals with the influence of small initial perturbations on the temporal development of the solution. The technique is based on the definition of stability in the sense of Liapunov for dynamical systems (see e.g. Jordan & Smith (1987)).

Let $X_0(t)$ be a solution of the dynamical system $\partial_t X = v(X, t)$. Then the solution $X_0(t)$ is said to be *Liapunov-stable* for $t > 0$, if for all $\epsilon > 0$ there exists a $\delta(\epsilon) > 0$ such that

$$||X(0) - X_0(0)|| < \delta \quad \Longrightarrow \quad ||X(t) - X_0(t)|| < \epsilon, \qquad (5.1)$$

for all $t \geq 0$, where $||...||$ is a norm. Otherwise, $X_0(t)$ is said to be *Liapunov-unstable*. Essentially, this means that the basic solution $X_0(t)$ is stable if all initially small perturbations remain sufficiently small for all times. If all perturbations vanish for $t \to \infty$ ($\lim_{t \to \infty} ||X(t) - X_0(t)|| = 0$), the basic solution is said to be *asymptotically stable*.

Now let $X_0(r) = [u_0(r), p_0(r), \theta_0(r)]$ be a stationary solution[1] of the Navier-Stokes equations for given boundary conditions. In addition, let the solution due to a small (infinitesimal) perturbation at time $t = 0$ be given by $X_0(r) + \epsilon X(r, t)$ for $t > 0$, where $\epsilon X = \epsilon(u, p, \Theta)$ is the deviation from the basic solution and $\epsilon \ll 1$ a small parameter. The Oberbeck-Boussinesq equations (2.14–2.16) for the evolution of the deviations $X(r, t)$ then read

$$\partial_t u(t) + \text{Re}\,[u(t) \cdot \nabla u_0 + u_0 \cdot \nabla u(t)] = -\nabla p(t) + \Delta u(t) + \frac{\text{Gr}}{\text{Re}} \Theta(t) e_z, \qquad (5.2)$$

$$\partial_t \Theta(t) + \text{Re}\,[u(t) \cdot \nabla \theta_0 + u_0 \cdot \nabla \Theta(t)] = \frac{1}{\text{Pr}} \Delta \Theta(t), \qquad (5.3)$$

together with the continuity equation for u. The nonlinear terms have been neglected, because they are of lesser order of magnitude ($\mathcal{O}(\epsilon^2)$). Since the equations for the perturbations $X(r, t)$ are autonomous, the solutions of the linear system (5.2) and (5.3) depend exponentially on time, $X(r, t) = Y(r) e^{-i\omega t}$. The equations therefore reduce to the eigenvalue problem

$$-i\omega Y(r) = \mathcal{A} Y(r), \qquad (5.4)$$

where $\mathcal{A} = \mathcal{A}[\text{Re}, \text{Gr}, \text{Pr}, \text{Bi}, u_0(r), p_0(r), \theta_0(r)]$ is a linear operator. For systems with finite spatial extension and realistic boundary conditions, (5.4) must in general be solved numerically. As the solution, one obtains a discrete spectrum of complex eigenvalues $\{\omega_n\}$ with associated spatial modes (eigenfunctions). The growth rate of a mode, $\sigma_n = \Im(\omega_n)$, is given by the imaginary part of ω_n, while the real part, $\Re(\omega_n)$, represents the oscillation frequency. Here, the spectrum $\{\omega_n\}$ depends on the parameters Re, Gr, Pr, and Bi, as well as on the basic flow state X_0.

An arbitrary perturbation at time $t = 0$ may be any superposition of the eigenfunctions Y. Therefore, within the linear theory, the perturbed solution grows exponentially without bounds if at least one fundamental mode Y_n exists for which the eigenvalue has a positive imaginary part. The basic flow

---

[1] The stability of time-periodic basic flows can be treated by Floquet theory; see Joseph (1976).

is then linearly unstable. If the imaginary parts of all eigenvalues are negative, the basic flow is linearly asymptotically stable. A basic flow is said to be neutrally or marginally stable if the maximum growth rate over all modes vanishes, i.e. $\max_n \Im(\omega_n) = 0$. The curves or hypersurfaces in the parameter space on which $\Im(\omega_n) = 0$ are called neutral curves (or surfaces). The linear-stability boundary is the envelope curve on which $\max_n \Im(\omega_n) = 0$. This curve separates parameter regions in which the basic flow is linearly stable from those in which it is linearly unstable. Typically, one is interested in the stability of a basic solution X_0 when a parameter (control parameter) is increased. That value of the control parameter, e.g. Re, for which $\max_n \Im(\omega_n) = 0$ is called the *critical* value (e.g. Re_c).

5.1.1 Temporal and Spatial Stability

If the system under consideration is of infinite extent in one or more coordinate directions and the basic state is independent of the corresponding coordinate, any arbitrary perturbation vector $X(r, t)$ can be expressed as a Fourier integral. In the case of infinite extension in the x direction, for instance,

$$X(r, t) = \int_{-\infty}^{\infty} Y_k(y, z)\, e^{i(kx - \omega t)}\, dk. \tag{5.5}$$

The modes $\propto e^{ikx}$ are called *normal modes*. The wavenumber k represents a continuous index and the spectrum $\{\omega_n(k)\}$ now has continuous parts. For a determination of the critical control parameter (critical point) the growth rates must also be maximized with respect to the continuous wavenumber k, or the control parameter on the neutral curves must be minimized with respect to k. Using normal modes, the eigenvector $Y_k(y, z)$ does not depend on the coordinate in which the system has infinite extent. In this case only a system of linear differential equations of a reduced spatial dimension must be solved. If one is primarily interested in the stability boundaries, this is a decisive advantage of the normal-mode linear-stability analysis compared to a full numerical simulation.

The linear analysis loses its validity in the supercritical range $(Re > Re_c)$, after the initial exponential growth has led to amplitudes for which quadratic terms in ϵ are no longer negligible. The nonlinear terms lead to a saturation of the amplitudes at finite values. It is clear that linear stability analysis cannot give any information about the form or the amplitude of the solution in the supercritical parameter range. Quite often, however, the structure of the weakly nonlinear solution is similar to that of the neutral solution on the stability boundary. For an approximation of the Navier–Stokes equations in the weakly nonlinear range the so-called *amplitude equations* have proven valuable (Cross & Hohenberg (1993), Newell et al. (1993), and Sect. 7.7.2) owing to their generic form for many different systems.

In the above conventional analysis the temporal growth rate $\sigma(k) = \Im[\omega(k)]$ is considered as function of the real wavenumber $k \in \mathbf{R}$. The approach is, therefore, called temporal stability analysis. Since we have assumed that (5.2) and (5.3), including the boundary conditions, are homogeneous in both the space coordinate x and time, we could have likewise taken $\omega \in \mathbf{R}$ independently with normal modes $\propto e^{-i\omega t}$, and enquired about the corresponding spatial growth rate $\sigma(\omega) = -\Im(k)$. Such an analysis is called spatial stability analysis. It is useful if the length of the system in the x direction is very large but finite. We shall come back to this approach in Sect. 8.1.2, where the influence of side walls is studied.

5.1.2 Convective and Absolute Stability

As a generalization of temporal and spatial analyses, one can regard both ω and k as complex. The eigenmodes $\propto e^{i(kx-\omega t)}$ are thus waves of wavenumber $\Re(k)$ and frequency $\Re(\omega)$ which grow exponentially in both space and time, with spatial growth rate $-\Im(k)$ and temporal growth rate $\Im(\omega)$, respectively. If one takes k as the independent variable, the eigenvalue problem (5.4) leads to a dispersion relation for the complex frequency

$$D(\omega, k, \mathrm{Re}, ...) = 0 \,, \tag{5.6}$$

which depends on the governing parameters, such as the Reynolds number. The concepts of convective and absolute stability have been reviewed by Huerre & Monkewitz (1990) and Huerre & Rossi (1998). Priede & Gerbeth (1997b) applied these methods to thermocapillary flows in plane layers. We follow these authors to illustrate the basic ideas.

Consider the temporal evolution of a Gaussian wave packet of width $4a$ centered around the basic wavenumber k_c,

$$\psi(x, t{=}0) = e^{-(x/2a)^2} e^{ik_c x} \,. \tag{5.7}$$

where $k_c \in \mathbf{R}$ is the real wavenumber at which, for given parameters, the temporal growth rate $\Im(\omega)$ has its maximum value (the fastest-growing mode). The spectral amplitudes are given by the Fourier transform

$$\hat{\psi}(k, t{=}0) = \frac{1}{\sqrt{2\pi}} \int_{-\infty}^{\infty} \psi(x, t{=}0)\, e^{-ikx} \mathrm{d}x = \sqrt{2}\, a\, e^{-a^2(k-k_c)^2} \,. \tag{5.8}$$

The temporal evolution of the wave packet is obtained by advancing and superposing all modes:

$$\psi(x, t) = \frac{1}{\sqrt{2\pi}} \int_{-\infty}^{\infty} \hat{\psi}(k, t{=}0)\, e^{i(kx-\omega t)} \mathrm{d}k \,. \tag{5.9}$$

For a wave packet sufficiently wide in space the amplitudes $\hat{\psi}$ drop off very sharply from k_c and the dispersion relation (5.6) can be expanded as

$$\omega(k) = \omega_c + \omega_k(k - k_c) + \frac{1}{2}\omega_{kk}(k - k_c)^2 + \mathcal{O}(k - k_c)^3 \,, \tag{5.10}$$

where $\omega_c = \omega(k_c) \in \mathbb{C}$ and the group velocity is ω_k. Since the growth rate has a maximum at k_c, the imaginary part of the group velocity must vanish, i.e. $\Im(\omega_k) = 0$ and $\Im(\omega_{kk}) < 0$. Using this expansion one obtains

$$\psi(x,t) = \sqrt{\frac{2a^2}{2a^2 + i\omega_{kk}t}} \, \exp\left(-\frac{(x - \omega_k t)^2}{4a^2 + 2i\omega_{kk}t}\right) e^{i(k_c x - \omega_c t)}. \tag{5.11}$$

The real part of the exponential factor determines the growth of the amplitude. With $U = x/t$ being the velocity of the frame of reference in which the wave packet is observed, the exponential factor takes the following asymptotic form for long times $t \to \infty$:

$$\exp\left(-\frac{(x - \omega_k t)^2}{4a^2 + 2i\omega_{kk}t}\right) \stackrel{t \to \infty}{\longrightarrow} \exp\left(\frac{i\omega_{kk}^*(U - \omega_k)^2 t}{2|\omega_{kk}|^2}\right), \tag{5.12}$$

where the asterisk denotes the complex conjugate. From (5.11) we obtain the growth rate in the moving frame of reference:

$$\sigma = \Im(\omega_c) + \Re\left(\frac{i\omega_{kk}^*(U - \omega_k)^2}{2|\omega_{kk}|^2}\right) = \Im(\omega_c) + \frac{\Im(\omega_{kk})}{2|\omega_{kk}|^2}(U - \omega_k)^2. \tag{5.13}$$

Since $\Im(\omega_{kk}) < 0$, the growth rate σ is always less than the maximum temporal growth rate $\Im(\omega_c)$. Only in the frame of reference moving with the group velocity $U = \omega_k$ does the packet attain its maximum growth rate, which is identical to that obtained from the temporal analysis, up to the algebraic decay due to the dispersion.

Since the convective stability limit is defined by a zero growth rate in the frame of reference moving with the group velocity of the packet, the temporal stability limit is identical to the the convective stability limit. The absolute stability limit is determined by $\sigma(U = 0) = 0$. In this case the decay of the amplitude due to the propagation of the wave packet away from a fixed location is just compensated by the temporal growth. Thus the absolute stability limit is generally larger than the convective stability limit.

If the system has reflection symmetry for $x \to -x$ the dispersion relation must be symmetric with respect to k. For a Hopf bifurcation ($\Re(\omega_c) \neq 0$), left- and right-traveling wave packets centered around k_c and $-k_c$, respectively, appear with opposite group velocities. If the instability is stationary ($\Re(\omega_c) = 0$) the group velocity is zero ($\omega_k = 0$). In this case the growth rate vanishes in the stationary frame of reference, rendering the convective and absolute stability boundaries the same. This is realized, for example, in the Rayleigh–Bénard problem (Drazin & Reid (1981)).

Another type of instability is of interest when the system is extended but has a finite length L. Waves can then be reflected by the distant boundaries and may give rise to a state that occupies the system globally. The critical condition for such a global state with real frequency ω is that its amplitude is stationary, i.e. the temporal growth must vanish ($\Im(\omega) = 0$). In the limit $L \to \infty$ the spectrum of discrete wavenumbers that satisfy a constructive interference condition becomes continuous and there is no restriction on the frequency. Moreover, the amplitude reduction upon reflection becomes insignificant compared to the spatial damping or amplification over L. Thus the critical state for a global instability is realized when the spatial growth rate of the fastest-growing mode in one direction just compensates the spatial decay rate of the least-attenuated mode in the other direction, i.e.

$$\Im(k_+) - \Im(k_-) = 0, \tag{5.14}$$

where the wavenumbers k_+ and k_- belong to the modes having the largest spatial growth rate in the positive and negative x directions, respectively. Equation (5.14) defines the global stability boundary in extended systems (see also Priede & Gerbeth (1997b)).

5.2 Energy Analysis

If a flow is evolving dynamically, a knowledge of the energy transport from its injection to its dissipation may provide useful insight into the fluid mechanics. This is particularly true of hydrodynamic instabilities, where energy is fed from the basic state to the disturbance flow. For later use, we shall derive the energy equations for a cylindrical liquid bridge. Generalizations are straightforward.

5.2.1 Reynolds–Orr Equation

Consider the momentum equation (2.14), using the summation convention,

$$\partial_t u_i + \mathrm{Re}\, u_j \partial_j u_i = -\partial_i p + \partial_j^2 u_i + \mathrm{Bd}\, \theta\, \delta_{i3} . \tag{5.15}$$

Scalar multiplication by u_i and integration over the volume yields

$$\partial_t E_{\mathrm{kin}} = \frac{1}{2}\partial_t \langle u_i^2 \rangle = -\mathrm{Re}\langle u_i u_j \partial_j u_i \rangle - \langle u_i \partial_i p \rangle + \langle u_i \partial_j \partial_j u_i \rangle + \mathrm{Bd}\langle w\theta \rangle , \tag{5.16}$$

where $E_{\mathrm{kin}} = \frac{1}{2}\langle u_i^2 \rangle$ is the kinetic energy and $\langle ... \rangle = \int_V dV$. The pressure term $\langle u_i \partial_i p \rangle$ in (5.16) vanishes upon partial integration. The nonlinear term is

$$\langle u_i u_j \partial_j u_i \rangle \overset{\partial_j u_j = 0}{=} \langle u_i \partial_j u_j u_i \rangle$$
$$\overset{\mathrm{Int.}}{=} \underbrace{\langle u_i e_j u_j u_i \rangle_S}_{\to 0} - \langle (\partial_j u_i) u_j u_i \rangle , \tag{5.17}$$

where $\langle ... \rangle_S = \int_S dS$ is the integral over the surface of the volume and e_j the unit normal vector of the surface. Since the equation has the form $a = -a$, the nonlinear term $\langle u_i u_j \partial_j u_i \rangle = 0$ vanishes identically. Consider next the diffusive term

$$\langle u_i \partial_j \partial_j u_i \rangle \overset{\mathrm{Int.}}{=} \langle u_i e_j \partial_j u_i \rangle_S - \langle (\partial_j u_i)(\partial_j u_i) \rangle . \tag{5.18}$$

The second term on the right-hand side is just the negative of the positive rate of dissipation in the volume $\langle(\nabla\boldsymbol{u})^2\rangle$.[2] This term always contributes to a reduction of the kinetic energy. The first term is zero for all rigid surfaces. But it is nonzero on the free surface, where it describes the Marangoni forcing

$$\langle u_i e_j \partial_j u_i\rangle_S = \int_S (\mathrm{d}^2\boldsymbol{r}\cdot\nabla\boldsymbol{u})\cdot\boldsymbol{u} = \int_S (\partial_r\boldsymbol{u})\cdot\boldsymbol{u}\;\mathrm{d}S$$

$$= \int_S \left(\underbrace{u\partial_r u}_{\to 0} + v\partial_r v + w\partial_r w\right)\mathrm{d}S$$

$$= \int_S v\tilde{S}_{r\varphi}\mathrm{d}S + \int_S w\tilde{S}_{rz}\mathrm{d}S + \int_S \frac{v^2}{r}\mathrm{d}S$$

$$= M_\varphi + M_z + \int_S \frac{v^2}{r}\mathrm{d}S\,; \tag{5.19}$$

here M_φ and M_z denote the work done per time by external shear forces; in this case, by azimuthal and axial Marangoni forces. The latter relation is valid because the components of the viscous stress tensor on the free cylindrical surface are $\tilde{S}_{rz}=(\partial_r w+\partial_z u)_S=\partial_r w$ and $\tilde{S}_{r\varphi}=[\partial_r v+\frac{1}{r}(\partial_\varphi u-v)]_S=\partial_r v-v/r$. We are left with the Reynolds–Orr equation

$$\partial_t E_{\mathrm{kin}} = -D + M_\varphi + M_z + I_{\mathrm{Gr}}\,, \tag{5.20}$$

where $I_{\mathrm{Gr}}=\mathrm{Bd}\langle w\theta\rangle$ is the work done per unit time by buoyancy forces and

$$D = \langle(\nabla\boldsymbol{u})^2\rangle - \int_S \frac{v^2}{r}\mathrm{d}S = \langle(\nabla\times\boldsymbol{u})^2\rangle - 2\int_S \frac{v^2}{r}\mathrm{d}S \tag{5.21}$$

---

[2] The dissipation in the volume $\langle(\nabla\boldsymbol{u})^2\rangle$ differs from the dissipation $D = \langle(\partial_i u_j)^2\rangle + \langle\partial_i u_j\partial_j u_i\rangle$ by a term that can be expressed as a surface integral (see Raynal (1996)). For rigid nonrotating boundary conditions, $\boldsymbol{u}=0$, the additional term vanishes. The dissipation in the volume can also be expressed in terms of the integral over the enstrophy $(\nabla\times\boldsymbol{u})^2$. Consider

$$\langle(\nabla\times\boldsymbol{u})^2\rangle = \langle\epsilon_{ijk}\partial_j u_k\epsilon_{ilm}\partial_l u_m\rangle.$$

By noting $\epsilon_{ijk}\epsilon_{ilm}=\delta_{jl}\delta_{km}-\delta_{jm}\delta_{kl}$ this is equal to

$$\langle\epsilon_{ijk}\partial_j u_k\epsilon_{ilm}\partial_l u_m\rangle = \langle(\partial_j u_k)(\partial_j u_k)\rangle - \langle(\partial_j u_k)(\partial_k u_j)\rangle\,.$$

The first term is the negative rate of dissipation in the volume, while the second term gives, upon partial integration,

$$-\int_S \boldsymbol{u}\cdot\nabla\boldsymbol{u}\cdot\mathrm{d}^2\boldsymbol{r} = -\int_S \left(u\partial_r + \frac{v}{r}\partial_\varphi + w\partial_z\right)\boldsymbol{u}\cdot\boldsymbol{e}_r\,\mathrm{d}S$$

$$= -\int_S \left(u\partial_r u + \frac{v}{r}(\partial_\varphi - v) + w\partial_z u\right)\mathrm{d}S = \int_S \frac{v^2}{r}\mathrm{d}S\,.$$

We thus have

$$\langle(\nabla\boldsymbol{u})^2\rangle = \langle(\nabla\times\boldsymbol{u})^2\rangle - \int_S \frac{v^2}{r}\mathrm{d}S\,.$$

is the total dissipation. The dissipation is reduced by the presence of a nonzero azimuthal velocity on the free surface.[3] Thus energy is injected only by the surface forcing and by work done by buoyancy forces.

5.2.2 Energy Equations for Infinitesimal Perturbations

When the time evolution of an infinitesimal perturbation (u, p, Θ) superposed on some basic nonlinear flow $(u^{(0)}, p^{(0)}, \theta^{(0)})$ is investigated, we must consider the linearized momentum equation

$$\partial_t u_i + \mathrm{Re} \left(u_j^{(0)} \partial_j u_i + u_j \partial_j u_i^{(0)} \right) = -\partial_i p + \partial_j^2 u_i + \mathrm{Bd}\,\Theta \delta_{i3}\,, \qquad (5.22)$$

and the energy balance for the disturbances must be supplemented on the right-hand side of (5.20) by

$$\partial_t E_{\mathrm{kin}} = \dots - \mathrm{Re} \left(\langle u_i u_j^{(0)} \partial_j u_i \rangle + \langle u_i u_j \partial_j u_i^{(0)} \rangle \right)\,. \qquad (5.23)$$

The first term vanishes by the same arguments as for the nonlinear term (5.17) above. The second term describes the change of energy of the perturbation flow component u_i by convective transport due to the disturbance flow $u_j \partial_j$ of the basic-state velocity $u_i^{(0)}$. If the basic state is axisymmetric, $u^{(0)} = u_0 e_r + w_0 e_z$, five contributions I_{vi} remain:

$$- \mathrm{Re} \langle u_i u_j \partial_j u_i^{(0)} \rangle = I_v = \sum_{i=1}^{5} I_{vi} \qquad (5.24)$$

$$= -\mathrm{Re} \int_V \left(v^2 \frac{u_0}{r} + u^2 \frac{\partial u_0}{\partial r} + uw \frac{\partial u_0}{\partial z} + wu \frac{\partial w_0}{\partial r} + w^2 \frac{\partial w_0}{\partial z} \right) \mathrm{d}V\,,$$

and we obtain the rate of change of kinetic energy of the disturbance flow,

$$\partial_t E_{\mathrm{kin}} = -D + M_\varphi + M_z + I_{\mathrm{Gr}} + \sum_{i=1}^{5} I_{vi}\,, \qquad (5.25)$$

where i numbers the terms consecutively as they appear in (5.24). The terms D, M_φ, M_z, and I_{Gr} have the same form as those for the basic state.

A similar procedure leads to the balance for the *thermal energy*[4] of the disturbance flow, defined as $E_T = \frac{1}{2} \int_V \Theta^2\,\mathrm{d}V$:

$$\partial_t E_T = -D_T - H + \sum_{i=1}^{3} I_{Ti}\,, \qquad (5.26)$$

where

---

[3] This reduction becomes clear when considering the rigid-body rotation $u = \frac{1}{2}\omega r e_\varphi$. Since there is no relative fluid motion, the dissipation is zero. The surface integral exactly compensates the volume integral over the enstrophy.

[4] E_T must not be confused with the thermal energy in thermodynamics.

$$D_T = \frac{1}{\text{Pr}} \langle (\nabla \Theta)^2 \rangle, \qquad H = \frac{\text{Bi}}{\text{Pr}} \int_S \Theta^2 \, dS, \tag{5.27}$$

$$I_T = \sum_{i=1}^{3} I_{Ti} = -\text{Re} \int_V \Theta \left(u \frac{\partial \Theta_0}{\partial r} + w \frac{\partial \Theta_0}{\partial z} + w \right) dV, \tag{5.28}$$

and $\Theta_0 = \theta_0 - z$ is the deviation of the temperature from the conductive profile z (heating from above).

5.3 Energy Stability

From the conventional temporal linear stability analysis it can be concluded that the basic flow is unstable if at least one growth rate $\Im[\omega_n(k)]$ exists with a positive real part. The converse of this statement is not true, however, since linear stability analysis deals with the limited class of infinitesimal perturbations. There are systems, such as plane shear flows and systems with perturbed bifurcations, which are linearly stable in some parameter ranges, but unstable to large-amplitude perturbations. To determine the parameter range within which the stability of the basic flow with respect to *all* perturbations is guaranteed, the energy method has been devised. The method can be traced back to Reynolds (1895), Orr (1907), and Serrin (1959) and has been further improved by Joseph (1976).[5]

In order to apply the energy method the kinetic energy of the flow field or some other generalized functional in the sense of Liapunov (Jordan & Smith (1987)) is considered. The only requirement of the generalized energy functional $E(t)$ is that it is a positive measure of the velocity, temperature, and, eventually, other perturbation fields. The basic flow is said to be asymptotically stable in the mean if and only if

$$\lim_{t \to \infty} \frac{E(t)}{E(0)} \longrightarrow 0, \tag{5.29}$$

for all perturbations. If, at $t=0$, the energy $E(t)$ decays ($\partial_t E|_{t=0} < 0$) for *all* admitted initial conditions, the basic flow is also said to be monotonically stable.

To find a condition for guaranteed stability we consider the generalized energy

$$E(t) = E_{\text{kin}}(t) + \lambda E_T(t), \tag{5.30}$$

where E_{kin} is the kinetic energy of the flow field of the perturbation and E_T a measure of the deviation of the temperature field from the basic-state temperature field. Here λ is the so-called coupling parameter. It can be selected arbitrarily. In particular, λ can be chosen to maximize the parameter range

---

[5] Further applications of the energy method to hydrodynamic problems can be found in Straughan (1992).

for which the basic flow is asymptotically stable (Joseph (1976)).[6] From the rates of change of energy ((5.25) and (5.26)), the advantage of using of the energy as the Liapunov functional becomes clear: the advective and convective terms vanish identically upon integration over the volume, as they are energy-conserving. Therefore, the energy functional $E(t)$ is merely quadratic in the perturbations.

In order that the basic flow is monotonically stable the rate of change of energy at time $t=0$ must be negative or at most zero for all perturbations. This condition leads to

$$
\frac{1}{E} \frac{dE}{dt} \bigg|_{t=0} \leq c = \max_{\tilde{H}} \mathcal{F}[\boldsymbol{u}, \Theta]_{t=0}
$$
$$
= \max_{\tilde{H}} \left(\frac{-D + M_z + M_\varphi + I_{\mathrm{Gr}} + I_v + \lambda(-D_T - H + I_T)}{E} \right)_{t=0},
$$
$$(5.31)$$

where $\tilde{H}$ denotes the space of kinematically admissible functions for which the maximum of $\mathcal{F}$ must be sought. Kinematically admissible functions are those that satisfy the boundary conditions and the incompressibility constraint

$$
\tilde{H} = \left\{ \boldsymbol{u}, \Theta \,\middle|\, \nabla \cdot \boldsymbol{u} = 0, \quad (\boldsymbol{u}, \Theta)_{z \pm 1/2} = 0, \quad \text{and} \quad (2.34) \right\} . \qquad (5.32)
$$

The parameter values for which the relative rate of change of energy is $c=0$ determine the range for which the basic flow is monotonically stable. Often, this range is characterized by a maximum Reynolds number Re_E, which is called the *energy limit*. Since the basic flow could possibly be stable in a nonmonotonic manner, Re_E is only an estimate of the stable range. The basic flow may still be asymptotically stable for $\mathrm{Re} > \mathrm{Re}_E$. Moreover, it is possible that the basic flow is unstable only to perturbations of a certain amplitude but stable otherwise. The flow is then said to be conditionally stable. From this it is clear that the energy limit Re_E is only a lower bound on the true stability limit. Hence, the conventional linear-stability boundary Re_c (instability for $\mathrm{Re} > \mathrm{Re}_c$) is generally larger than the energy limit Re_E (stability for $\mathrm{Re} < \mathrm{Re}_E$). Only in some exceptional cases do both boundaries coincide.[7]

To determine Re_E, the functional $\mathcal{F}$ has to be maximized. As a necessary condition the first variation of $\mathcal{F}$ over the space of kinematically admissible functions must vanish

$$
\delta_{\tilde{H}} \mathcal{F} = 0. \qquad (5.33)
$$

---

[6] The optimization of λ can be crucial for the determination of the stability range (Shen et al. (1990)).

[7] This is the case, for example, for the onset of axisymmetric thermal convection in rigid cylinders heated from below; see Charlson & Sani (1970) and Yamaguchi et al. (1984).

Frequently, the incompressibility condition $\nabla \cdot \boldsymbol{u} = 0$ is included into the functional by means of a Lagrange multiplier which takes the role of the pressure in the resulting Euler–Lagrange equations. Then the variation need only be carried out for perturbations that satisfy the boundary conditions (Straughan (1992)). Instead of deriving the classical Euler–Lagrange equations, which are linear owing to the quadratic form of the functional $\mathcal{F}$, the variation can also be carried out directly by numerical methods (cf. Sect. 7.8 and Shen et al. (1990)).

Since the energy stability limit Re_E yields an estimate of the stable parameter range, the energy method is usually employed only if subcritical instabilities occur. An indication of a subcritical instability would be the experimental observation of hysteresis during variation of the Reynolds number.

5.4 A Posteriori Energy Balances

In the preceding section the rate of change of the energy of an arbitrary perturbation has been employed to find parameter ranges of guaranteed stability. But the energy balance equations (5.25) and (5.26) can also provide useful information in combination with a linear stability analysis. Since the nonlinear terms $\boldsymbol{u} \cdot \nabla \boldsymbol{u}$ and $\boldsymbol{u} \cdot \nabla \Theta$ do not enter the energy equations, the equations are equally valid for the neutral modes of a linear analysis. If the linear modes are inserted into the energy balances, the sign and the absolute value of the individual contributions to the energy growth can be determined. Because each term in the equations is associated with a specific energy transfer process, it is possible to identify those properties of the basic flow that are responsible for the instability in question. In the framework of a linear stability analysis the integrands in the rates of change of energy are explicitly known and one can calculate the spatial distribution of the energy transfer rates. The knowledge of these distributions does not only help understand the physical instability processes, it could also be valuable for control of the flow and the temperature field by external forces if, for instance, the onset of a particular instability is to be suppressed.[8]

Another advantage is that the energy balance equations allow one to check the energy conservation of solutions, which are usually obtained numerically. This test is independent of the method used to calculate the solution, but it depends on how the derivatives and integrals appearing in (5.25) and (5.26) are evaluated. To obtain an error measure we define the error in the kinetic energy balance as the normalized residual

$$\delta_{\mathrm{kin}} = \frac{\left| -\partial_t E_{\mathrm{kin}} - D + M_z + M_\varphi + I_{\mathrm{Gr}} + I_v \right|}{\max \left\{ D, |M_z|, |M_\varphi|, |I_{\mathrm{Gr}}|, |I_{v1}|, ..., |I_{v5}| \right\}}. \tag{5.34}$$

---

[8] The control of thermocapillary flows, aimed at preventing the onset of oscillatory convection, has become important recently; see Petrov et al. (1996) and Benz et al. (1998).

The error in the thermal energy balance is defined in the same way:

$$\delta_T = \frac{|-\partial_t E_T - D_T - H + I_T|}{\max\{D_T, H, |I_{T1}|, |I_{T2}|, |I_{T3}|\}} . \tag{5.35}$$

Since the dissipation rate D and the heat diffusion rate D_T are always positive, both processes lead to a decay of the generalized energy, i.e. both processes have a stabilizing effect. It can be seen, moreover, that the heat exchange of the perturbation with the ambient medium $(Bi \neq 0)$ has a stabilizing effect, since $H \geq 0$ (5.27). It must be kept in mind, however, that the basic state depends on the Biot number as well. Therefore, an increase of Bi could also have a destabilizing effect.

In the case of instability both the kinetic energy E_{kin} and the thermal energy E_T will grow in time. As the coupling parameter λ is undetermined, the relative importance of the two types of energy transfer process cannot be assessed a priori. By a careful consideration of the limits of small or large Prandtl numbers, however, thermal or inertial effects, respectively, may be excluded as a cause of the instability (see Sect. 7.4.1.).

6. Plane Thermocapillary Layers

For basic investigations of thermocapillary flow instabilities it is useful to consider a simplified model that only contains the essential features. Such a model is provided by an infinite plane liquid layer bounded from below by a rigid bottom and from above by a planar, free surface supporting thermocapillary stresses. As described in Sect. 5.1, stability analysis for infinite systems is facilitated by normal modes. Boundary effects that are present in all real, finite systems are not present in this model. But a comparison with realistic systems (Sect. 7.4.3) shows that the convective instability mechanisms in plane thermocapillary layers are also operative in finite-size systems.[1]

6.1 Convective Stability of the Return Flow

The convective stability of one-dimensional thermocapillary flows was first considered by Smith & Davis (1983a).[2] They calculated the two- and three-dimensional linear instabilities of flows driven by an imposed, constant, tangential temperature gradient on the free surface, while neglecting free-surface deformations (Ca→0). The physical mechanisms of the convective instabilities in plane thermocapillary layers have been discussed by Smith (1986a). He also calculated the nonlinear stability of supercritical thermocapillary convection (Smith (1988)) by means of amplitude equations (Cross & Hohenberg (1993), Newell & Whitehead (1969), Segel (1969)). In a second paper, Smith & Davis (1983b) investigated the stability of the same one-dimensional basic flows with respect to two-dimensional surface waves. The temperature field does not play an active role in the surface-wave instability. It is merely required to maintain the basic flow field. The results of Smith & Davis (1983a,b), among others, have been summarized in a review article by Davis (1987).

---

[1] In finite-size systems other types of instabilities may also arise, in particular those that are caused by the presence of no-penetration boundaries (Sect. 7.4.1).

[2] The stability of combined thermocapillary–buoyant convection was investigated by Parmentier et al. (1993), Mercier & Normand (1996), and Priede & Gerbeth (1997b). Garr-Peters (1992a,b) considered the stability of the flow for both directions of gravity, the thermocapillary layer being either the right way up or upside down.

Owing to its importance, the problem of thermocapillary flow in plane layers is revisited here. However, a method different from the one of Smith & Davis (1983a) will be employed.

Consider a liquid layer extended infinitely in the (x, y) direction and having a depth d. Let the liquid be bounded from below by a rigid wall at $z = -d/2$ and from above by a free surface at $z = d/2$. If the medium above the liquid layer is a passive gas at a temperature depending linearly on the x coordinate, $T = T_0 - bx$ with $T_0 = \mathrm{const.}$, a constant surface stress in the x direction is induced via the thermocapillary effect. Analogously to Sect. 2.3, we select the scales d, d^2/ν, $\gamma bd/\rho\nu$, γb, and bd for length, time, velocity, pressure, and temperature $(\theta = (T - T_0)/bd)$, and the volume equations (2.14–2.16) are considered for $\mathrm{Gr} = 0$. The boundary conditions for the basic state $(\boldsymbol{u}_0, \theta_0)$ are

$$\boldsymbol{u}_0 = \partial_z \theta_0 = 0, \qquad\qquad \text{on} \quad z = -\frac{1}{2}, \tag{6.1}$$

$$w_0 = \partial_z v_0 = \partial_z u_0 - 1 = \partial_z \theta_0 = 0, \qquad \text{on} \quad z = \frac{1}{2}, \tag{6.2}$$

where the free surface has been assumed to be thermally insulating. The basic equations allow for a set of solutions

$$\boldsymbol{u}_0 = -g''(z)\, \boldsymbol{e}_x, \tag{6.3}$$
$$\theta_0 = -x + \mathrm{RePr}\, g(z), \tag{6.4}$$

where the Reynolds number is given by

$$\mathrm{Re} = \frac{\gamma bd^2}{\rho\nu^2}. \tag{6.5}$$

Since most real systems are laterally bounded by solid walls, the solution with vanishing lateral through flow $(\int_{-1/2}^{1/2} u_0 dz = 0)$ is of particular interest. It is called the *return flow* solution.[3] The solution is easily found to be

$$g(z) = -\frac{1}{64}\left(4z^4 + \frac{8}{3}z^3 - 2z^2 - 2z + \frac{11}{12}\right). \tag{6.6}$$

The profiles of the basic-state velocity and temperature are shown in Fig. 6.1.

For the present laterally unbounded system, the solutions of the linearized disturbance equations (5.2–5.3) and the continuity equation can be written as normal modes in the x and y directions. The neutral solutions of a temporal stability analysis are plane waves propagating at an angle α with respect to the negative temperature gradient, i.e. the direction of the basic flow at the free surface. In a coordinate system in which the x axis is parallel to the $\boldsymbol{k}$ vector (see Fig. 6.2), the basic state reads

---

[3] For the stability of thermocapillary plane Couette flow, see Smith & Davis (1983a).

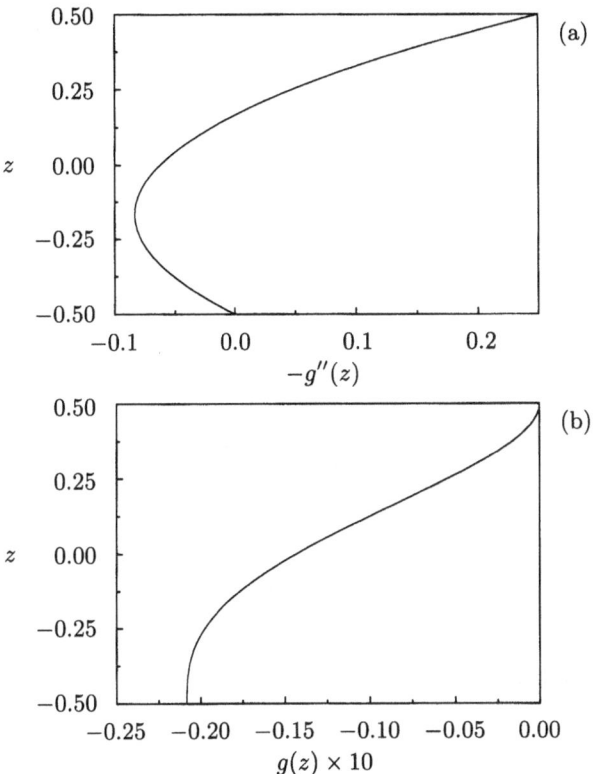

Fig. 6.1. Profiles of the basic-state velocity $-g''(z)$ (**a**) and temperature $g(z)$ (**b**) in an adiabatic, planar, thermocapillary layer without through-flow

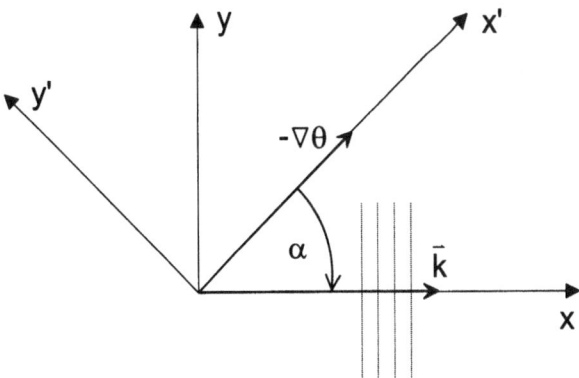

Fig. 6.2. Orientation of $-\nabla\theta_a$ and k relative to the coordinate system

$$u_0 = -g''(z) \cos \alpha \,, \tag{6.7}$$
$$v_0 = -g''(z) \sin \alpha \,, \tag{6.8}$$
$$w_0 = 0 \,, \tag{6.9}$$
$$\theta_0 = -(x \cos \alpha + y \sin \alpha) + \mathrm{RePr}\, g(z) \,. \tag{6.10}$$

Eliminating the pressure and introducing the stream function, where ψ

$$u = \partial_z \psi \,, \qquad w = -\partial_x \psi \,, \tag{6.11}$$

one obtains the nonlinear disturbance equations for the deviations (ψ, v, θ) from the basic state:

$$(\partial_t - \Delta)\, \Delta \psi = -\mathrm{Re}\left[(\psi_z \partial_x - \psi_x \partial_z - cg'' \partial_x)\, \Delta \psi + cg'''' \psi_x\right] \,, \tag{6.12}$$
$$(\partial_t - \Delta)\, v = -\mathrm{Re}\left[(\psi_z \partial_x - \psi_x \partial_z - cg'' \partial_x)\, v + sg''' \psi_x\right] \,, \tag{6.13}$$
$$\left(\partial_t - \frac{\Delta}{\mathrm{Pr}}\right) \theta = -\mathrm{Re}\left[(\psi_z \partial_x - \psi_x \partial_z - cg'' \partial_x)\, \theta - c\psi_z - sv\right]$$
$$+\mathrm{Re}^2 \mathrm{Pr}\, g' \psi_x \,, \tag{6.14}$$

where the abbreviations $c = \cos \alpha$ and $s = \sin \alpha$ have been used. The boundary conditions are

$$\partial_z \psi = v = \partial_x \psi = \partial_z \theta = 0 \,, \qquad \text{on} \quad z = -\frac{1}{2} \,, \tag{6.15}$$
$$\partial_x \psi = (\partial_z^2 \psi + \partial_x \theta) = \partial_z v = \partial_z \theta = 0 \,, \qquad \text{on} \quad z = \frac{1}{2} \,. \tag{6.16}$$

By a suitable choice of the coordinate system the unknowns $\phi \in \{\psi, v, \theta\}$ no longer depend on y, and the solution of the linearized problem can be written as

$$\phi(x, z, t) = \tilde{\phi}(z) e^{ikx} e^{(\sigma - i\omega)t} + \text{c.c.} \tag{6.17}$$

One obtains the linear system

$$\left[(\sigma - i\omega - \partial_z^2 + k^2 - ikc\mathrm{Re}\, g'')(\partial_z^2 - k^2) + ikc\mathrm{Re}\, g''''\right] \tilde{\psi} = 0 \,, \tag{6.18}$$
$$(\sigma - i\omega - \partial_z^2 + k^2 - ikc\mathrm{Re}\, g'')\, \tilde{v} + iks\mathrm{Re}\, g''' \tilde{\psi} = 0 \,, \tag{6.19}$$
$$\left(\sigma - i\omega - \frac{\partial_z^2 - k^2}{\mathrm{Pr}} - ikc\mathrm{Re}\, g''\right) \tilde{\theta} - s\mathrm{Re}\, \tilde{v} - \mathrm{Re}\,(c\partial_z + ik\mathrm{RePr}\, g')\, \tilde{\psi} = 0 \,,$$
$$\tag{6.20}$$

for the functions $\tilde{\phi}(z)$. The above equations are invariant under the transformations

$$\begin{pmatrix} \alpha \\ s \\ \tilde{\theta} \\ \tilde{\psi} \end{pmatrix} \longrightarrow \begin{pmatrix} -\alpha \\ -s \\ -\tilde{\theta} \\ -\tilde{\psi} \end{pmatrix} \qquad \text{and} \qquad \begin{pmatrix} \alpha \\ c \\ \omega \\ \tilde{\psi} \end{pmatrix} \longrightarrow \begin{pmatrix} \pi - \alpha \\ -c \\ -\omega \\ -\tilde{\psi} \end{pmatrix} \,, \tag{6.21}$$

where, for the latter transformation, the complex conjugate of (6.18–6.20) has to be taken. Thus it is sufficient to consider the quadrant $\alpha \in [0, \pi/2]$ only.

The system of equations (6.18–6.20) must be solved numerically. We employ a Chebyshev-τ method (Canuto et al. (1988), see also Priede & Gerbeth (1997b,c)). To this end the variable $z \longrightarrow 2z$ is stretched and all quantities $\phi(z)$ are expanded into Chebyshev polynomials $T_n(z) = \cos[n \arccos(z)]$ with $z \in [-1, 1]$:

$$\left\{ \tilde{\psi}, \tilde{v}, \tilde{\theta} \right\}(z) = \sum_{n=0}^{N} \left\{ \hat{\psi}_n, \hat{v}_n, \hat{\theta}_n \right\} T_n(z), \tag{6.22}$$

where the infinite sum has been truncated at order N. This ansatz is inserted into (6.18–6.20), and the resulting equations are projected onto the orthogonal polynomials $T_m(z)$ by means of a suitable scalar product (Sect. A.4). In this procedure, (6.18) for $\tilde{\psi}$ is projected onto $T_m(z)$ with $0 \leq m \leq (N-4)$, while the remaining two equations are projected onto the modes $0 \leq m \leq (N-2)$. Inserting the ansatz into the boundary conditions (6.15–6.16) yields $4(N+1) + 2 \times 2(N+1)$ additional equations. This way one obtains a linear algebraic system of equations of the form

$$\begin{bmatrix} A^{(\Re)} & -A^{(\Im)} \\ A^{(\Im)} & A^{(\Re)} \end{bmatrix} \begin{pmatrix} \psi^{(\Re)} \\ \psi^{(\Im)} \end{pmatrix} = 0. \tag{6.23}$$

This is described in detail in Sect. A.4. Here, $\psi = \psi^{(\Re)} + i\psi^{(\Im)} = (\hat{\psi}_n, \hat{v}_n, \hat{\theta}_n)$ denotes the $3(N+1)$-dimensional vector of the complex field amplitudes $((\psi^{(\Re)}, \psi^{(\Im)}) \in \mathbb{R}^{6(N+1)})$. Considering neutral stability (growth rate $\sigma = 0$), the complex coefficient matrix $A = A^{(\Re)} + iA^{(\Im)}$ depends on the parameters α, k, ω, Pr, and Re, and eventually on Bi. For the existence of a nontrivial solution, the coefficient determinant in (6.23) must vanish. Owing to the structure of A it is easy to show that $\det(A) \geq 0$. Thus the zero (ω, Re) of $\det(A)$ is also a minimum. It can be calculated numerically for given parameters $(\text{Pr}, \text{Bi}, \alpha, k)$ by a Newton method. Re is then minimized with respect to α and k using a gradient method (Press et al. (1989)) to find the critical values Re_c, ω_c, k_c, and α_c.

The critical mode consists of a pair of plane waves that propagate at the angles $\pm|\alpha|$ with respect to the negative temperature gradient. For these waves, Smith & Davis (1983a) have coined the term *hydrothermal waves*. The critical Marangoni number Ma_c, the wavenumber k_c, the propagation angle α_c, and the phase velocity ω_c/k_c are shown in Fig. 6.3 for $N = 10$ as functions of the Prandtl number, for $\text{Bi} = 0$ (full curves). To the accuracy of the lines' thickness the curves for $N = 10$ agree with those for $N = 20$. The calculated critical Reynolds numbers agree well with those of Smith & Davis (1983a) (dashed lines). The values for the critical angle α_c and for the critical phase speed ω_c/k_c deviate somewhat for very small and very large Prandtl num-

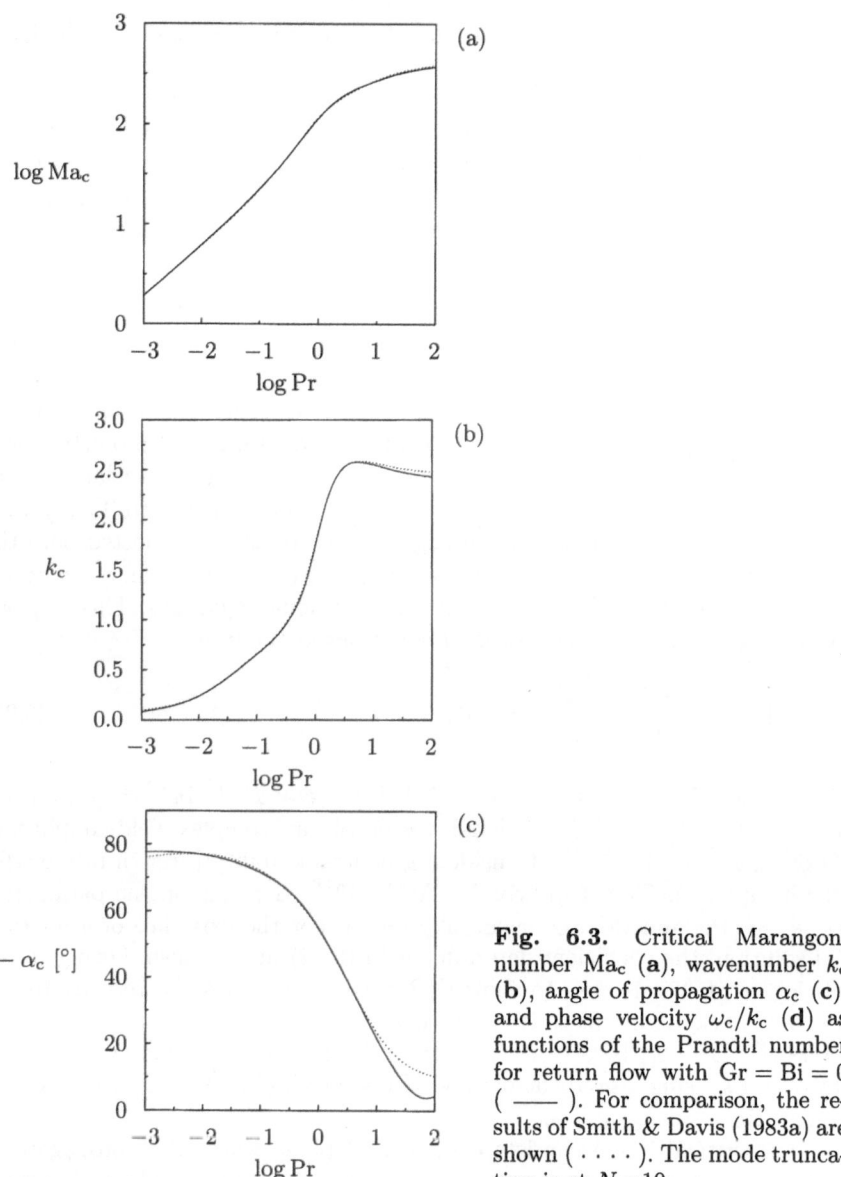

Fig. 6.3. Critical Marangoni number Ma_c (**a**), wavenumber k_c (**b**), angle of propagation α_c (**c**), and phase velocity ω_c/k_c (**d**) as functions of the Prandtl number for return flow with $Gr = Bi = 0$ (——). For comparison, the results of Smith & Davis (1983a) are shown ($\cdots$). The mode truncation is at $N = 10$

bers.[4] For high Prandtl numbers the hydrothermal waves propagate with a wavenumber $k \approx 2.5$ nearly parallel to the temperature gradient and opposite to the surface flow, whereas for small Prandtl numbers, they propagate nearly perpendicular to the applied temperature gradient and have

---

[4] The differences are presumably due to inaccuracies of the numerical method used by Smith & Davis (1983a) for extreme Prandtl numbers.

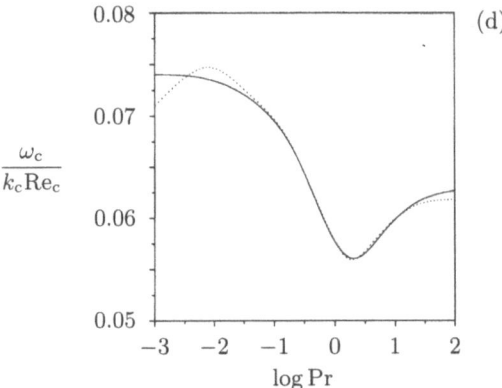

Fig. 6.3. (Continued)

a very small wavenumber. Further calculations show that a nonzero Biot number does not change the character of the instability. It can be shown that the basic flow (6.3–6.4) does not depend on the Biot number. Hence, $\mathrm{Re_c(Bi>0)>Re_c(Bi=0)}$ (cf. the energy term H in (5.27)), and the critical Reynolds numbers for $\mathrm{Bi>0}$ are monotonically shifted to higher values (see also Smith & Davis (1983a) and Priede & Gerbeth (1997a)).

Using an energy analysis Smith (1982, 1986a) showed that the hydrothermal waves at small Prandtl numbers draw their energy from the basic velocity field. By means of the classical Marangoni effect (Pearson (1958)) the x-momentum of the disturbance field is amplified through a supply of x-momentum from the basic flow. At high Prandtl numbers the hydrothermal waves exhibit strong temperature extrema in the bulk of the liquid layer (see Fig. 6.4). Convective transport of the basic-state temperature field owing to the disturbance flow maintains the internal temperature extrema. The pattern propagates, since the weak temperature extrema on the free surface are phase-shifted with respect to the corresponding strong temperature extrema in the bulk. The same mechanism can also be found in thermocapillary liquid bridges and is discussed in detail in Sect. 7.4.2.

By a perturbation technique Sen & Davis (1982) have shown that the thermocapillary flow in finite-length containers with lengths $L \gg d$, far away from the lateral end walls, can be well described by the above Couette–Poiseuille profile (6.3–6.4).[5] This has been confirmed experimentally by Saedeleer et al. (1996) for small Marangoni numbers. In view of the simplifications inherent in the model, however, it is difficult to experimentally observe the ideal hydrothermal waves that appear only at higher Marangoni numbers. For instance, the temperature distribution on the free surface cannot simply be enforced.

---

[5] In order that no dynamic surface deformations occur, the relation Ca $<$ $\mathcal{O}[\mathrm{Re}(d/L)^4]$ must be satisfied in the limit $d/L \to 0$.

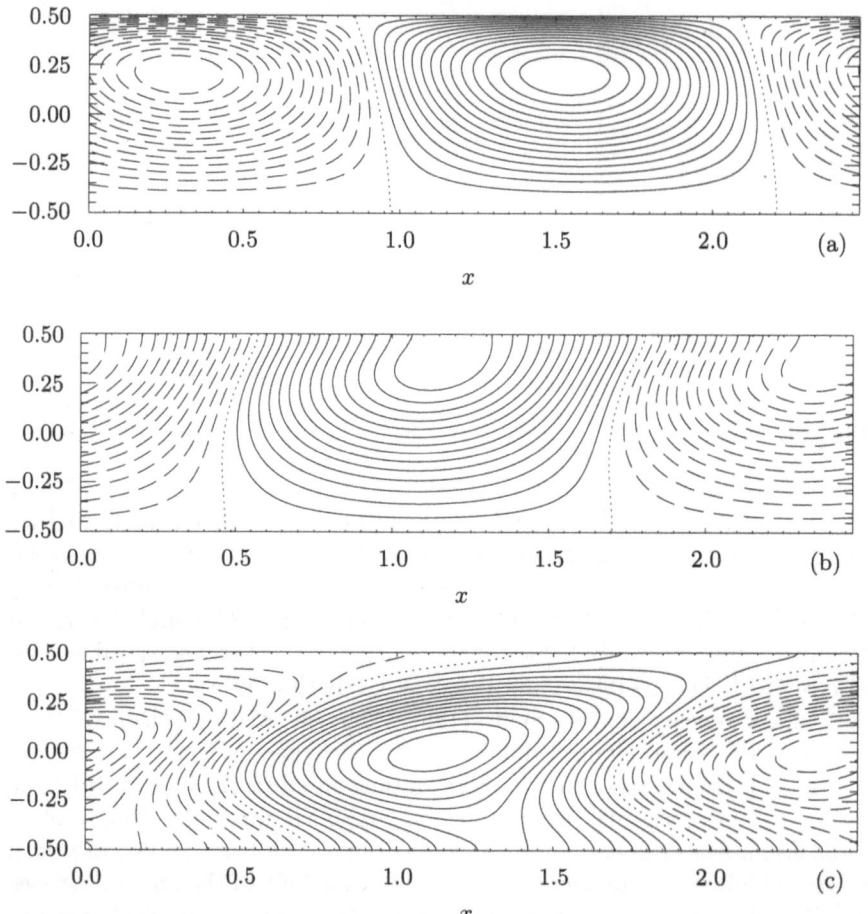

Fig. 6.4. Critical mode for $Pr = 13.9$ and $Bi = Gr = 0$. Isolines of the stream function ψ (**a**), transverse velocity v (**b**), and perturbation temperature θ (**c**) are shown. Positive values are indicated by *full lines*, negative values are *dashed*. The vertical axis is the z axis. The critical data are $Re_c = 21.2$, $k_c = 2.54$, $\omega_c = 3.26$. The wave propagates upstream against the surface flow (here to the *left*) at an angle $\pi - \alpha_c = 16.8°$. The truncation order is $N = 15$

Moreover, the Poiseuille–Couette profile is spatially unstable (Laure et al. (1990) and Priede & Gerbeth (1997b); see also Villers & Platten (1992)), and stationary two-dimensional vortices with equal senses of rotation resembling Kelvin's cat's eyes (Kelvin (1880)) can occupy parts or the whole of the liquid layer[6] (see e.g. Saedeleer et al. (1996)). These stationary corotating vortices are induced by the perturbation of the parallel flow by the presence of the

---

[6] Such cat's-eye flows have also been observed by Schwabe et al. (1992) and Schneider (1995) in a shallow thermocapillary annular gap.

cold and the hot side walls. In the limit of low Prandtl numbers ($Pr \to 0$), for instance, this type of stationary perturbation can grow spatially towards the cold wall to reach a saturated amplitude if $Re \gtrsim 400$ (cf. Sect. 8.1.2). Since the critical Reynolds number for hydrothermal waves diverges like $Re_c \sim Pr^{-1/2}$ in the limit $Pr \to 0$ (cf. Sect. 6.2), the small-Prandtl-number basic state will become two-dimensional on an increase of the Reynolds number beyond $Re \approx 400$, before hydrothermal waves can appear.

Natural convection cannot always be neglected. For a comparatively large Prandtl number $Pr = 13.9$ and for modified Bond numbers $Bd \leq 0.22$ (shallow layers) Riley & Neitzel (1998) found a direct transition from a quasi-one-dimensional thermocapillary flow to a single hydrothermal wave. The critical Reynolds number extrapolated to $Bd = 0$ was found to be in very good agreement with the theoretical prediction of Smith & Davis (1983a). For higher values of Bd, cat's-eye flows were observed to occur first. These results have been compared successfully with different types of stability analyses for plane layers by Priede & Gerbeth (1997b). Waves in shallow liquid layers of high Prandtl number have also been studied by Ezersky et al. (1993). A comparison with the theoretical predictions of the infinite-layer theory turned out difficult, since the observed waves propagated in the opposite direction. Experiments of Garcimartín et al. (1997) for a high Prandtl number fluid in the presence of gravity showed that these latter waves are emitted from a large-amplitude recirculation roll near the hot wall, and these authors speculated that the origin of the waves is due to an instability of the hot-wall thermal boundary layer (Schlichting (1968)).

In addition to a modification of hydrothermal waves by buoyancy effects (see e.g. Parmentier et al. (1993)), surface waves may appear (see also Sect. 9.1.2.). According to Smith & Davis (1983b) the critical wavelength of the most dangerous surface wave diverges for the return flow ($\lambda_c \to \infty$) and the associated critical Marangoni number becomes small. Since it scales like $Ma_c \sim Pr$ for $Pr \to 0$, the critical Reynolds number remains $\mathcal{O}(1)$. Therefore, the return flow in plane layers with a deformable free surface will first become unstable with respect to surface waves when the Prandtl number is small.[7] Owing to the long wavelength of the most dangerous surface waves, a small lateral extension of the system has a stabilizing effect by restricting the wavelengths of surface waves from above. For high Prandtl numbers, on the contrary, hydrothermal waves appear before surface waves on an increase of the Reynolds number.

---

[7] The analysis of Smith & Davis (1983b) requires the capillary number to be small too.

6.2 Asymptotics of the Critical Values for Pr → 0

The scaling of the critical parameters for small Prandtl numbers can be deduced analytically. To this end, the governing equations are considered in orders of magnitude in the limit Pr → 0. The following analysis is due to Priede & Gerbeth (1997a).

From the numerical analysis described in Sect. 6.1 we have seen that the critical wavenumber tends to zero as Pr → 0 and that the most dangerous mode propagates nearly perpendicular to the applied temperature gradient. For simplicity we investigate, therefore, longitudinal waves in the long-wave limit. For longitudinal waves ($c=0, s=1$) and on the stability threshold $\sigma=0$, the bulk equations are ($i \to -i$)

$$\left[(i\omega - \partial_z^2)(\partial_z^2 - k^2)\right]\tilde{\psi} = 0, \tag{6.24}$$

$$\left(i\omega - \partial_z^2 + k^2\right)\tilde{v} - ik\operatorname{Re} g'''\tilde{\psi} = 0, \tag{6.25}$$

$$\left(i\omega - \frac{\partial_z^2 - k^2}{\operatorname{Pr}}\right)\tilde{\theta} - \operatorname{Re}\tilde{v} + ik\operatorname{Re}^2\operatorname{Pr} g'\tilde{\psi} = 0. \tag{6.26}$$

Consider now a long-wavelength perturbation ψ of the return flow. When the oscillation frequency is low, the flow perturbation will vary on a vertical length scale given by the layer's thickness, i.e. on a length of $\mathcal{O}(1)$, so that $\partial_z^2\tilde{\psi} \sim \tilde{\psi}$. If the frequency is high, the perturbation will vary on a length scale given by the depth of the skin layer $\delta^{-2} \sim i\omega$, which is determined by the balance between viscous and time-dependent terms in the $\tilde{\psi}$ equation (6.24).[8] Therefore, we write

$$\partial_z^2\tilde{\psi} \sim (1 + i\omega)\tilde{\psi}, \tag{6.27}$$

in the sense that we consider the leading-order term, depending on whether ω is small or large. From the longitudinal-velocity equation (6.25) we get, to leading order ($k \to 0$, $\partial_z^2 \to (1+i\omega)$, $g'''=\mathcal{O}(1)$),

$$(1 + i\omega)\tilde{v} \sim ik\operatorname{Re}\tilde{\psi}. \tag{6.28}$$

It is important to notice the following property of insulating boundary conditions. The thermal diffusion time over a length l is l^2/κ. In the present scaling, the vertical heat diffusion time is given by Pr. Thus a thermal perturbation will relax locally to some average temperature within a time $\mathcal{O}(\operatorname{Pr})$. Since the Prandtl number is assumed small, the relaxation will be fast and the local temperature will be practically indepent of the vertical coordinate. It will, however, depend on the lateral location, since lateral temperature variations relax only on a timescale associated with the wavelength, i.e. on a timescale of order $\mathcal{O}(\operatorname{Pr}/k^2)$.[9] Therefore, the relevant term in the temperature equation (6.26) is $k^2\operatorname{Pr}^{-1}\tilde{\theta}$ and not $\operatorname{Pr}^{-1}\partial_z^2\tilde{\theta}$. Hence we get the thermal balance ($k \to 0$)

---

[8] Formally, δ is complex here to describe spatial oscillations.

[9] This time scale is not present for conducting boundaries, since the fluctuations will decay to the same value (zero) everywhere within the time $\mathcal{O}(\operatorname{Pr})$.

$$\left(i\omega + \frac{k^2}{\mathrm{Pr}}\right)\tilde{\theta} \sim \mathrm{Re}\,\tilde{v}\,. \tag{6.29}$$

Finally, the shear stress condition on the free surface (6.16) yields $(\partial_z^2\tilde{\psi} \to (1+i\omega)\tilde{\psi})$

$$(1+i\omega)\tilde{\psi} \sim ik\tilde{\theta}\,. \tag{6.30}$$

When $\tilde{\psi}, \tilde{v}, \tilde{\theta}$ are eliminated from these estimates one obtains the dispersion relation

$$(1+i\omega)^2\left(i\omega + \frac{k^2}{\mathrm{Pr}}\right) \sim k^2\mathrm{Re}^2\,. \tag{6.31}$$

The imaginary part yields

$$\omega \sim \left(1 + \frac{2k^2}{\mathrm{Pr}}\right)^{1/2} \xrightarrow{k\to 0} \sim 1\,, \tag{6.32}$$

i.e. the frequency is determined by the viscous diffusion time over the height of the layer (~ 1) as long as the thermal diffusion time over the wavelength ($\sim \mathrm{Pr}/k^2$) is very large. The real part of (6.31) results in

$$\mathrm{Re} \sim \frac{1}{k}\left[2\omega^2 + (1-\omega^2)\frac{k^2}{\mathrm{Pr}}\right]^{1/2} \xrightarrow{k\to 0} \sim k^{-1}\,, \tag{6.33}$$

meaning that the neutral (k-dependent) Reynolds number decreases $\sim k^{-1}$ when the wavenumber is small. In this case the heat diffusion term $\sim \mathrm{Pr}/k^2$ drops out (long-wavelength limit). The decrease of the neutral Reynolds number with increasing k is due to the fact that the effective shear stress $(1+i\omega)\tilde{\psi}$ increases $\sim k$ (see (6.16)), providing a more efficient driving of the shorter waves. However, when heat diffusion becomes important the neutral Reynolds number tends to a constant value. If we had not neglected higher-order terms in k, the Reynolds number would have reached a minimum, in fact. It is the crossover when $\mathrm{Pr}/k^2 \sim 1$ that determines the critical wavenumber. We thus have

$$k_c \sim \mathrm{Pr}^{1/2}\,, \tag{6.34}$$

while the associated Reynolds number scales like

$$\mathrm{Re}_c \sim \mathrm{Pr}^{-1/2}\,. \tag{6.35}$$

This is the observed asymptotic dependence of the full system for Pr → 0, even though the waves are not exactly longitudinal.

Priede & Gerbeth (1997a) have calculated the proportionality factors and obtained $k_c = 2.599\mathrm{Pr}^{1/2}$, $\mathrm{Re}_c = 65.566\mathrm{Pr}^{-1/2}$, and $\omega = 14.087$. They also derived the asymptotic scaling for the conducting-bottom boundary condition, which is different. The formulas yield good results, even for the full problem, as long as $\mathrm{Pr} < 0.01$.

7. Convection in Cylindrical Geometry

The model of a thermocapillary liquid bridge, or half-zone (see Fig. 2.1), includes the main features of the technological float-zone method. These are the axially symmetric geometry, the rigid boundaries at constant temperatures, and the lateral thermocapillary free surface. However, some other aspects are not always taken into account by half-zone models. Among these are deviations from the cylindrical free-surface shape due to the high Bond numbers encountered in practice. Moreover, the model consists of only half a zone and thus exhibits a monotonic variation of the surface temperature θ_a, in contrast to the full floating zone, for which the surface temperature reaches a maximum somewhere between the supporting cylindrical rods. None of these latter features neglected in the cylindrical half-zone model seems, however, to have a significant qualitative influence on the two-dimensional laminar convection structures.

In this chapter, therefore, thermocapillary and natural convection in a circular cylindrical volume with a free lateral surface are treated. The emphasis is on the linear stability of stationary axisymmetric flows. In Sect. 7.3 three-dimensional instabilities of pure buoyant convection (Re=0) are considered, while we shall be concerned with pure thermocapillary convection (Gr=0) in Sect. 7.4. The instabilities of mixed convection flows will be investigated in Sect. 7.5, followed by a comparison with experimental results. Section 7.7 is devoted to finite-amplitude three-dimensional thermocapillary flows in liquid bridges. The chapter will be concluded with a brief exposition of the results of energy stability analyses in Sect. 7.8.

7.1 Axisymmetric Convection in Liquid Bridges

The assumption of a small Reynolds number made in Chap. 4 is usually not satisfied for crystal growth from the melt. The nonlinear terms in (2.14) and (2.16) cannot be neglected and the two-dimensional convection must be calculated numerically. The qualitative vortex structures, i.e. the main toroidal vortex and the weak secondary vortices in the interior if $\Gamma \ll 1$, prevail at all Reynolds numbers. In the following the most important properties of two-dimensional nonlinear convection in liquid bridges will be addressed.

Among others, Shen (1989) has carried out various two-dimensional calculations, which nowadays are a standard task.

Semiconductor melts processed by the floating-zone or the Czochralski method (Hurle & Cockayne (1994)) have a low Prandtl number. The Prandtl number of silicon, for instance, is $\mathrm{Pr}\,(\mathrm{Si}) \approx 0.02$ (see Table A.1). According to Sect. 3.2.2 (conductive inertial limit), it is expected that thermo-convective effects are small in the range $\mathrm{Re} \ll \mathrm{Pr}^{-3/2} \approx 350$ for silicon, while they should become important for $\mathrm{Re} \gg \mathrm{Pr}^{-3/2} \approx 350$. For small Reynolds numbers, therefore, the strength of the flow will be constant on the scale given in Table 2.1. When inertial effects become important, the scaled amplitude will decay like $\mathrm{Re}^{-1/3}$ and eventually become independent of Re, being $\sim \mathrm{Pr}^{1/2}$.[1] Since the inertia terms violate the mirror symmetry of (2.14–2.16) with respect to $z=0$, the streamlines become increasingly asymmetric and the center of the vortex is shifted towards the cold corner if the Prandtl number is small. For $\mathrm{Pr}=0.02$ and $\mathrm{Re}=2000$ this is illustrated in Fig. 7.1, in which the thermoconvective

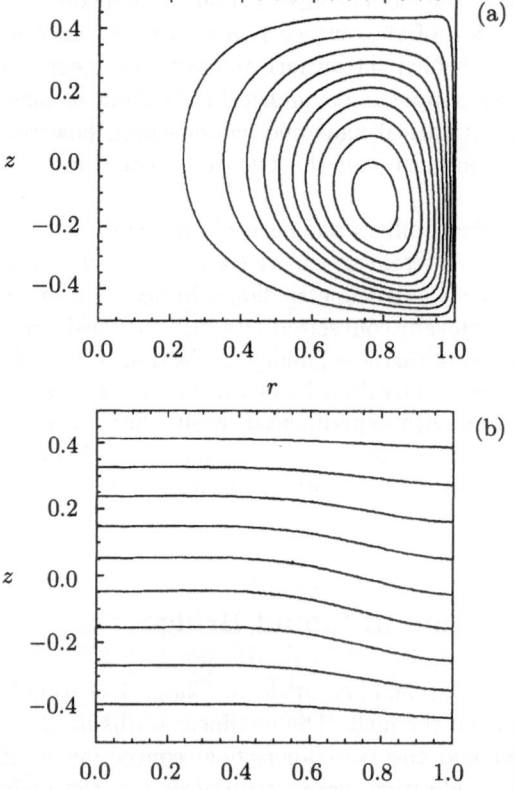

Fig. 7.1. Stream function (a) and isotherms (b) in a liquid bridge for $\varGamma=1$, $\mathrm{Pr}=0.02$, $\mathrm{Re}=2000$, and $\mathrm{Bi}=\mathrm{Gr}=0$

---

[1] The range within which the velocity scales like $\mathrm{Pr}^{1/2}$ has not yet been observed; see Sect. 3.2.2.

effects are weak. They are, in fact, weaker than expected from Table 3.1. The flow still exhibits the conductive inertial scaling. When the Reynolds number is increased further, the vortex develops a nearly inviscid core (Prandtl–Batchelor theorem, see e.g. Acheson (1990), Carpenter & Homsy (1990)) in which the vorticity ω varies linearly with r ($\omega/r = $ const., Batchelor (1956)) and which is surrounded by viscous boundary layers.

Unlike the case for liquid metals, thermoconvective effects are more pronounced in transparent liquids, which are commonly used for experimental modelling. Typical Prandtl numbers range from $\mathrm{Pr} \approx 1$ for KCl melts (Velten et al. (1991)) up to $\mathrm{Pr} > 100$ for some silicone oils. For $\mathrm{Pr} = 2$ and $\mathrm{Re} = 2000$ an example of the two-dimensional flow is shown in Fig. 7.2. The not yet fully developed thermal boundary layer near $z = 0.5$ is clearly

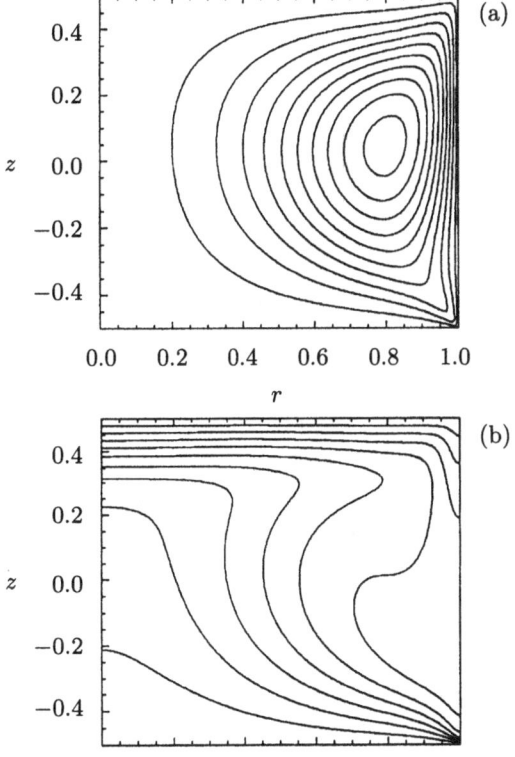

Fig. 7.2. Stream function (a) and isotherms (b) in a liquid bridge for $\Gamma = 1$, $\mathrm{Pr} = 2$, $\mathrm{Re} = 2000$, and $\mathrm{Bi} = \mathrm{Gr} = 0$

visible, as well as the compression of the isotherms near the cold corner ($z = -0.5$, $r = 1$) together with the associated crowding of streamlines. Unlike the behavior at small Prandtl numbers, the center of the vortex is located closer to the hot corner now, because of its increased importance for

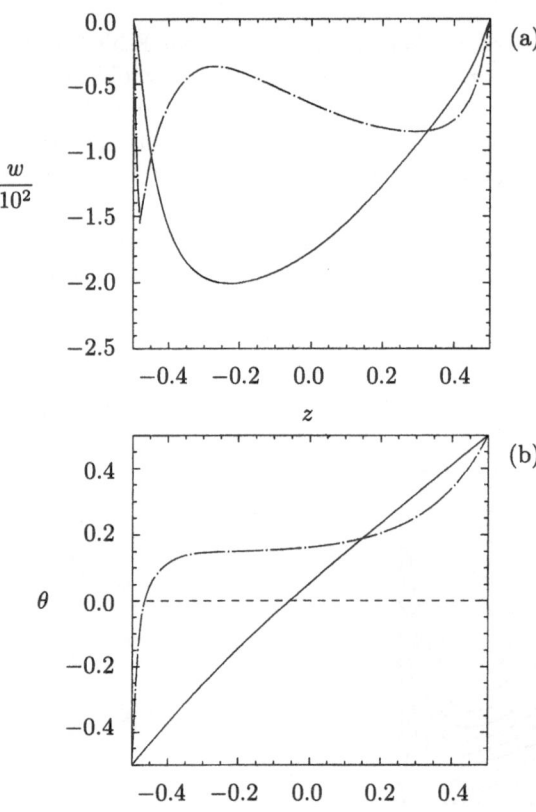

Fig. 7.3. Axial velocity $w(z)$ in units of ν/d (**a**) and temperature $\theta(z)$ (**b**) on the free surface at $r = 1/\Gamma$ for $\Gamma = 1$, Re=2000, and Bi=Gr=0. (——) Pr=0.02; (– · – · –) Pr=2

the driving of the flow (cf. Sect. 3.2.3). The axial velocity and temperature distributions along the free surface are shown in Fig. 7.3. The high gradients near $z \approx -0.5$ (for higher Marangoni numbers, also near $z \approx 0.5$) make numerical computations for high Marangoni numbers increasingly costly. The strong localization of the axial thermal gradient close to the cold corner leads to a heat transfer to the solid wall at $z = -0.5$ that is mainly concentrated in a narrow annular region. This behavior makes difficult a precise calculation of the Nußelt number Nu $= J_{\text{total}}/J_{\text{conductive}}$, which is a measure of the heat current J and which is frequently employed to check the energy conservation.

The above properties of the two-dimensional thermocapillary convection essentially agree with the experimental findings (Preisser et al. (1983)). The observed deviations from the results of model calculations are mainly caused by noncylindrical surface deformations, additional buoyancy forces, contamination of the free surface, temperature-dependent material parameters, and momentum and heat transfer to the ambient gas. These effects will be discussed in Chap. 9. In particular, Newton's heat-transfer law may often be a crude approximation. To account for the heat transport in the ambient gas Fu & Ostrach (1985) have calculated the steady two-dimensional flow in the liquid bridge as well as in a limited exterior volume, the rigid boundary of which is kept at the cold-wall temperature. The authors found that the flow in a liquid bridge without exterior convection but with a heat transfer given by (2.26), and with $\theta_a = -0.5$ and $Bi = 0.3$ is nearly identical to that with zero heat transfer ($Bi = 0$) but where convection in the exterior gas is taken into account. For this reason, these modified thermal conditions have been employed in later investigations by Shen (1989), Shen et al. (1990), and Neitzel et al. (1991). Frequently, however, the real conditions at the free surface deviate considerably from the assumptions made by Fu & Ostrach (1985) (see e.g. Velten et al. (1991)).

Another important driving force for two-dimensional convection is buoyancy. For $Re = 0$ the conductive state ($u = 0$) is stable when the bridge is heated from above and when no radial temperature gradients exist. When the heating is from below, the liquid-bridge problem is similar to the Rayleigh–Bénard problem (Charlson & Sani (1970, 1971)) and the conductive state may be unstable (Sect. 7.3). For small nonzero Rayleigh and Reynolds numbers the flow is two-dimensional and stationary. Since thermocapillary convection is always accompanied by radial temperature variations, the flow is modified by buoyancy forces, independent of whether the acceleration due to gravity g is parallel or antiparallel to e_z. The relative strength of thermocapillary forces compared to buoyancy forces depends on the dynamical Bond number $Bd = Gr/Re = \rho g \beta d^2 / \gamma$. Therefore, buoyant convection can be neglected when the length scale tends to zero ($d \to 0$).

When the heating is from below, the two-dimensional combined buoyant–thermocapillary flow is not always unique. One example is given in Fig. 7.4 for $Ra = 3200$ (heating from below), $Re = -5$, and $Pr = 4$. This flow is mainly driven by buoyancy. Since $|Re|$ is small, there exist two flow solutions, each characterized by a single main vortex of nearly equal strength of circulation but with different sense of rotation (Charlson & Sani (1975)). Depending on the sense of rotation the weak thermocapillary surface forces, acting always in the positive z direction, either enhance (Fig. 7.4a) or suppress the main vortex motion (Fig. 7.4b). The main vortex rotates counterclockwise in Fig. 7.4a, while it rotates clockwise in Figs. 7.4b,c. In the two latter cases separated vortices with opposite senses of rotation may arise that are either restricted to a small area close to the hot and the cold corner (Fig. 7.4b) or extended all along the free surface (Fig. 7.4c). If buoyant and thermocapillary forces favor the same sense of rotation (Fig. 7.4a) the Nußelt number is a maximum, while it is minimum for the situation corresponding to Fig. 7.4c, since the convective heat transport is significantly reduced by the mutual suppression of the inner buoyant and the outer thermocapillary vortex.

The question remains as to which of the different two-dimensional stationary states can be realized in an experiment. This depends on the initial conditions and on the stability of the corresponding solutions. An efficient method to determine the parameter ranges of instability is linear stability

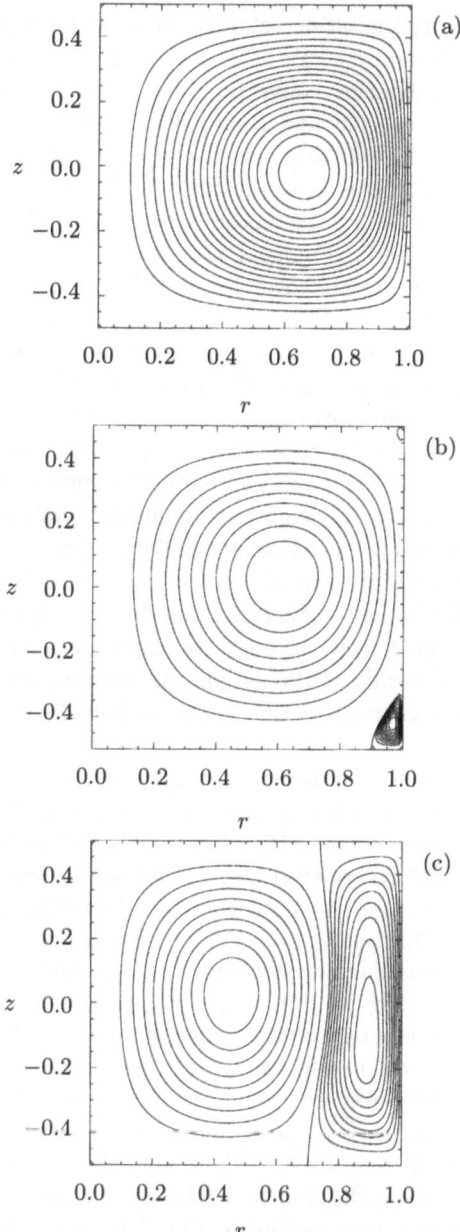

Fig. 7.4. Axisymmetric flow in a liquid bridge heated from below (Stokes stream function) for $Pr = 4$, $Ra = -PrGr = 3200$, $Re = -5$, $\Gamma = 1$, and $Bi = 0$. Three different solutions, (**a**), (**b**), and (**c**), exist. The increments between contour lines are constant within each separated vortex

analysis. This will be carried out for a number of cases in the following sections.

7.2 Linear Stability Analysis

For a theoretical treatment of the flows in liquid bridges, one relies on numerical methods. Several techniques can be employed. Here we use a mixed method of finite differences in the axial direction and a Chebyshev collocation method radially, making a compromise between rapid radial convergence and minimization of the Gibbs oscillations due to the discontinuities of the boundary conditions. Details of the method can be found in Wanschura (1996). The advantages of mixed methods have been demonstrated by application to the Taylor–Couette problem (Jones (1985)).

The starting point for the analysis is the stationary axisymmetric basic state described in Sect. 7.1. To calculate this basic state a stream-function-vorticity formulation (ψ_0, ω_0) is employed, where ψ_0 and $\omega_0 = \tilde{L}\psi_0 = (DD_* + \partial_z^2)\psi_0$ are chosen as in (4.2–4.3). The bulk equations for the basic state are given by (4.32–4.33), supplemented by the buoyancy term $\mathrm{Gr}D\Theta_0$ on the right-hand side of (4.32). The unknowns $(\psi_0, \omega_0, \Theta_0)$ are calculated on a grid consisting of $M+1$ equidistant finite difference points in the axial direction and of $N+1$ radial Gauß-Lobatto points (Canuto et al. (1988))

$$\{\psi_0, \omega_0, \Theta_0\}(x, z) \longrightarrow \{\psi_0, \omega_0, \Theta_0\}_{ij} , \tag{7.1}$$

where

$$x_i = \cos\left(\frac{i\pi}{N}\right) , \qquad i \in [0, N], \tag{7.2}$$

$$z_j = -\frac{1}{2} + \frac{j}{M} , \qquad j \in [0, M], \tag{7.3}$$

and the radial coordinate has been transformed as $x = 2\Gamma r - 1$. Within the discretization, the axial differential operators in (4.32–4.33) must be replaced by the appropriate finite-difference matrices of second order. The radial derivatives are obtained by multiplying the vector of unknowns at the Gauß–Lobatto points with the Chebyshev collocation derivative matrix, which can be obtained by differentiation of the Lagrange polynomials (Lanczos (1956))

$$L_i(x) = \frac{1}{F'(x_i)} \frac{F(x)}{x - x_i} , \qquad \text{with} \quad F = T'_N(x), \tag{7.4}$$

for the Gauß–Lobatto points (7.2). This is explicitly given in Canuto et al. (1988). The nonlinear equations discretized in this way are solved with a Newton–Raphson method.

For the stability analysis of the discretized basic state, small perturbations are written as normal modes

$$\{u, v, w, p, \Theta\}(r, z, \varphi, t) = \left\{\hat{u}, \hat{v}, \hat{w}, \hat{p}, \hat{\Theta}\right\}(r, z)\, e^{\gamma t + im\varphi} + \text{c.c.}, \tag{7.5}$$

where m is an azimuthal wavenumber and γ the complex growth rate. For a solution of the resulting linear stability equations (5.2–5.3), the azimuthal velocity is eliminated by use of the continuity equation[2]

$$\hat{v} = -\frac{r}{im}\left(D_*\hat{u} + \partial_z\hat{w}\right) . \tag{7.6}$$

The equations to be solved consist of radial and axial momentum balances, the temperature equation, and the Poisson equation for the pressure, which is obtained from $\nabla\cdot(5.2)$. The linearized equations read

$$\gamma\hat{u} + \mathrm{Re}\left(\boldsymbol{u}_0 \cdot \nabla\hat{u} + \hat{\boldsymbol{u}} \cdot \nabla u_0\right) = -D\hat{p} + \left(\Delta - \frac{1}{r^2}\right)\hat{u}$$

$$+ \frac{2}{r}\left(D_*\hat{u} + \partial_z\hat{w}\right) , \tag{7.7}$$

$$\gamma\hat{w} + \mathrm{Re}\left(\boldsymbol{u}_0 \cdot \nabla\hat{w} + \hat{\boldsymbol{u}} \cdot \nabla w_0\right) = -\partial_z\hat{p} + \Delta\hat{w} + \mathrm{Bd}\,\hat{\Theta} , \tag{7.8}$$

$$\gamma\hat{\Theta} + \mathrm{Re}\left(\boldsymbol{u}_0 \cdot \nabla\hat{\Theta} + \hat{\boldsymbol{u}} \cdot \nabla\Theta_0 + \hat{w}\right) = \frac{1}{\mathrm{Pr}}\Delta\hat{\Theta} , \tag{7.9}$$

$$-2\mathrm{Re}\nabla \cdot (\hat{\boldsymbol{u}} \cdot \nabla u_0) = \Delta\hat{p} , \tag{7.10}$$

where ∂_φ is to be replaced by im in the differential operators. The boundary conditions require

$$\hat{u} = \hat{w} = \partial_z\hat{w} = \hat{\Theta} = 0 , \qquad \text{on} \quad z = \pm\frac{1}{2} , \tag{7.11}$$

as well as

$$\left.\begin{array}{r} \hat{u} = 0 , \\ D\hat{w} + \partial_z\hat{\Theta} = 0 , \\ r^2(D_*D\hat{u} + \partial_z D\hat{w}) + m^2\hat{\Theta} = 0 , \\ (D + \mathrm{Bi})\hat{\Theta} = 0 , \end{array}\right\} \quad \text{on} \quad r = \frac{1}{\Gamma} , \tag{7.12}$$

$$\left.\begin{array}{r} \hat{u} = D\hat{w} = D\hat{p} = D\hat{\Theta} = 0 , \text{ if } m = 0 , \\ D\hat{u} = \hat{w} = \hat{p} = \hat{\Theta} = 0 , \text{ if } m = 1 , \\ \hat{u} = \hat{w} = \hat{p} = \hat{\Theta} = 0 , \text{ if } m > 1 , \end{array}\right\} \quad \text{on} \quad r = 0 . \tag{7.13}$$

The discretization of the perturbation equations is carried out in analogy to the basic-state discretization

$$\left\{\hat{u}, \hat{w}, \hat{p}, \hat{\Theta}\right\}(r, z) \longrightarrow \left\{\hat{u}, \hat{w}, \hat{p}, \hat{\Theta}\right\}_{ij} , \tag{7.14}$$

leading to the generalized eigenvalue problem

$$\gamma\,\mathsf{A} \cdot \boldsymbol{x} = \mathsf{B} \cdot \boldsymbol{x} , \tag{7.15}$$

with $\boldsymbol{x} = (\hat{u}, \hat{w}, \hat{p}, \hat{\Theta})_{ij}$. The matrix B has a block-diagonal structure owing to the particular method of discretization, if the index i of the Chebyshev polynomials is selected as the fast (inner) index (Fig. 7.5). A is diagonal and

---

[2] For $m=0$, a stream-function-vorticity formulation as for the basic state is used.

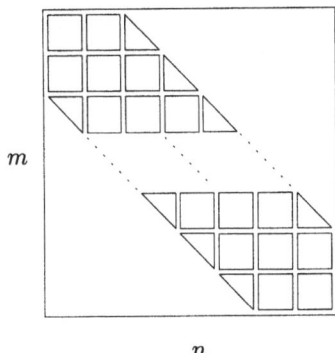

m

n

Fig. 7.5. Schematic representation of the pentadiagonal block structure of the matrix B in (7.15). The size $4(N+1) \times 4(N+1)$ of the blocks is determined by the number of unknowns $(\hat{u}, \hat{w}, \hat{p}, \hat{\Theta})_i$ in the radial direction

singular. The system (7.15) was solved either for all eigenvalues and eigenvectors, using mathematical subroutine libraries (Visual Numerics (1994)), or for the eigenvector and eigenvalue with the largest real part, using inverse iteration (Gollub & Loan (1989)). The Brent method (Press et al. (1989)) was employed to trace the real part of the eigenvalue of the most dangerous mode to zero.

The results of the basic-state and linear-stability calculations have been verified by comparison with the well-established results of Fu & Ostrach (1985), Shen (1989), Kuhlmann & Rath (1993a), Charlson & Sani (1971), and Xu & Davis (1984).[3] The remaining errors in the energy balances, δ_{kin} and δ_T, were typically smaller than 1%. In a few cases only, a maximum energy error of 5% was tolerated.

In experiments, the applied temperature difference is usually selected as the control parameter. Therefore, both thermocapillary and natural convection arise simultaneously. To investigate the influence of both driving forces on the pattern formation, the natural convection in cylinders when thermocapillary forces are absent (Re=0) is considered first (Sect. 7.3). Then, pure thermocapillary instabilities (Gr=0) and the critical flow patterns are analyzed (Sect. 7.4). Finally, the stability of combined buoyant–thermocapillary convection is investigated in Sect. 7.5.

7.3 Buoyant Instability

The buoyant convection in a liquid layer heated from below is a paradigm of hydrodynamic pattern formation. It is known as the *Rayleigh–Bénard problem* (Chandrasekhar (1961)). After the early work of Bénard (1900) and Rayleigh (1916), Malkus & Veronis (1958) in particular contributed to the understanding of the system. Reviews of the Rayleigh–Bénard problem have been given

---

[3] For comparison with the onset of three-dimensional flows in infinite thermocapillary cylinders (Xu & Davis (1984)), (7.11) was replaced by periodic boundary conditions.

by Busse (1978), Koschmieder (1993), and other authors. In the absence of thermocapillary effects one may use the scales ν/d and $\rho\nu^2/d^2$ for the velocity and the pressure (compare Table 2.1), and consider a cylindrical volume heated from below.[4] The modified scaling corresponds to $u \to u/\mathrm{Re}$ and $p \to p/\mathrm{Re}$, and the governing equations read

$$\partial_t u + u \cdot \nabla u = -\nabla p + \Delta u + \mathrm{Gr}\,\theta e_z \,, \tag{7.16}$$

$$\nabla \cdot u = 0 \,, \tag{7.17}$$

$$\partial_t \theta + u \cdot \nabla \theta = \frac{1}{\mathrm{Pr}} \Delta \theta \,, \tag{7.18}$$

with the rigid-wall, constant-temperature boundary conditions

$$u = 0, \quad \theta = \mp\frac{1}{2}, \qquad \text{on} \quad z = \pm\frac{1}{2}. \tag{7.19}$$

The stability equations (7.7–7.13) must be rescaled correspondingly. If lateral temperature gradients are absent, as is the case for adiabatic or perfectly conducting lateral boundaries, the basic state is given by the conductive temperature profile $\theta_0 = -z$ and $u_0 = 0$. The onset of convection in the cylindrical Rayleigh–Bénard problem is always stationary[5] and it is three-dimensional for most aspect ratios. Since stationary solutions of the linearized equations (7.16–7.19) do not depend on the Prandtl number, the critical Rayleigh number $\mathrm{Ra_c}$ is independent of Pr. For cylinders, $\mathrm{Ra_c}$ was first calculated by Charlson & Sani (1970, 1971), and later also by Buell & Catton (1983).

All previous stability analyses have been carried out either for rigid lateral boundary conditions

$$u = 0, \qquad \text{on} \quad r = \frac{1}{\Gamma}\,, \tag{7.20}$$

for idealized lateral boundary conditions (Rosenblat (1982))

$$u = \partial_r w = \left(\partial_r + \frac{1}{r}\right) v = 0, \qquad \text{on} \quad r = \frac{1}{\Gamma}\,, \tag{7.21}$$

or for a linear combination of the two types of boundary conditions (Chen (1992a,b)). The idealized boundary conditions (7.21) require the vertical component of the vorticity to vanish on the lateral surface. In this case a separation of variables is possible (Zierep (1963), Charlson & Sani (1970), Rosenblat (1982)). On the other hand, the stress-free boundary conditions

$$u = \partial_r w = \left(\partial_r - \frac{1}{r}\right) v = 0, \qquad \text{on} \quad r = \frac{1}{\Gamma}\,, \tag{7.22}$$

do not allow a separation of variables, and the stability of the conductive state has only recently been investigated for these boundary conditions (Wanschura et al. (1996)).

---

[4] The scales and the definition of the temperature difference (heated from below) apply only to this section (Sect. 7.3).

[5] The linear stability problem is self-adjoint; see Drazin & Reid (1981).

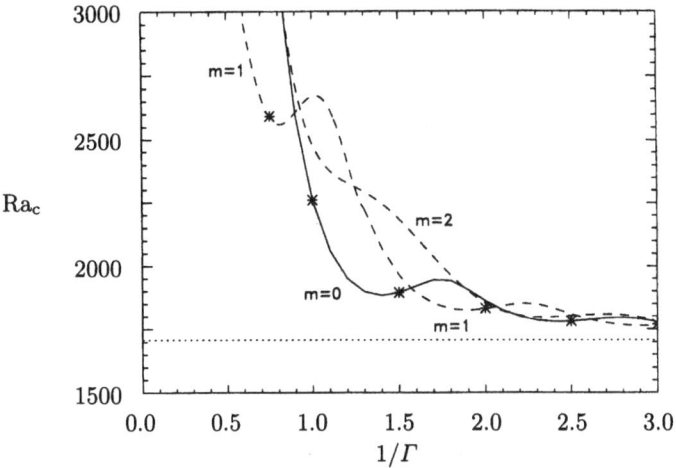

Fig. 7.6. Critical Rayleigh number as a function of the aspect ratio $1/\Gamma = R/d$ for a rigid cylinder heated from below with adiabatic lateral walls. (——) neutral curve for $m=0$; (- - -) Neutral curves for $m=1$ and $m=2$; (· · · ·) asymptotic value $\mathrm{Ra}_c^\infty = 1708$; (∗) values from Hardin & Sani (1993)

The stability boundaries for the onset of convection in cylinders with adiabatic ($\partial_r \theta = 0$) and rigid side walls (7.20) are shown in Fig. 7.6. In the limit of a shallow layer $\Gamma \to 0$ the asymptotic value $\mathrm{Ra}_c^\infty = 1708$ is reached (Chandrasekhar (1961)). The critical curve approaches Ra_c^∞ from above.[6] Figure 7.7 displays the stability boundaries for stress-free conditions (7.22). Within the range of aspect ratios considered, $1/\Gamma < 2.5$, the stability boundaries are strictly ordered according to the azimuthal wavenumber m. Moreover, the asymptotic value Ra_c^∞ is approached from below. Thus the conductive state in a cylindrical volume with, for example $1/\Gamma = 0.57$, becomes unstable for $\mathrm{Ra}_c(m=1) = 1527$. This value is 11% less than Ra_c^∞. In agreement with the theory of Rosenblat (1982) and Chen (1992a), the local minima of the neutral curves for $m=0$ take exactly the value Ra_c^∞. Modes with $m=0$ are never critical for liquid bridges heated from below, i.e. convection always sets in three-dimensionally.

Other than for stress-free liquid bridges, there exist ranges of the aspect ratio Γ for rigid adiabatic boundaries within which axisymmetric convection is possible (see Fig. 7.6). Only the range $0.9 < 1/\Gamma < 1.57$ is of interest here, since the other Γ intervals are very narrow. For Γ in this range there exist only a few, partly contradictory results about the sequence of instabilities that occur on an increase of the Rayleigh number. For $\mathrm{Pr} = 6.7$ (water) and $\mathrm{Ra} = 2800$ Müller et al. (1984) and Neumann (1990) found by numerical simulations that two different flow states can exist, one with $m=0$ and another one with $m=2$, depending on the initial conditions. Only the axisymmetric flow state with $m=0$, however, was observed in an experiment conducted

---

[6] See also Chen (1992a,b).

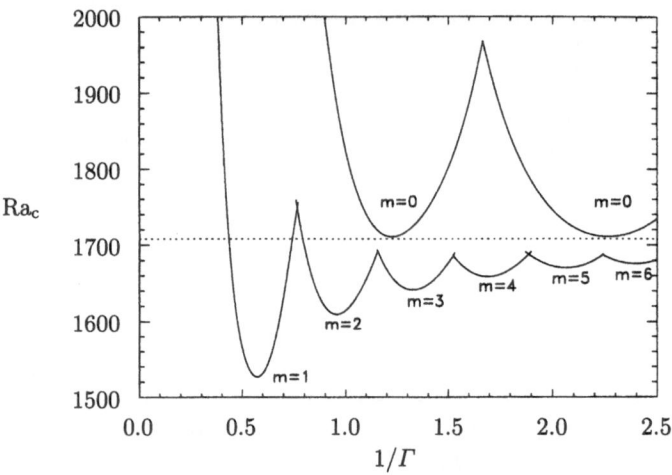

Fig. 7.7. Critical Rayleigh number as a function of the aspect ratio $1/\Gamma = R/d$ for an adiabatic cylindrical liquid bridge heated from below. ($\cdots$) Asymptotic value $Ra_c^\infty = 1708$. The critical curve is the envelope of all neutral curves for wavenumbers $m \geq 1$. Axisymmetric modes ($m=0$) are never critical

in parallel (Müller et al. (1984)). The axisymmetric state turned out to be stable up to $Ra \approx 10Ra_c$. On the other hand, Wagner et al. (1994), in their numerical simulation, found that the three-dimensional flow with $m = 2$ is unstable for $Ra = 2800$ and slowly decays to the axisymmetric state. Finally, Hardin & Sani (1993) predicted a linear instability of the axisymmetric flow with respect to a mode with $m = 2$ for $Ra = 2850$, in contradiction to the experimental findings.

It is clear that the neutral curves for $m=1$ and $m=2$ (dashed in Fig. 7.6) do not mark the transition Rayleigh number for three-dimensional convection, since the conductive state is not realized for $Ra > Ra_c (m=0)$. In the following, the instability of the axisymmetric convection state will be investigated in the range $0.9 < 1/\Gamma < 1.57$. As representative Prandtl numbers, $Pr = 0.02$ (liquid metals) and $Pr = 1$ (transparent fluids) will be considered for adiabatic boundary conditions ($Bi = 0$). The only stability analysis of the axisymmetric state performed to date is due to Charlson & Sani (1975), who made use of a full spectral Galerkin method (Finlayson (1972)) but did not take into account a sufficient number of modes.[7]

The relation between the Rayleigh–Bénard problem with rigid boundaries and liquid bridges is not obvious. In crystal growth, high-purity materials are used and the liquid–gas interfaces are nearly fully mobile. A mobile interface cannot, however, always be realized. In some experiments with liquid gallium Hardy (1985), for instance, found that the liquid may be covered with a thin,

---

[7] Hardin & Sani (1993) also used a severe mode truncation.

homogeneous, solid layer enforcing a rigid boundary condition at the surface.[8] Convection in rigid cylinders can thus be considered as the limiting case in which the surface becomes completely immobile. Moreover, rigid boundaries are present in the Bridgman crystal-growth technique (Müller et al. (1984), Monberg (1994)). The problem is also of fundamental interest for the bifurcation sequences in the Rayleigh–Bénard problem, and it is important for a comparison with experiments (Stork & Müller (1975)), in which radial temperature gradients can hardly be avoided. These gradients lead to an additional driving of the flow (Koschmieder (1966)) and can substantially influence the pattern formation.

7.3.1 Small Prandtl Numbers: Pr = 0.02

The stability boundaries for Pr = 0.02 are shown in Fig. 7.8. The dash-dotted line between the vertical dashed lines marks the critical Rayleigh number for the onset of axisymmetric convection. The axisymmetric flow is not unique. There exist two equivalent two-dimensional flows that can be transformed one into the other by the symmetry operation (see Liang et al. (1969) and Charlson & Sani (1975))

$$(z, u_0, w_0, \Theta_0) \longrightarrow (-z, u_0, -w_0, -\Theta_0). \tag{7.23}$$

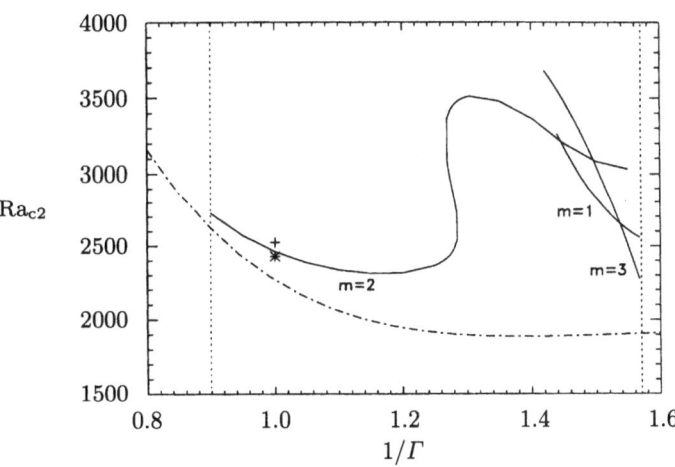

Fig. 7.8. Critical Rayleigh numbers for Pr = 0.02 as a function of the aspect ratio $1/\Gamma$ for a rigid cylinder with adiabatic side walls heated from below. Within the *dashed vertical lines* convection sets in axisymmetrically. ($- \cdot - \cdot$) $\mathrm{Ra_c}(m = 0)$; (——) $\mathrm{Ra_{c2}}$; (+) Neumann (1990); (*) Hardin & Sani (1993)

---

[8] A mathematical modelling of partially mobile interfaces and a comparison with experiments have been carried out by Mansell et al. (1994).

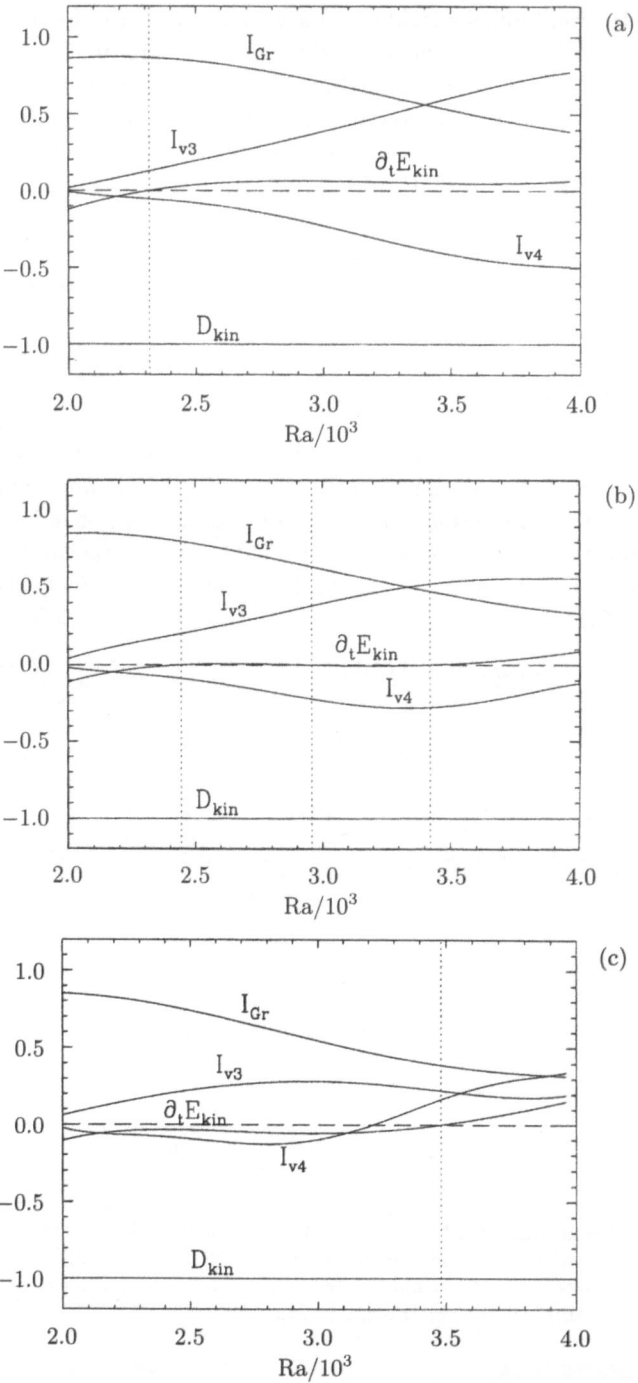

Fig. 7.9. Kinetic-energy balance (5.25) for $Pr = 0.02$ and $m = 2$ as a function of the Rayleigh number Ra. (a) $1/\Gamma = 1.2$; (b) $1/\Gamma = 1.275$; (c) $1/\Gamma = 1.35$. All terms are normalized by D

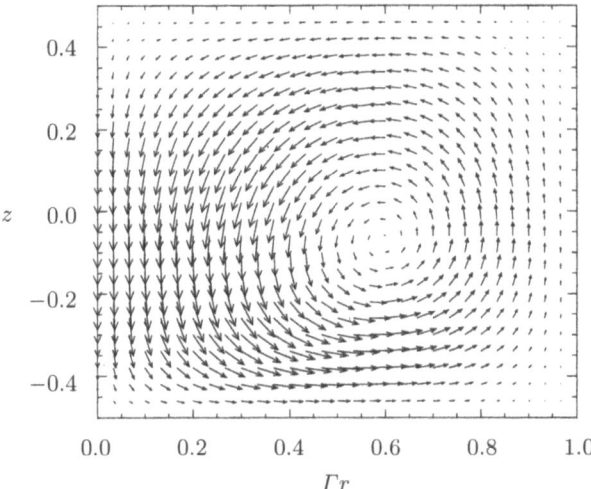

Fig. 7.10. Basic flow for $1/\Gamma=1.2$, $Pr=0.02$, and $Ra=Ra_{c2}(m=2)=2315$

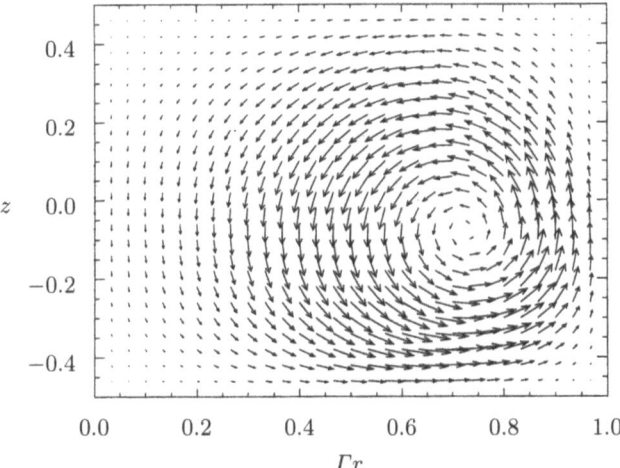

Fig. 7.11. Basic flow for $1/\Gamma=1.35$, $Pr=0.02$, and $Ra=Ra_{c2}(m=2)=3480$

Since the neutral modes of the corresponding linear stability problem possess the same symmetries, the neutral curves for both axisymmetric basic states are the same. They are given by the solid lines in Fig. 7.8. The envelope is the linear-stability boundary Ra_{c2}. Up to a narrow range around $1/\Gamma \approx 1.5$ the critical azimuthal wavenumber is $m=2$. The onset of three-dimensional convection is stationary throughout the entire interval, i.e. $\omega_{c2}=0$.

These results agree well with previous ones. By numerical simulation Neumann (1990) found $Ra_{c2}(\Gamma=1)=2525$ with $m=2$. This value is only

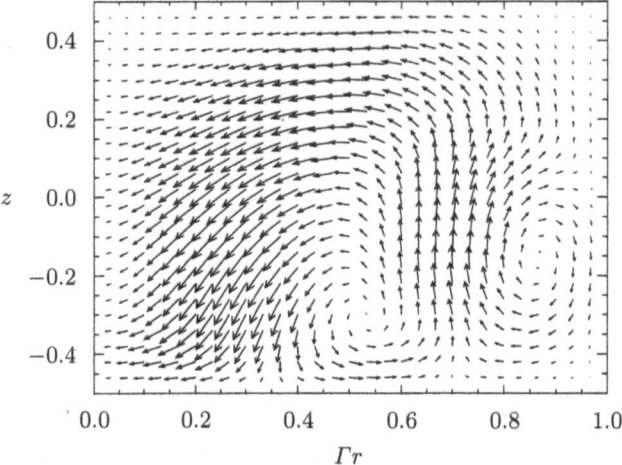

Fig. 7.12. Neutral velocity field for $1/\Gamma = 1.35$, $\mathrm{Pr} = 0.02$, and $\mathrm{Ra} = \mathrm{Ra}_{c2}(m = 2) = 3480$. A section is shown at $\varphi = 0$ where $v = 0$

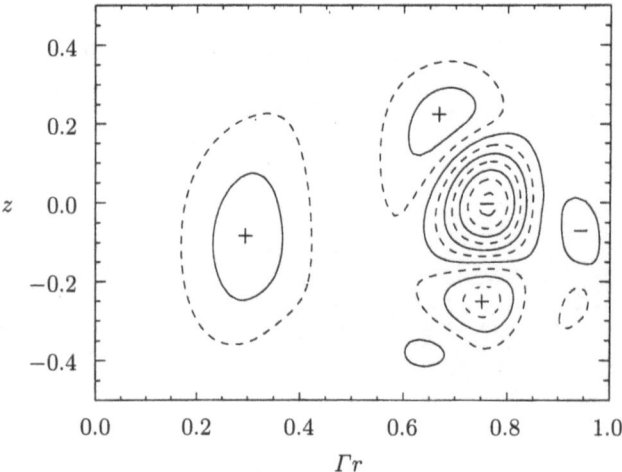

Fig. 7.13. Isolines of $-wu\partial_r w_0$ (integrand of I_{v4}) in the (r, z) plane at $\varphi = 0$ for $1/\Gamma = 1.35$, $\mathrm{Pr} - 0.02$, $m = 2$, and $\mathrm{Ra} = \mathrm{Ra}_{c2}(m - 2) - 3480$

2% higher than the actual value $\mathrm{Ra}_{c2} = 2463$. The corresponding value of Hardin & Sani (1993) is $\mathrm{Ra}_{c2} = 2430$, which is only 1.3% less.

The instability mechanisms can be elucidated by considering a posteriori energy balances (cf. Sect. 5.4). To this end, the different contributions to the rate of change of energy are normalized with respect to the rate of dissipation D and the rate of heat diffusion D_T, respectively, and plotted as functions of the Rayleigh number.

The integrals of the thermal-energy balance for $m = 2$ show little variation in the range $2000 < \text{Ra} < 4000$. Nearly all thermal energy is transferred to the perturbation temperature Θ by the vertical convective transport $(w\partial_z)$ of temperature of the basic conductive profile $(\theta_0 = \Theta_0 - z \approx -z)$, i.e. $I_{T3} = \int_V \Theta w_0 dV$ is balanced by D_T. Therefore, the thermal-energy balance is unimportant for the instability mechanism. The mechanism must be inertial.

The kinetic-energy balances are shown in Fig. 7.9. The linear-stability boundaries of the axisymmetric state are indicated by dotted vertical lines. Near $1/\Gamma = 1.275$ the critical Rayleigh number varies strongly by $\mathcal{O}(10^3)$ (cf. Fig. 7.8). This behavior is caused by a change of the instability mechanism. While the conventional buoyancy term I_{Gr} always has a destabilizing effect, although its relative importance decreases as Ra increases, I_{v3} is destabilizing, and I_{v4} is stabilizing for $1/\Gamma \lesssim 1.275$. On the other hand, for $1/\Gamma \gtrsim 1.275$, I_{v4} has a destabilizing effect, while the relative importance of I_{v3} diminishes.

This change of the instability mechanism can be traced back to a change of the basic flow (Figs. 7.10 and 7.11). I_{v3} is a measure of the amplification of radial velocity perturbations (u) by axial transport $(w\partial_z)$ of radial basic flow (u_0). Similarly, I_{v4} is a measure of the amplification of axial velocity perturbations (w) by radial transport $(u\partial_r)$ of axial basic flow (w_0). Unlike the axisymmetric vortex for $1/\Gamma = 1.2$, the vertical flow near the axis is very weak for $1/\Gamma = 1.35$. On a further increase of $1/\Gamma$ a second axisymmetric vortex with the opposite sense of rotation would form there. For this reason the vertical basic flow has strong radial gradients near $\Gamma r \approx 0.25$. The process described by I_{v4} acts on these gradients. The significant radial (u) and axial (w) perturbation flows at $(\Gamma r \approx 0.25, z \approx -0.2)$ (Fig. 7.12) cause the high absolute value of the integrand of I_{v4}. The associated maximum of the local energy transfer rate which determines I_{v4} is well developed (see Fig. 7.13).

7.3.2 Large Prandtl Numbers: Pr=1

The stability diagram for $\text{Pr} = 1$ and $\text{Bi} = 0$ is shown in Fig. 7.14. For high Prandtl numbers axisymmetric convection can be linearly stable up to very high Rayleigh numbers. In contrast to $\text{Pr} = 0.02$, it becomes unstable to a stationary $(\omega_{c2} = 0)$ mode with $m = 1$ for most aspect ratios. The $m = 2$ mode is stationary too. Only for $1/\Gamma \approx 1.5$ does the instability of the basic flow appear in the form of oscillatory $(\omega_{c2} \neq 0)$ modes with $m = 3$ or $m = 4$. The neutral curve for $m = 2$ exhibits a *nose*. Thus, for $1.05 < 1/\Gamma < 1.089$, there exists a range of Rayleigh numbers within which the two-dimensional basic flow is linearly stable, and from which both an increase and a decrease of the Rayleigh number Ra will lead to three-dimensional convection. By a numerical simulation Neumann (1990) predicted the critical value $\text{Ra}_{c2} > 4100$ for $\Gamma = 1$ and wavenumber $m = 1$. This value is only 2.9% less than the actual neutral value $\text{Ra}_{c2} = 4224$ for $m = 1$. The critical linear-stability boundary $(\text{Ra}_{c2} = 3016)$ is, however, much smaller and occurs for the wavenumber $m = 2$.

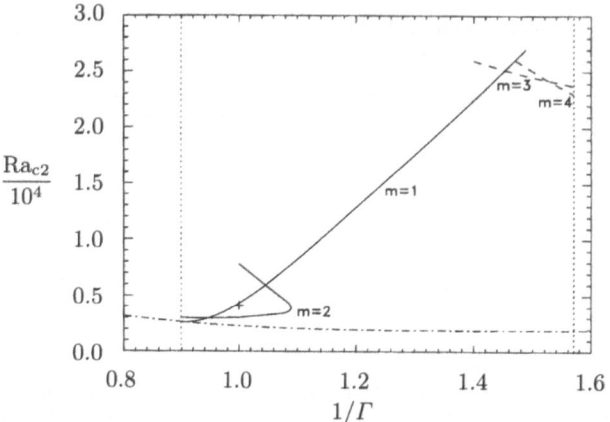

Fig. 7.14. Critical Rayleigh numbers for $Pr = 1$ as a function of the aspect ratio $1/\Gamma = R/d$ for a rigid adiabatic cylinder heated from below. Within the *dashed vertical lines* convection sets in axisymmetrically. ($- \cdot - \cdot$) $Ra_c(m=0)$; (——) Ra_{c2} for stationary instabilities ($\omega_{c2} = 0$); (- - -) secondary oscillatory instabilities ($\omega_{c2} \neq 0$); (+) Neumann (1990)

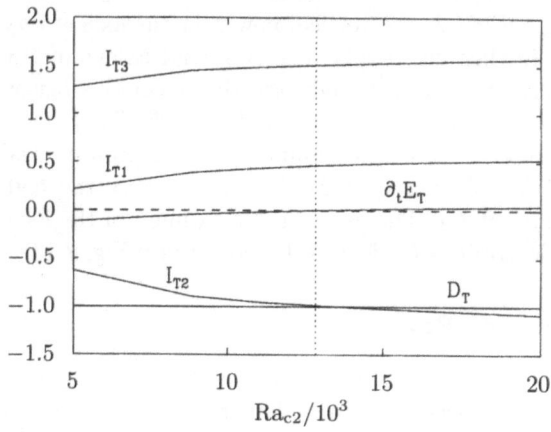

Fig. 7.15. Rates of change of the contributions to the thermal energy (5.26) of the neutral mode as functions of Ra for $1/\Gamma = 1.2$, $Pr = 1$, $Bi = 0$, and $m = 1$

For high Prandtl numbers, inertia effects do not play a significant role. The thermal-energy balance is important. It is exemplified in Fig. 7.15 for $1/\Gamma = 1.2$. By far the largest energy transfer to the thermal field results from I_{T3}, the amplification of temperature perturbations (Θ) by vertical transport ($w\partial_z$) of the temperature of the conductive profile ($-z$). The local contributions to the energy growth are shown in Fig. 7.16 and the corresponding neutral mode ($m = 1$) is given in Fig. 7.17 for $\varphi = 0$. Owing to the conductive background temperature field the velocity field of the critical mode induces a

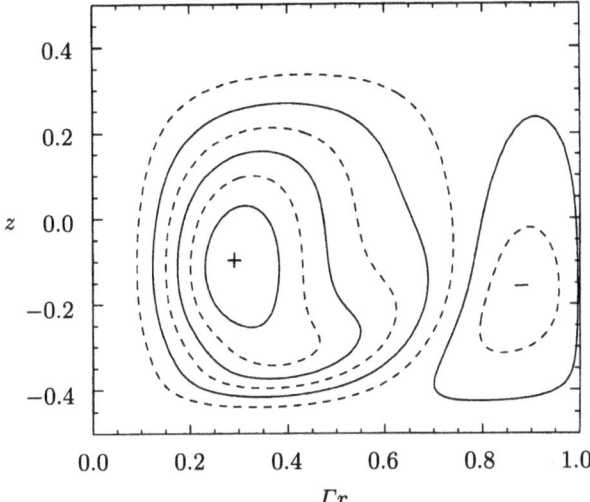

Fig. 7.16. Local rate of energy transfer $-\Theta w$ (integrand of I_{T3}) at $\varphi = 0$ for $1/\Gamma = 1.2$, $\text{Pr} = 1$, $\text{Bi} = 0$, and $\text{Ra} = \text{Ra}_{c2}(m=1) = 12\,821$

hot spot near $(z \approx 0, \Gamma r \approx 0.25)$.[9] The horizontal temperature gradient associated with the hot spot induces additional buoyancy forces that enhance the existing perturbation velocity field. Although this is the mechanism leading also to the first instability (Ra_c), the three-dimensional stability boundary $\text{Ra}_{c2}(m=1) = 12\,821$ is much larger than the critical threshold for the onset of axisymmetric convection. It is also higher than the neutral value $\text{Ra}_c(m=1)$. This strong stabilization is due to the nontrivial basic-state temperature field Θ_0. The contribution I_{T2} (see Fig. 7.15) describes a process similar to I_{T3}. For I_{T2} the perturbation velocity w couples, however, to the vertical temperature gradient $\partial_z \Theta_0$. Owing to the form of the basic-state temperature field, I_{T2} is negative.

The same stabilizing (D_T, I_{T2}) and destabilizing (I_{T3}) mechanisms are at work in the oscillatory instabilities around $1/\Gamma \approx 1.5$. As an example, the critical $m=3$ mode for $1/\Gamma = 1.47$ is shown in Fig. 7.18. The structures of the neutral fields within one azimuthal period are very similar to those for $1/\Gamma = 1.2$. Owing to subtle details of the instability, the oscillatory modes rotate, however, about the axis $r = 0$. During the instability process competing forces arise that lead to a certain *frustration*. The frustration is partially overcome by an azimuthal phase shift between the temperature and the velocity field of the neutral mode. It is the phase shift that makes the pattern rotate.[10] The mechanism responsible for the appearance of oscillations due to a frustration will be explained in Sect. 7.4.2 for hydrothermal waves in liquid bridges.

---

[9] A corresponding cold spot is generated at $\varphi = \pi$.
[10] For details, see Wanschura et al. (1996).

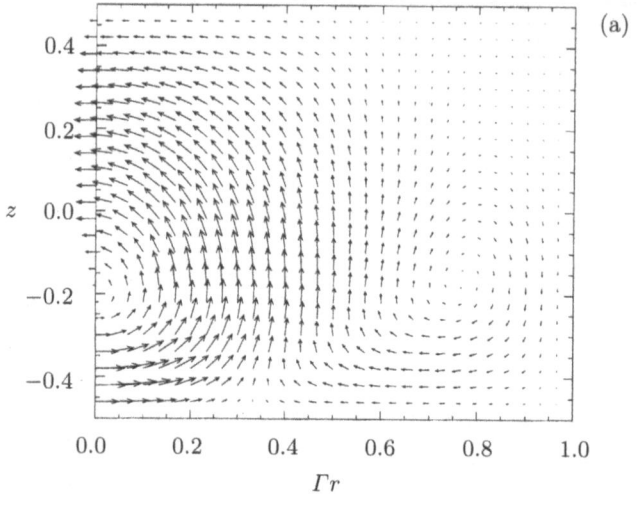

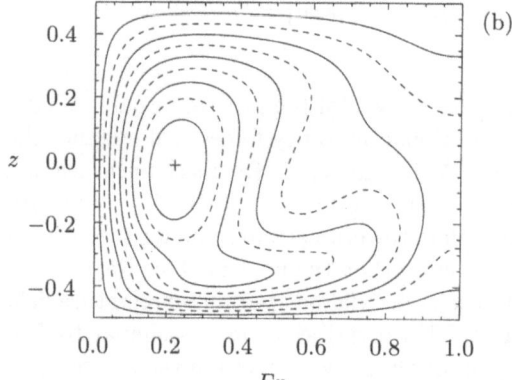

Fig. 7.17. Velocity field (**a**) and temperature field (**b**) of the neutral solution shown at $\varphi = 0$ for $1/\Gamma = 1.2$, $\text{Pr} = 1$, $\text{Bi} = 0$, and $\text{Ra} = \text{Ra}_{c2}(m = 1) = 12\,821$

The stability diagram for $\text{Pr} = 6.7$ (water) has not been calculated in detail. It is very likely that the structures of the diagrams are similar for both $\text{Pr} = 6.7$ and $\text{Pr} = 1$. For $\text{Pr} = 6.7$ and $\Gamma = 1$ Hardin & Sani (1993) found a stationary bifurcation of the axisymmetric flow to a mode with $m = 2$ at $\text{Ra}_{c2} = 2850$. A linear stability analysis for $\text{Pr} = 6.7$ shows, however, that this instability does not occur for $\Gamma = 1$. The growth rate for the $m = 2$ mode merely has a local maximum near $\text{Ra} = \mathcal{O}(3000)$ – but at a negative value (see Fig. 7.19). This behavior suggests that the neutral curve for $m = 2$ must possess a similar nose to that for $\text{Pr} = 1$, the tip of it being located at $1/\Gamma < 1$. Therefore, the axisymmetric flow for $\Gamma = 1$ becomes unstable first for $\text{Ra}_{c2} = 10\,134$ in the form of an $m = 1$ mode. This result is consistent with the calculations of Wagner et al. (1994) and the experimental

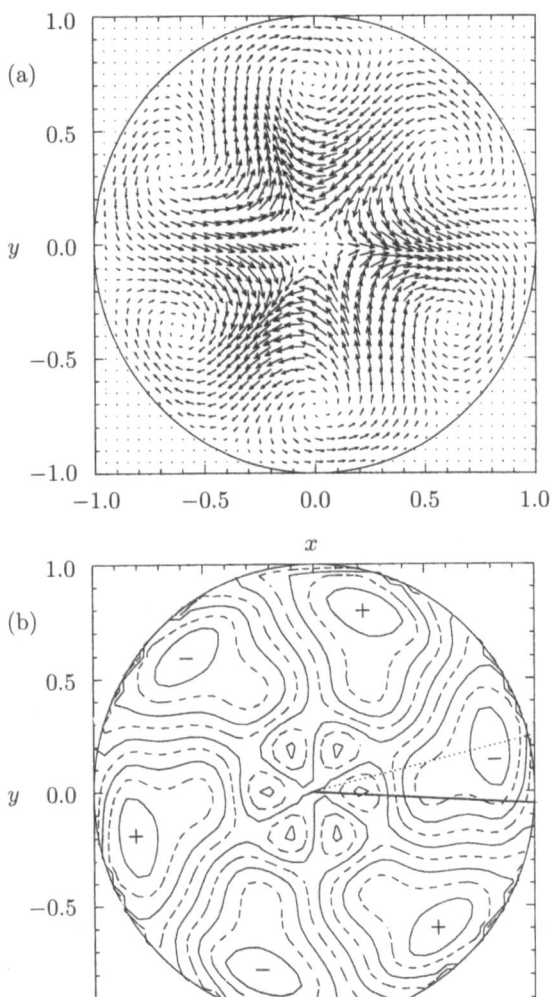

Fig. 7.18. Velocity field (**a**) and temperature field (**b**) of the neutral solution for $1/\Gamma = 1.47$, $Pr = 1$, $Bi = 0$, and $Ra = Ra_{c2}(m = 3) = 24\,928$ in a horizontal section at $z = -0.3$. The phase shift between the velocity and temperature fields is an indication of pattern rotation. The *full line* in (**b**) indicates the angle for which the radial outflow of the neutral mode is maximum. The angle at which the neutral temperature field attains its minimum is shown by a *dashed line*. The pattern rotates clockwise

data of Müller et al. (1984), who found $Ra_{c2}(m = 1) \approx 22\,600$.[11] The numerical results of Müller et al. (1984) and Neumann (1990) do not agree with the stability analysis. Even though it cannot be precluded in the framework of linear stability analysis that $m = 2$ modes with a *finite* amplitude are possible in the vicinity of $Ra = 2800$, their existence is very unlikely, since

---

[11] Since ideal adiabatic boundary conditions cannot be realized experimentally, the excessively high experimental value of $Ra_{c2}(m = 1) = 22\,600$ may be caused by radial temperature gradients near the lateral wall. Owing to its symmetry, such an additional drive favors axisymmetric flows.

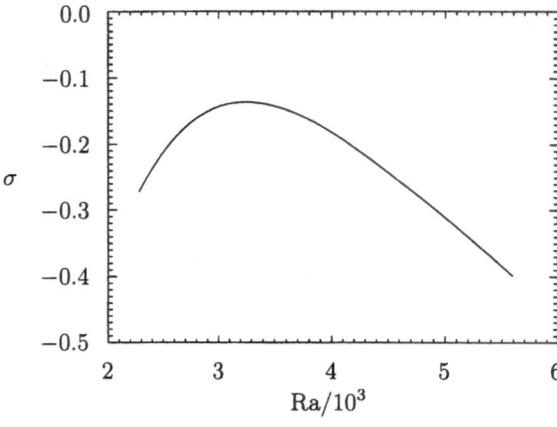

Fig. 7.19. Linear growth rate σ of the $m=2$ mode as function of the Rayleigh number for $\mathrm{Pr}=6.7$ and $\Gamma=1$

Wagner et al. (1994) found numerically an exponential decay of the finite-amplitude mode with wavenumber $m=2$.

The axisymmetric buoyant convection at high Prandtl numbers is very stable in the range $0.9 < 1/\Gamma < 1.56$ (up to $\mathrm{Ra}_{c2} = \mathcal{O}(10 \times \mathrm{Ra}_c)$). The ensuing instabilities are caused by buoyancy forces. Small-Prandtl-number flows for such high Rayleigh numbers would possess considerable inertia. Since the secondary instabilities for small-Prandtl-number fluids are caused by inertia effects, the critical Rayleigh numbers Ra_{c2} for $\mathrm{Pr} \ll 1$ are much smaller than the ones for $\mathrm{Pr} = \mathcal{O}(1)$. It can be concluded that there must exist a transition range between the two Prandtl numbers considered here in which the instability mechanism changes from advective (momentum transport) at low Prandtl numbers to convective (heat transport) at high Prandtl numbers.

7.4 Thermocapillary Instability

Hydrothermal waves are important convection structures in thermocapillary plane layers. It has been speculated for a long time that the flow oscillations in thermocapillary liquid bridges are also caused by hydrothermal waves (Preisser et al. (1983), Xu & Davis (1984), Velten et al. (1991)). Despite many experiments this question has not been settled satisfactorily. Moreover, the dependence of the stability boundary on the Prandtl number has been largely unknown. In this section, therefore, convective instabilities in liquid bridges will be treated and the fluid-dynamical mechanisms responsible for the instability will be analyzed.

The early publications by Chang & Wilcox (1975, 1976) on the modeling of thermocapillary flows in melt zones stimulated further investigations. Important contributions are due to Schwabe et al. (1978), Schwabe & Scharmann (1979), Chun & Wuest (1979), Schwabe et al. (1982), Preisser et al. (1983), Kamotani et al. (1984), Fu & Ostrach (1985), Velten et al. (1989, 1991), and other authors. Numerical work regarding hydrodynamic instabilities in liquid bridges has been published by Shen et al. (1990), Mittelmann (1990), Neitzel et al. (1991, 1993), Kuhlmann & Rath (1993a,b), Mittelmann (1994), and Wanschura et al. (1995a). A number of contributions resulted from a controversy about the physical mechanisms causing the observed thermocapillary instabilities (Chun (1980), Kamotani et al. (1984), Ostrach et al. (1985), Hu & Tang (1990), Velten et al. (1991)). This question will be addressed here by a linear stability analysis of axisymmetric stationary thermocapillary convection in the absence of buoyancy (Gr = 0). The neutral modes and the physical instability mechanisms will be discussed following Wanschura et al. (1995a) on the basis of energy transfer rates.

The stability boundary depends strongly on the Prandtl number. The typical behavior as a function of Pr can be demonstrated for the aspect ratio $\Gamma = 1$ and an adiabatic free surface (Bi = 0).[12] The linear stability analysis yields a critical azimuthal wavenumber $m = 2$ for $\Gamma = 1$ and Bi = 0, independent of Pr.

The dependence of the critical Reynolds number Re_c on Pr is shown in Figs. 7.20 and 7.21. For very small Prandtl numbers the critical Reynolds

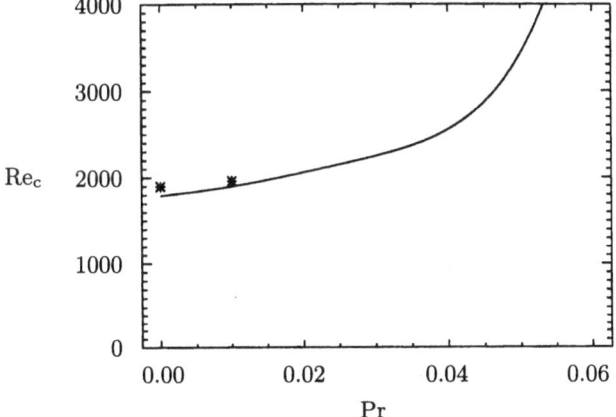

Fig. 7.20. Critical Reynolds number Re_c for $\Gamma = 1$, Bi = 0, and small Prandtl numbers; (∗) Levenstam (1994)

number increases moderately from the limiting value $Re_c = 1793$ at Pr = 0. For Prandtl numbers larger than Pr ≳ 0.05, however, the Reynolds number increases very rapidly, and this branch of the stability boundary has not been traced to higher Prandtl numbers, because the numerical error in the calculated critical values becomes too large.[13] In the range of intermediate Prandtl

---

[12] Additional calculations reveal that the critical Reynolds number increases upon an increase of Bi if Gr=0.

[13] The calculation of the basic state becomes increasingly demanding because of the appearance of boundary layers and the associated large local gradients in

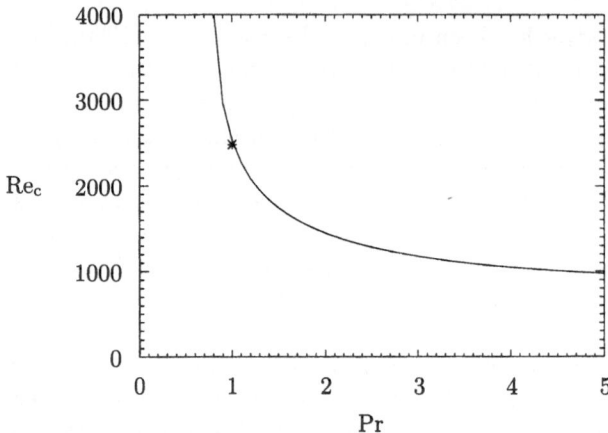

Fig. 7.21. Critical Reynolds number Re_c for $\Gamma = 1$, $Bi = 0$, and large Prandtl numbers; (∗) Neitzel et al. (1993)

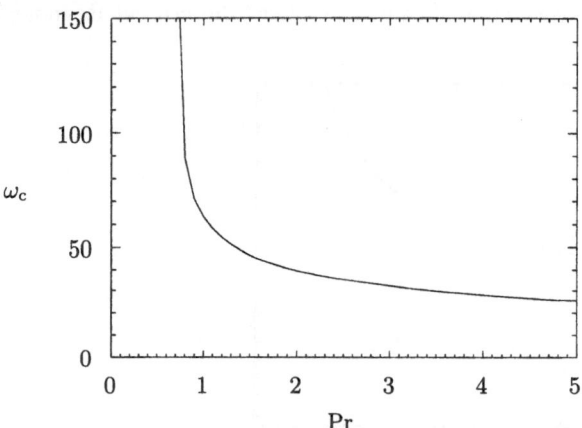

Fig. 7.22. Oscillation frequency ω_c of the critical mode ($m=2$) for high Prandtl numbers ($\Gamma=1$, $Bi=0$)

numbers ($0.05 \lesssim Pr \lesssim 0.7$) the basic flow is linearly stable for $Re \lesssim 5000$, not only for $m=2$ but also for other wavenumbers. The relatively stable range is followed by a range of Prandtl numbers for which the first instability has a different character. The critical mode also has $m=2$, but it is oscillatory. The oscillation frequencies are shown in Fig. 7.22. Since the character of the critical mode does not change along either of the two individual critical curves, the instabilities are now analyzed in more detail for two representative high and low Prandtl numbers.

---

the velocity and temperature fields. Chen et al. (1998) carried out a linear stability analysis for intermediate Prandtl numbers without, however, specifying the numerical error.

7.4.1 Small Prandtl Numbers: Pr=0.02

The stationary character of the thermocapillary instability was first predicted by Rupp et al. (1989).[14] The authors did not, however, specify the stability boundary nor did they clarify the cause of the stationary instability. A grid resolution of $M = 25$ and $N = 80$ employed in the present work for the basic state as well as for the neutral mode yields well-converged stability boundaries.[15] The neutral mode for $Pr = 0.02$ ($Re_c = 2062$) is shown in Fig. 7.23. The pattern consists of four equal convection cells, each occupying a

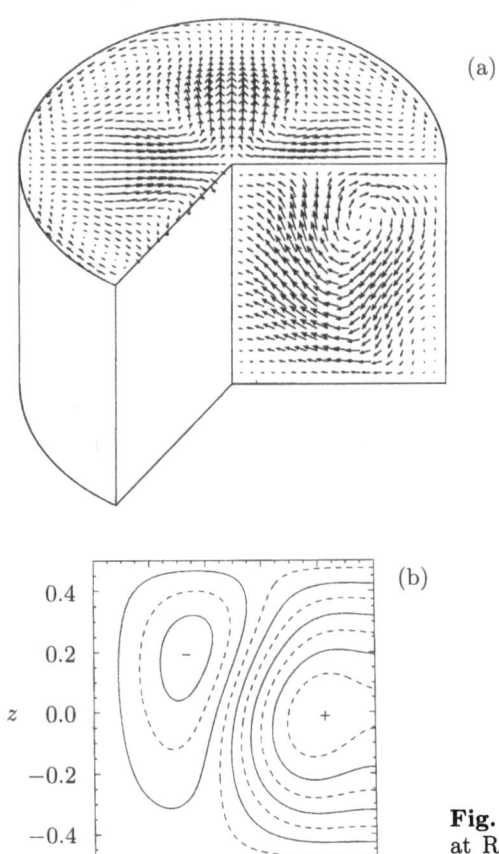

Fig. 7.23. Critical mode for $Pr = 0.02$ at $Re_c = 2062$ ($m = 2$, $\Gamma = 1$, $Bi = 0$). Velocity field at $\varphi = 0$ (**a**) and its projection onto $z = -0.25$; temperature field Θ at $\varphi = 0$ (**b**)

radial section $\Delta\varphi = \pi/2$. The $m = 2$ symmetry is readily identified by the regions of alternating radial in- and outflow visible in a horizontal section

---

[14] Rupp et al. (1989) simulated the flow for $\Gamma = 1.2$ (see Sect. 7.7).
[15] The error in the energy balances for $Gr = 0$ was less than 5% for E_{kin} and less than 1% for E_T.

(Fig. 7.23a). The azimuthal velocity vanishes on the vertical cell boundaries at constant φ, and the neutral velocity field at $\varphi=0$ appears as a single vortex similar to the basic flow. Owing to the small Prandtl number the convective temperature transport is comparatively weak. The temperature field of the neutral mode is mainly determined by the convective transport $(w\partial_z)$ of the conductive basic-state temperature $(\theta\approx z)$.[16]

In the limit $\mathrm{Pr}\to 0$ the temperature deviations from the conductive profile vanish, and the critical Reynolds number takes the finite value $\mathrm{Re_c}(\mathrm{Pr}\to 0)=1793$. It is seen from (4.33) that the basic-state temperature field must be $\Theta_0=\mathcal{O}(\mathrm{Pr})$ for $\mathrm{Re}=\mathcal{O}(1)$. Since $\partial_t\Theta=0$ on the stability boundary $(\omega_c=0)$, the temperature field of the disturbance also scales like $\Theta=\mathcal{O}(\mathrm{Pr})$ (see (7.9)). It follows that $\Delta\Theta\sim\mathrm{RePr}\,w=\mathcal{O}(\mathrm{Pr})$, i.e. the convective heat transport from the conducting temperature field is compensated by heat conduction. The balance between $\Delta\Theta$ and $\mathrm{RePr}\,w$ corresponds to a balance between the rates of change of energy D_T and I_{T3} of the neutral mode. It is seen from Fig. 7.24 that this balance is satisfied to a good approximation for $\mathrm{Pr}=0.02$. Since the temperature field of the basic state Θ_0 only enters I_{T1} and I_{T2}, it

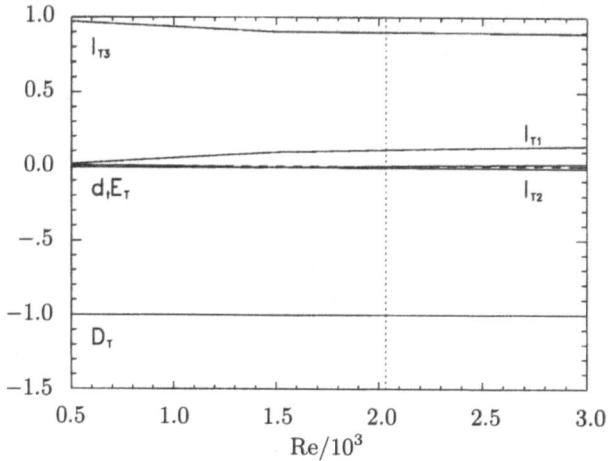

Fig. 7.24. Thermal-energy balance as a function of the Reynolds number for $\mathrm{Pr}=0.02$, $\Gamma=1$, and $\mathrm{Bi}=0$

is insignificant for the instability at low Prandtl numbers. The ratio of the two total rates of change of energy scales like $E_{\mathrm{kin}}/E_T=\mathcal{O}(\mathrm{Pr}^{-1})$. Hence, the instability mechanism for $\mathrm{Pr}\to 0$ must be purely inertial.

The normalized rates of change of kinetic energy are shown in Fig. 7.25 for the most dangerous mode. With increasing Reynolds number the interaction term I_{v4} increases and dominates the whole balance for $\mathrm{Re}\gtrsim 2500$. On the stability threshold $\mathrm{Re_c}$, D, and I_{v4} nearly compensate each other. All other

---

[16] The heating is from the upper side (at $z=0.5$).

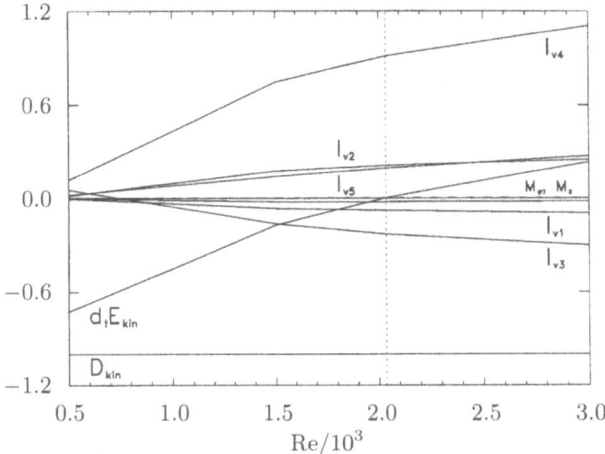

Fig. 7.25. Kinetic-energy balance as a function of the Reynolds number for $Pr=$ 0.02, $\Gamma=1$, and $Bi=0$

contributions to the kinetic-energy balance are much smaller, even locally. The integral $I_{v4} = -\mathrm{Re} \int w\, u\, \partial_r w_0\, dV$ is a measure of the transport of axial basic-state momentum (w_0) to the disturbance (w) by the radial velocity field of the perturbation (u). The efficiency of this process is proportional to the magnitude of the radial gradient of the axial basic-state velocity $(\partial_r w_0)$, i.e. proportional to the strain rate of the axial basic flow. The high absolute value of the shear gradient $(\partial_r w_0 < 0)$ of the basic flow near the free surface is caused by the driving Marangoni forces. They are visible in Fig. 7.1a for conditions close to criticality. For $\varphi = 0$, the radial velocity of the neutral mode close to the free surface is essentially negative (Fig. 7.23a). It transports negative axial momentum of the basic flow from the free surface into the interior of the liquid bridge. Since the axial velocity w of the neutral mode is also negative for $r \gtrsim 0.7$ at $\varphi = 0$, it is enhanced there. Similar considerations apply to the amplification for $r \lesssim 0.5$ and explain the two amplification maxima of the local energy transfer rate $-\mathrm{Re}\, w\, u\, \partial_r w_0$ for $\varphi = 0$, shown in Fig. 7.26 as a function of r and z. The peak close to the free surface makes by far the largest contribution to the energy gain. Since the critical perturbations for $m = 2$ are π-periodic, the same energy transfer mechanisms are operative at $\varphi = \pi$. The explanation is equally valid at $\varphi = \pi/2$ and $\varphi = 3\pi/2$, but with inverted signs of u and w. Although the shear flow is driven by the thermocapillary effect, the rates of change of energy due to the Marangoni forces M_φ and M_z, induced by the the neutral temperature field, do not play any role in the instability mechanism (Fig. 7.25). The low-Prandtl-number instability is purely inertial and is caused by the thermocapillary-induced strain of the basic flow.

In order that the amplification mechanism discussed above leads to instability, a feedback from the amplified velocity w to u is required. Here,

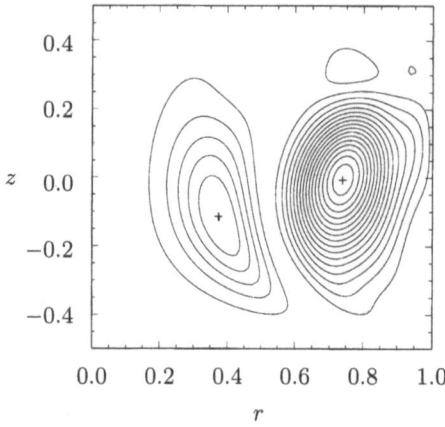

z

r

Fig. 7.26. Local transfer rate of kinetic energy (integrand of I_{v4}) at $\varphi = 0$ for Pr = 0.02, Re$_c$ = 2062, $m=2$, $\Gamma=1$, and Bi$=0$

this feedback is provided by the rigid boundaries at $z = \pm 1/2$ together with the continuity of the flow. Bounding rigid walls are not present in the infinitely long liquid-bridge model (Xu & Davis (1984)). Therefore, the feedback is lacking in the infinite system, preventing the stationary instability. This instability is suppressed for $\Gamma \to \infty$.[17]

The question arises as to why the critical Reynolds number increases for larger Prandtl numbers. By artificially neglecting the azimuthal Marangoni forces in the boundary condition for the neutral mode, Prange (1997) has shown that azimuthal Marangoni effects are the dominant cause of the flow stabilization.[18] With increasing Prandtl number the radial outward disturbance flow creates ever-colder surface-temperature spots, since the basic-state temperature difference between the bulk and the free surface increases. Azimuthal Marangoni forces are thus induced that oppose the azimuthal disturbance surface flows arising from continuity. This mechanism corresponds to the stable situation in the classical Marangoni problem (Pearson (1958)), where a liquid layer with an upper free surface is heated from above.

The influence of the aspect ratio on the instability for Pr$=0.02$ is shown in Fig. 7.27. With increasing normalized radius $1/\Gamma$ the critical azimuthal wavenumber increases because of a crossing of the neutral curves for different values of m. The structure of the critical modes within a single azimuthal convection cell $0 \le \varphi \le \pi/m$ is, however, quite similar. Thus the instability mechanism discussed above is operative all along the critical curve shown in Fig. 7.27. The wavenumbers are well ordered in the interval of Γ considered, similar to the situation for thermal convection in liquid bridges (Fig. 7.7) and for the thermocapillary instability at high Prandtl numbers (see Sect. 7.4.2 and Preisser et al. (1983)). The azimuthal wavelength on the free surface,

---

[17] Instead, the first instability in infinitely long liquid bridges is oscillatory; see also Sect. 7.4.3.

[18] This mechanism of stabilization was also suggested by Kuhlmann & Rath (1993a).

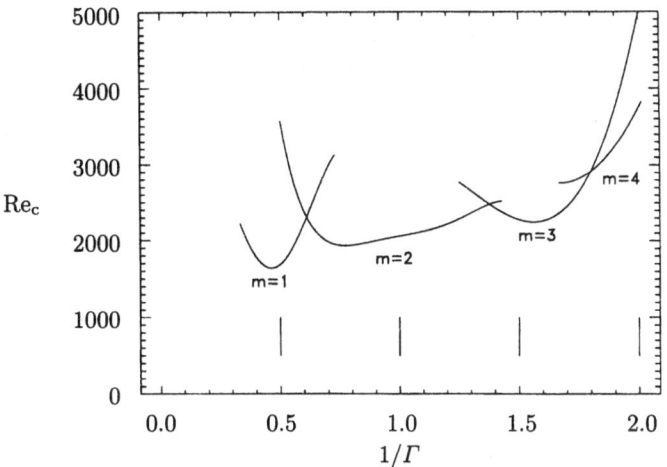

Fig. 7.27. Critical Reynolds number as a function of the aspect ratio for $Pr=0.02$ and $Bi=0$. The *vertical marks* indicate the aspect ratios for which $m=2/\Gamma$

$2\pi R/m \approx \pi d$, thus scales with the height of the liquid bridge.[19] Some neutral values are given in Table 7.1.

Table 7.1. Dependence of the neutral Reynolds number and wavenumber on the aspect ratio for $Pr=0.02$ and $Bi=0$

Γ	Re_c	m
3.00	2219	1
2.00	1689	1
2.00	3565	2
1.50	2056	2
1.00	2062	2
0.60	2358	3
0.50	4957	3
0.50	3744	4

7.4.2 Large Prandtl Numbers: $Pr=4$

When the Prandtl number is high, large temperature gradients develop near the corners (Sect. 7.1). The temperature gradient near the cold corner ($z = -0.5$) scales with the Marangoni number $\sim Ma$ in the viscous convective limit

---

[19] The scaling of the wavelengths of symmetry-breaking convection patterns with the macroscopic characteristic length is observed in many systems (see e.g. Koschmieder (1993) and Cross & Hohenberg (1993)).

(Sect. 3.2.2). Moreover, a thermal boundary layer develops on the hot wall at $z = 0.5$, leading to a large surface temperature gradient $\sim \mathrm{Ma}^{2/7}$ in the hot corner (Cowley & Davis (1983)). A typical temperature field is shown in Fig. 7.2b. The surface-velocity peak of w_0 resulting from the temperature gradient in the cold corner for high Marangoni numbers can only be resolved with a very fine numerical mesh. Here, the driving basic-state shear stress in (2.29) is regularized, as in Shen (1991):

$$
\boldsymbol{e}_z \cdot \mathsf{S} \cdot \boldsymbol{e}_r
$$
$$
= -\left\{
\begin{array}{cc}
\frac{1}{4}\left\{1 - \cos\left[10\pi\left(\frac{1}{2} + z\right)\right]\right\}^2, & z \leq -0.4 \\
1, & -0.4 < z < 0.4 \\
\frac{1}{4}\left\{1 - \cos\left[10\pi\left(\frac{1}{2} - z\right)\right]\right\}^2, & z \geq 0.4
\end{array}
\right\} \boldsymbol{e}_z \cdot \nabla(\Theta_0 + z).
$$
$$
\tag{7.24}
$$

In this way high gradients in the upper and lower 10% of the free surface are smoothed (see Fig. 7.28). For $\Gamma = 1$, $\mathrm{Pr} = 4$, and the grid resolution

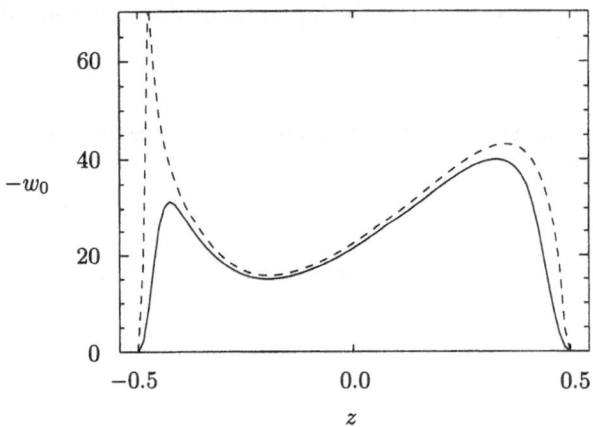

Fig. 7.28. Surface velocity w_0 of the basic state in units of ν/d for $\mathrm{Pr}=4$, $\mathrm{Re}=1047$, $\Gamma = 1$, and $\mathrm{Bi} = 0$. (- - -) Unregularized; (——) regularized according to (7.24)

employed, this regularization leads to an increase of the critical Reynolds number by $\approx 5\%$. This small value indicates that the corner flow is not of prime importance for the instability.[20] For this reason the regularization is kept throughout for $0.5 < \mathrm{Pr} < 5$.[21]

---

[20] The large-scale convection structures of the basic state depend only weakly on the details of the corner flows. For a similar problem, see Koplik & Banavar (1995).

[21] For higher Prandtl numbers, when the surface temperature gradients are confined to very narrow corner regions, this regularization is not justified and may yield erroneous results, since it would substantially alter the strength of the driving

The instability mechanism at high Prandtl numbers is fundamentally different from the one at small Pr. At the critical threshold Re_c (see Fig. 7.21) a pair of azimuthally traveling waves with frequencies $\pm\omega_c$ becomes unstable through a Hopf bifurcation (Guckenheimer & Holmes (1983)). The oscillation frequency ω_c (Fig. 7.22) has a similar dependence on Pr as Re_c, and both decrease with increasing Prandtl number. Within the interval $0.9 < Pr < 5$ the critical wavenumber is given by $m_c = 2$.

In the following, the instability mechanism is considered for the mode with $\omega_c > 0$. For $Pr = 4$, $\Gamma = 1$, and $Bi = 0$ the critical Reynolds number is $Re_c = 1047$ with Hopf frequency $\omega_c = 27.9$.[22]

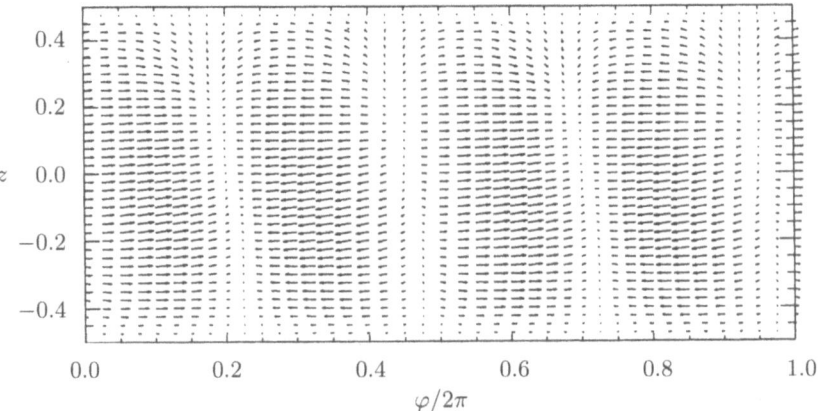

Fig. 7.29. Tangential flow on the free surface at $r = 1/\Gamma$ for the neutral mode with Pr=4, $Re_c = 1047$, $m=2$, and Bi=0

Figure 7.29 shows the flow field of the neutral mode on the free surface. The azimuthal velocity components are by far larger than the vertical ones. The small axial variations of the flow field suggest a weak influence of the rigid walls at $z = \pm 1/2$ on the instability. This hypothesis is confirmed by an a posteriori energy analysis. Those terms in (5.26) that include axial variations are comparatively small. The kinetic-energy balance is shown in Fig. 7.30. The inertia terms I_v are very small compared to the dissipation. Kinetic energy is mainly produced by the work of azimuthal Marangoni forces M_φ. Thus the perturbation flow is driven by the azimuthal Marangoni effect induced by the temperature field of the neutral mode. This is different from

---

forces, which become concentrated near the heated corners. Any change of the driving strength, however, would necessarily alter the bulk flow and thus its stability.

[22] The behavior of the mode with $\omega_c < 0$ is analogous. At Re_c both modes become unstable simultaneously and the neutral solution is a superposition of both waves. In the framework of a linear stability analysis the modal amplitudes for $Re > Re_c$ cannot be determined (see Sects. 7.6 and 7.7).

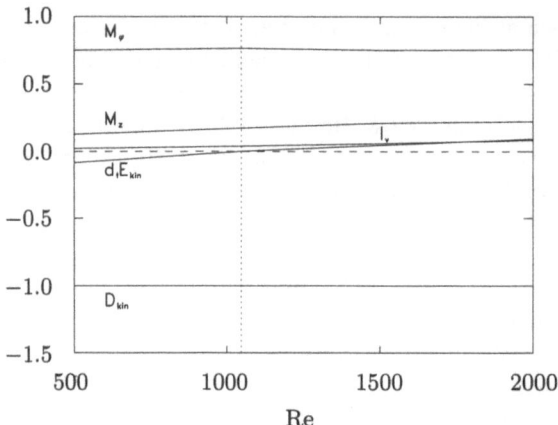

Fig. 7.30. Kinetic-energy balance as a function of the Reynolds number for $Pr=4$, $\varGamma=1$, $m=2$, and $Bi=0$

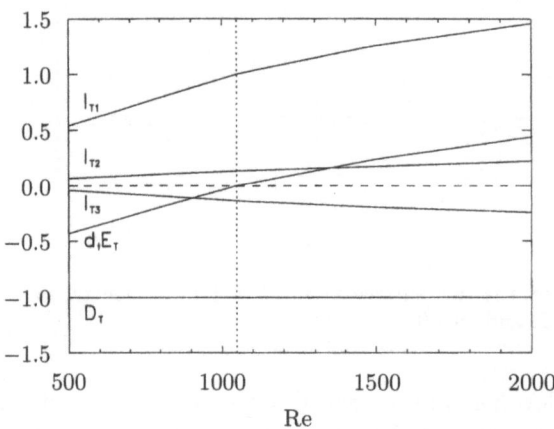

Fig. 7.31. Thermal-energy balance as a function of the Reynolds number for $Pr=4$, $\varGamma=1$, $m=2$, and $Bi=0$

the low-Prandtl-number ($Pr \ll 1$) instability, which is driven by a mechanical interaction with the basic flow. Inspecting the thermal-energy balance in Fig. 7.31, it is seen that I_{T3} is always stabilizing and nearly compensates I_{T2}. Thermal energy is mainly generated through the process described by I_{T1}, a term which in fact dominates the energy balance for $Re > 1000$. I_{T1} is the rate of change of thermal energy due to radial transport ($u\partial_r$) of the basic-state temperature ($\varTheta_0$) to the temperature field ($\varTheta$) of the perturbation. This process will now be explained in detail.

Since the axial terms in the energy equations are very small, the energetics can be discussed by considering the fields in a horizontal section at $z=0$. The temperature field of the neutral mode has pronounced extrema in the

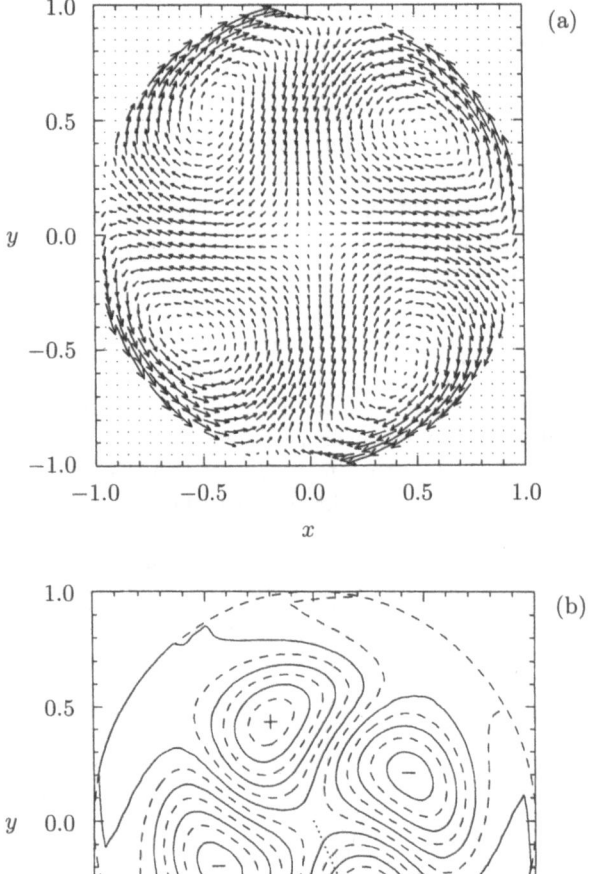

(a)

(b)

Fig. 7.32. Flow (**a**) and temperature field (**b**) of the critical mode for Pr = 4, $Re_c = 1047$, $m = 2$, $\Gamma = 1$, and Bi = 0 in a horizontal section at $z = 0$. The *dotted line* indicates the instantaneous angle of a temperature maximum (cf. Fig. 7.33). The pattern rotates clockwise

interior of the liquid bridge (Fig. 7.32b). An internal disturbance temperature maximum heats the free surface conductively and causes a weak hot spot on the surface. The latter generates Marangoni flows that are directed from the interior to the hot spot on the free surface and along the free surface away from the hot surface spot. In the present case the interior disturbance temperature extrema are only infinitesimal perturbations of the basic temperature, which is lower in the interior than on the free surface. Therefore, the hot free-surface spot is cooled by the radial outward convection. This situation is stable in the sense of the classical Marangoni effect. It establishes, however, a frustration owing to competing processes: conductive heating and convective cooling of the free-surface hot spot. Clearly, the convective cooling becomes more efficient with increasing Reynolds number.

The frustration is resolved if both competing processes are phase-shifted in the azimuthal direction. This, in fact, happens as can be seen from Fig. 7.32b, where the phase shift between the internal disturbance temperature extrema and those on the free surface is clearly visible. The flow at the free surface is directed azimuthally away from the surface temperature maxima at $\varphi \approx 0$ and π and it is directed towards the corresponding minima at $\varphi \approx \pi/2$ and $3\pi/2$. Owing to continuity, the azimuthal Marangoni flows are connected with radial flows that transport colder fluid (basic state) from the interior to the free surface. This way, the internal temperature minima are amplified. The location of maximum amplification, however, is slightly phase-shifted azimuthally. Similarly, the internal temperature maxima are amplified by the corresponding hot inward flows. The phase shift between the internal temperature extrema and the locations of their maximum amplification leads to a rotation of the pattern as a whole in the negative azimuthal direction. The spatial distribution of the local amplification rate confirms this explanation. The integrand $-\mathrm{Re}\,\Theta\,u\,\partial_r\Theta_0$ of I_{T1} is shown in Fig. 7.33 at $z=0$. It can be

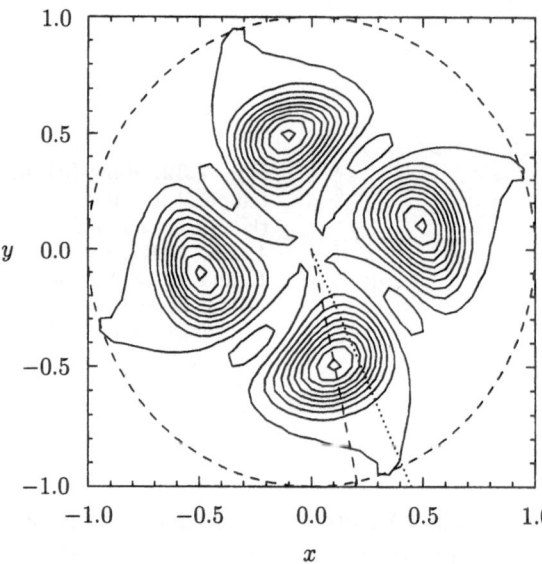

Fig. 7.33. Isolines of the local transfer rate of thermal energy $-\mathrm{Re}\,\Theta\,u\,\partial_r\Theta_0$ (integrand of I_{T1}) at $z = 0$ for $\mathrm{Pr} = 4$, $\mathrm{Re_c} = 1047$, $m = 2$, $\Gamma = 1$, and $\mathrm{Bi} = 0$. The *dashed line* indicates the angle at which the local amplification rate has its maximum value. The *dotted line* corresponds to the angle for the associated temperature maximum (compare Fig. 7.32)

seen that the locations of maximum amplification are in advance of those of the corresponding temperature extrema in the negative azimuthal direction.

The dependence of the linear-stability boundary on the aspect ratio Γ is shown in Fig. 7.34 for $\mathrm{Pr}=4$. Similarly to the case for small Prandtl numbers, an approximate scaling law $m_c \sim 2/\Gamma$ is valid. The instability mechanism

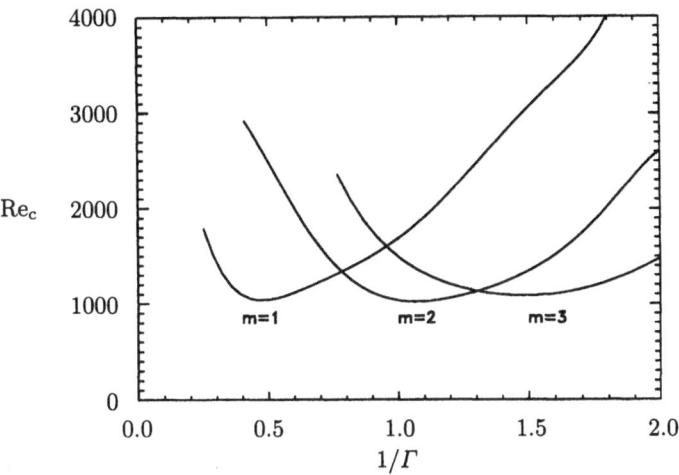

Fig. 7.34. Neutral stability boundaries as a function of the inverse aspect ratio $1/\Gamma$ for $\mathrm{Pr}=4$ and $\mathrm{Bi}=0$

and the modal structure (within one azimuthal wavelength) do not change qualitatively in the range of Γ considered.

7.4.3 Remarks on the Stability of Pure Thermocapillary Flows

The analyses carried out in Sects. 7.4.1 and 7.4.2 have shown that there exist at least two different types of instability in thermocapillary liquid bridges. For the range of medium Prandtl numbers $\mathrm{Pr}<0.5$ and for $\Gamma=1.2$, Rupp et al. (1989) predicted the first instability of the axisymmetric flow to be stationary. This is in contrast to recent results of Chen et al. (1998), who found oscillatory instabilities with $m=3$ for $0.1 \leq \mathrm{Pr} \leq 0.8$ at very high Reynolds numbers ($\Gamma=1$). Owing to the high Reynolds numbers these results must, however, be considered with care.

The instability mechanism for high-Prandtl-number liquid bridges is very similar to the one for plane thermocapillary liquid layers (cf. Chap. 6, Smith & Davis (1983a), and Smith (1986a)). Therefore, the critical modes in high-Prandtl-number liquid bridges may also be termed *hydrothermal waves*. Xu & Davis (1984) have investigated the stability of a long thermocapillary liquid bridge in the limit $\Gamma \to \infty$. They found a critical mode with $m=1$ which is consistent with the present analysis. Notwithstanding the different azimuthal wavenumbers for finite zones, the critical modes are quite similar in both systems. Both modes are slightly corkscrew-twisted (see Figs. 7.29 and 7.46, as well as Muehlner et al. (1997)). This corresponds to a small axial component of the wave vector which is directed streamwise of the free-surface flow in both cases. The relative strength of the axial wave-vector component becomes smaller for decreasing Prandtl number ($\mathrm{Pr}<5$), for $\Gamma=1$ as well as for $\Gamma \to \infty$. As can be seen from Table 7.2 the critical Reynolds number for

Table 7.2. Some neutral Reynolds numbers and oscillation frequencies as functions of Pr for $\Gamma = 1$ and Gr = Bi = 0. The wavenumber is $m = 2$ throughout. This is the critical wavenumber, except for Pr = 0.5, for which $m_c = 3$

Pr	Re_c	ω_c
10^{-8}	1793	0
0.02	2062	0
0.05	3434	0
0.5	10 252	245.2
1.0	2539	63.2
2.0	1449	39.5
4.0	1047	27.9

hydrothermal waves increases sharply when the Prandtl number is decreased beyond Pr = 1.

For Pr $\lesssim$ 1 significant differences exist between the two systems. The critical Reynolds number for inertia-induced hydrothermal waves diverges like Pr^{-1} (Xu & Davis (1984)). This is consistent with the linear stability of the Hagen–Poiseuille flow (Drazin & Reid (1981)). However, a finite critical Reynolds number for a stationary mode exists in finite-length liquid bridges (see Figs. 7.20 and 7.27).

The marks in Fig. 7.27 indicate those aspect ratios for which

$$m = \frac{2}{\Gamma} . \tag{7.25}$$

This empirical correlation is well satisfied in the interval of Γ considered, for both low and high Prandtl numbers. Similar correlations were found by Levenstam & Amberg (1995) (low Pr) and Leypoldt et al. (1998). The former authors suggested the similarity of the present low-Pr instability to the instability of ring vortices in an ideal fluid (Widnall & Tsai (1977)). The unstable modes have a similar structure in both cases and show a similar azimuthal-wavenumber dependence ($\sim \Gamma^{-1}$). The instability of ring vortices is caused by a *self-induced* straining motion owing to the curvature effect. Whether this or the *externally applied* shear force is responsible for the present instability can only be clarified by an analysis of a thermocapillary-driven vortex in a rectangular trench ($\Gamma' \to 0$). In any case, the unstable mode draws its energy from the basic state by the same process as that in which the perturbation flow is directed parallel to the principal strain direction of the basic flow (at an angle of 45 degrees with respect to the shear flow direction).

In the past, several authors have tried to explain the convective instabilities in liquid bridges in terms of the deformability (Ca $\neq$ 0) of the free surface (Kazarinoff & Wilkowski (1989)) or in terms of buoyancy effects (Hu & Tang (1990), Hu et al. (1994)). The detailed analysis presented above shows, however, that these effects cannot be responsible for the instabilities found, because the results from model experiments (see Sects. 7.5.1 and 7.6) agree qualitatively and semiquantitatively with the instabilities discussed in this section.

7.5 Instability of Mixed Convection

Usually, ΔT is selected as the control parameter in experiments. Then Gr and Re are no longer independent of each other. Their relative magnitude, measured by the dynamic Bond number $\mathrm{Bd} = \mathrm{Gr}/\mathrm{Re}$, may vary. In the experiments of Velten et al. (1991) with liquid bridges of lengths $\mathcal{O}(3\ \mathrm{mm})$ the Rayleigh numbers at the critical onset of oscillations were of the order of magnitude $|\mathrm{Ra}| \approx 1300$ for $\mathrm{Pr} = 1$, $|\mathrm{Ra}| \approx 9000$ for $\mathrm{Pr} = 7$, and $|\mathrm{Ra}| \approx 11\,000$ for $\mathrm{Pr} = 49$. Since the latter two of these values are significantly larger than the critical Rayleigh number for the onset of natural convection (Sect. 7.3), buoyancy effects cannot be a priori disregarded, even for liquid volumes as small as those employed by Velten et al. (1991). When investigating the combined buoyant-thermocapillary flow it is useful to distinguish between small and large dynamic Bond numbers. For small Bd the stationary two-dimensional flows and their instabilities will be dominated by thermocapillarity, while for sufficiently large values of Bd natural convection will dominate, modified by small thermocapillary effects. The influence of a second driving cannot be assessed in an ad hoc manner, since both the basic state and the neutral mode depend nontrivially on Bd. Therefore, we shall investigate the Bond-number dependence separately for small and for large Bd, keeping $\Gamma = 1$ and $\mathrm{Bi} = 0$ fixed and considering a representative case of small Prandtl number $(\mathrm{Pr} = 0.02)$ and one of a typical large Prandtl number $(\mathrm{Pr} = 4)$. The reader is also referred to Wanschura et al. (1997a).

The Oberbeck–Boussinesq equations (2.14–2.16) and the boundary conditions (2.34) are invariant under the transformation

$$(\mathrm{Re}, z, u, v, p, \Theta) \longrightarrow (-\mathrm{Re}, -z, -u, -v, -p, -\Theta)\ . \qquad (7.26)$$

It follows that the two-dimensional basic states for mixed convection $(\mathrm{Bd} \neq 0)$, as well as their stability properties, are symmetric with respect to $\mathrm{Re} = 0$. The above symmetry operation corresponds to a reflection of the streamlines at $z = 0$ and an inversion of the vorticity. The following discussion is based on a liquid bridge heated from above. A change from a stable thermal stratification $(\mathrm{Gr} > 0,\ \mathrm{Ra} < 0,\ \boldsymbol{g} \downarrow\uparrow \nabla T_{\mathrm{a}})$ to an unstable one $(\mathrm{Gr} < 0,\ \mathrm{Ra} > 0,\ \boldsymbol{g} \uparrow\uparrow \nabla T_{\mathrm{a}})$ can be imagined as an inversion of the $\boldsymbol{g}$ vector.

7.5.1 Thermocapillarity Dominating

Pr $= 0.02$. For $\mathrm{Pr} = 0.02$ the critical mode is stationary with wavenumber $m = 2$ as in the case of pure thermocapillary flow $(\mathrm{Ra} = 0)$. The stability boundaries are shown in Fig. 7.35 for $|\mathrm{Ra}| < 500$. In the range considered,

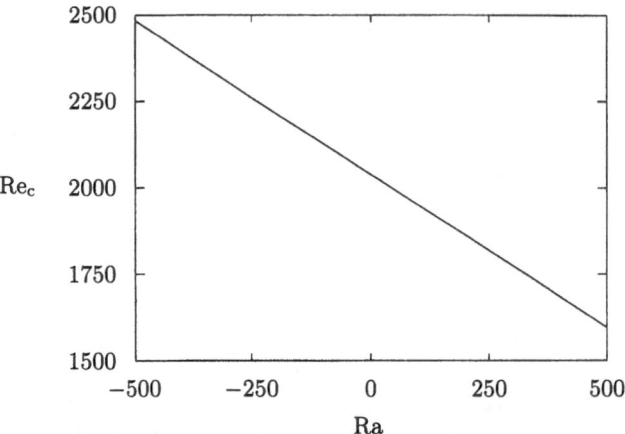

Fig. 7.35. Stability boundary Re_c as a function of the Rayleigh number Ra for $\Gamma=1$, Pr$=0.02$, and Bi$=0$. The neutral mode has an azimuthal wavenumber $m=2$

the critical curve is almost linearly dependent on Ra. For an unstable density stratification (Gr < 0, Ra > 0), the basic state is destabilized, while it is stabilized for a stable density stratification (Gr > 0, Ra < 0). This behavior is expected, since the temperature field does not deviate much from the conducting profile. However, the result is nontrivial. The thermal-energy balance does not vary much in the range of Rayleigh numbers considered and cannot be invoked to explain the threshold shift. The kinetic-energy balance for Re $= 2000$ as a function of the Rayleigh number is shown in Fig. 7.36. Similarly to the pure thermocapillary case (see Sect. 7.4.1), the contribution of I_{v4} is dominant and the instability mechanism is inertial. The work per unit time done by buoyancy forces $I_{\mathrm{Gr}} = \mathrm{Bd} \int_V w\Theta \, dV$ is insignificant, since the neutral temperature field Θ is very small. The basic-state temperature field for Re≈ 2000, however, deviates sizably from the pure conduction profile.

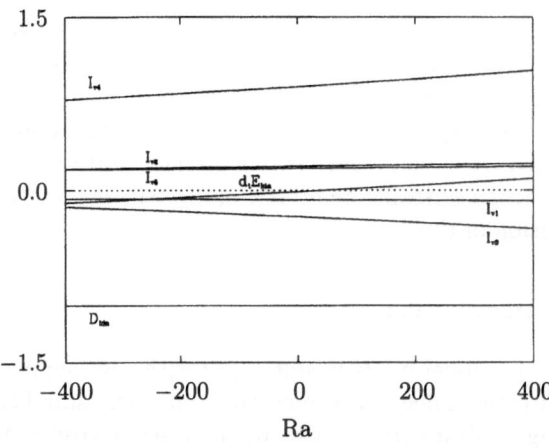

Fig. 7.36. Kinetic-energy balance as a function of the Rayleigh number Ra for Re$=2000$, $\Gamma=1$, Pr$=0.02$, and Bi$=0$

Thus buoyancy forces act to modify the basic state. In this way, the energy-providing shear gradient $(\partial_r w_0)$ is reduced for $Gr > 0$ just at the location of maximum amplification (cf. Fig. 7.26), reducing the destabilizing energy term I_{v4}. The influence of Ra on the shear gradient is shown in Fig. 7.37. For $Gr < 0$ the opposite holds true and I_{v4} is increased. These arguments are valid provided that the neutral mode remains unaltered. For the range of Ra considered the critical mode has in fact nearly the same form as for $Ra = 0$.

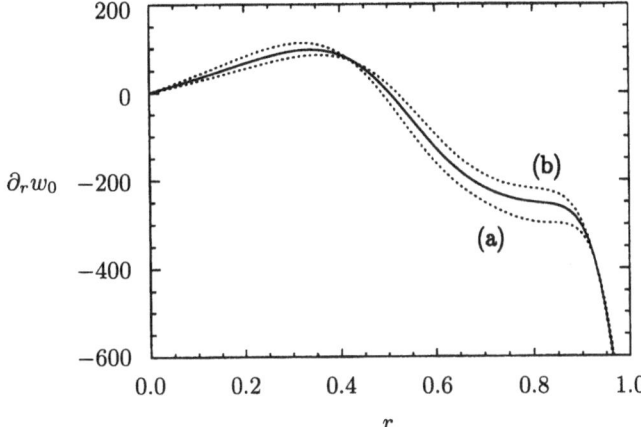

Fig. 7.37. Radial dependence of the shear rate $\partial_r w_0$ of the axisymmetric basic state at $z = 0$ for $Re = 2000$, $\Gamma = 1$, $Pr = 0.02$, and $Bi = 0$. (——) $Ra = 0$; (- - -) $Ra = 400$ (a) and $Ra = -400$ (b).

Pr = 4. The stability properties are different for high Prandtl numbers. As for $Ra = 0$ the critical mode is oscillatory with $m = 2$ and it develops continuously out of the pure hydrothermal wave. Hence the critical mode also exhibits strong internal temperature extrema. The stability boundary of the basic state is shown in Fig. 7.38. Except for a small range of Rayleigh numbers (the critical curve has a minimum at $Ra = -360 \pm 20$), buoyancy leads to a stabilization, independent of the sign of the Rayleigh number. Similarly to pure thermocapillary flow, the kinetic-energy balance does not change much with a variation of the parameters (here Ra) in the range under consideration. For $Pr = 4$, the thermal-energy balance is decisive. It is plotted in Fig. 7.39 for $Re = 1200$. For $Ra \approx 0$ the contributions due to I_{T2} and I_{T3} are relatively small. For $|Ra| = 10^4$ they reach, however, the order of magnitude of the energy term I_{T1} which is responsible for the appearance of hydrothermal waves. The neutral fields of a pure hydrothermal wave ($Ra = 0$) vary nearly exclusively in the r and φ directions and the thermal energy (I_{T1}) is transported mainly radially. But for $Ra = \mathcal{O}(10^4)$ a significant amount of energy (I_{T2}, I_{T3}) is convectively transported vertically. This vertical transport is carried by a strong w component of the neutral mode and it couples

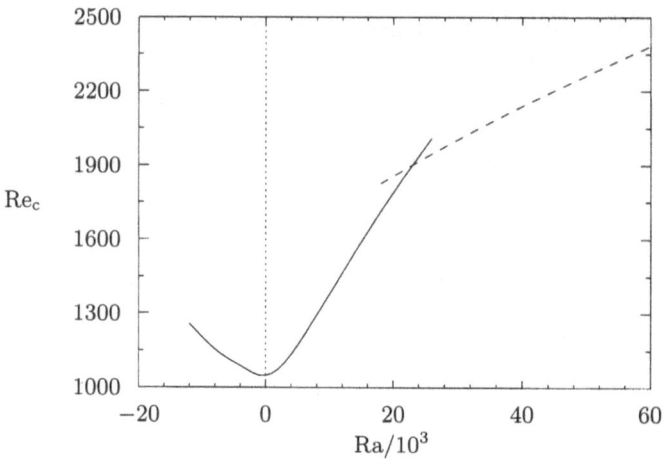

Fig. 7.38. Stability boundary Re_c as a function of Ra for $\Gamma=1$, Pr$=4$, and Bi$=0$. For small Rayleigh numbers the critical mode has the azimuthal wavenumber $m=2$ (——); for larger Rayleigh numbers (Ra $\gtrsim$ 22 000) the wavenumber is $m=1$ (- - -). Both modes are oscillatory

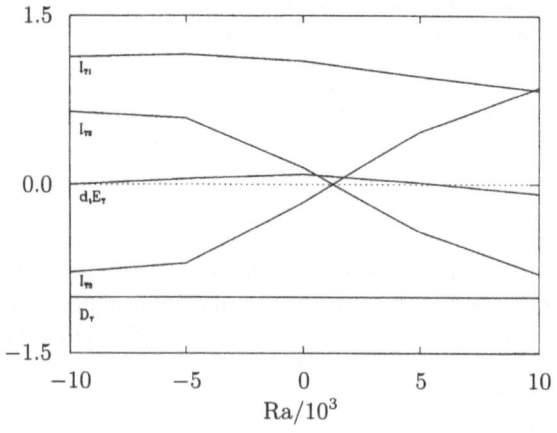

Fig. 7.39. Thermal-energy balance as a function of the Rayleigh number Ra for Re$=1200$, $\Gamma=1$, Pr$=4$, and Bi$=0$

to the basic-state temperature field. The coupling to the conductive profile (I_{T3}) is, as expected, stabilizing for Ra <0 and destabilizing for Ra >0 (see Fig. 7.39). The coupling to Θ_0 (I_{T2}), though, has a stabilizing effect as in the case of pure buoyant convection.[23] Both effects are strongly developed for Ra$=\mathcal{O}(10^4)$, but they nearly compensate each other. In addition, I_{T1} be-

---

[23] It was shown in Sect. 7.3.2 that the deviation Θ_0 of the basic-state temperature from the conducting profile has a strongly stabilizing effect in a gravity field, if the heating is from below (Ra>0).

comes smaller for Ra $\neq$ 0. Thus there is a sensitive balance between strong, opposing effects. The total balance leads to the calculated stabilization for Ra > 0 as well as for Ra < 0.[24]

Since the amount of stabilization is less for negative Rayleigh numbers, it is expected that the critical Reynolds number measured in the earth's gravity field is smaller when the heating is from above as compared to the critical Reynolds number for equally strong heating from below. This was, in fact, observed by Velten et al. (1991) using fluids with Pr = 1 and Pr = 7 (see also Figs. 7.44 and 7.45). Thus the relative stabilization for heating from below can be understood on the basis of the present linear stability analyses. Accordingly, the measured critical Reynolds numbers under weightlessness (Ra = 0) should be smaller than those under gravity conditions (Ra $\neq$ 0) regardless of the heating direction, except possibly for a small range ($-700 \lesssim$ Ra < 0 for Pr = 4) around the minimum of $Re_c(Ra)$. Schwabe & Scharmann (1984, 1985) provided experimental evidence for this conclusion.[25]

7.5.2 Buoyancy Dominating

If buoyancy dominates, the stability diagram is more complicated than for thermocapillary-determined flows. The main reason is that up to three different basic states may exist. The typical branching structures will be discussed here for $\Gamma = 1$, Bi = 0, and Pr = 4.

Let us consider, for the moment, only *two-dimensional* flows, regardless of their stability with respect to *three-dimensional* perturbations. For Re = 0 the conductive state becomes unstable at $Ra_c(m = 0) = 1825$ through a perfect pitchfork bifurcation (cf. Fig. 7.7). The bifurcation diagram is shown in Fig. 7.40 (dotted lines). Apart from the unstable conductive state, two nontrivial supercritical-solution branches exist for Ra > $Ra_c(m = 0)$, which correspond to up- and downflow, respectively, in the center ($r = z = 0$). If Re $\neq$ 0, a structural instability (Drazin (1992)) occurs and the perfect bifurcation is perturbed. The perturbed bifurcation is shown as full lines in Fig. 7.40 for Re = +5 > 0. For a thermally unstable stratification (Ra > 0) upflow in the center ($r = z = 0$) is favored (strong state (a)) when Re > 0. The corresponding flow field consists of a single toroidal vortex whose sense of rotation is supported by both buoyant and thermocapillary forces (see Figs. 7.41 and 7.4a).[26] For Rayleigh numbers larger than Ra_* (the value $Ra_*(Re) \geq$

---

[24] A somewhat hand-waving interpretation is the following. Hydrothermal waves are essentially characterized by vertical vorticity, while natural convection favors horizontal vorticity. Therefore, the generic modal structures are not compatible with each other, resulting in a mutual suppression of the associated three-dimensional flow structures. This corresponds to a stabilization of the basic state.

[25] Sometimes a higher value of the critical Reynolds number under weightlessness as compared to the value on the ground is caused by liquid contaminations (Schwabe & Scharmann (1985)).

[26] Figure 7.4 applies to Re = $-5 < 0$. For Re > 0 the streamlines must be reflected at $z = 0$.

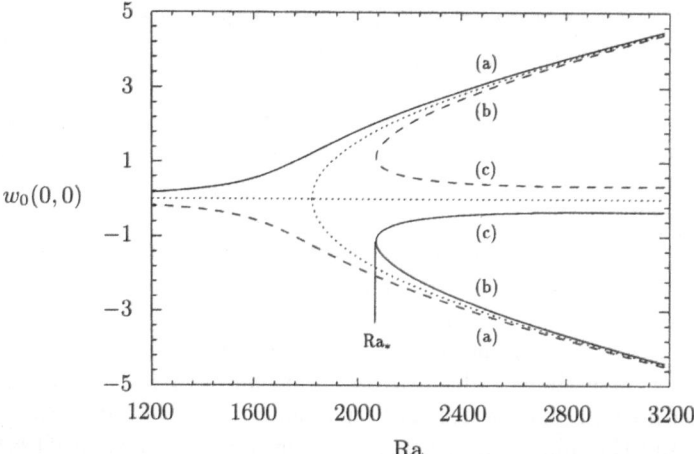

Fig. 7.40. Perturbation of the perfect bifurcation of the axisymmetric buoyant convection (· · · ·) by thermocapillary surface forcing. The axial velocity $w_0(r = 0, z = 0)$ is shown as a function of Ra for $\Gamma = 1$, Pr $= 4$, and Bi $= 0$. (———) Imperfect bifurcation for Re $= +5$; (a) denotes the strong state, (b) the weak state, and (c) the state that corresponds to the supercritical conducting solution. Ra$_*$ denotes the branching point of the perturbed bifurcation. (- - -) Imperfect bifurcation for Re $= -5$

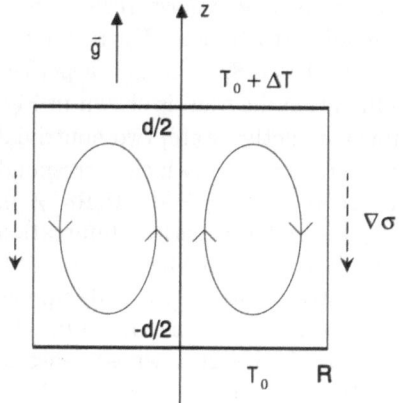

Fig. 7.41. Sketch of the geometry and the driving forces for Ra > 0 ($g \parallel e_z$)

Ra$_c$($m = 0$) depends on Re), two additional solutions exist, the weak state (b) and the state (c) (Fig. 7.40), both of which coincide for Ra $=$ Ra$_*$. In both latter states thermocapillary forces oppose buoyancy forces. This leads to the creation of additional small vortices of thermocapillary origin in the cold and hot corners, where thermocapillary forces are strong and buoyancy is weak. The sense of rotation of these small vortices is opposite to the rotation sense of the mainly buoyancy-driven bulk vortex. The weak state (b) (corresponding to Fig. 7.4b) develops out of the bifurcation point $Ra = Ra_*$. It is characterized by a strong toroidal vortex in the bulk whose sense of rotation (downflow at $r = z = 0$ for $Re > 0$) is determined by the buoyant convection. Owing to the strong buoyancy convection, the thermocapillary vortex is suppressed on an increase of Ra and only two small thermocapillary corner vortices remain.[27] Eventually both corner vortices vanish on a further increase of Ra.

In the presence of thermocapillary perturbation, state (c) (corresponding to Fig. 7.4c) evolves from the supercritical conducting state. Its temperature field, as compared to the strong (a) and weak (b) states, deviates least from the conducting profile and it is thus characterized by the smallest Nußelt number. On the other hand, the strong state (a) is associated with the largest Nußelt number. In state (c) both of the thermocapillary corner vortices grow in size on an increase of the Rayleigh number beyond Ra_* and finally merge to form a single thermocapillary vortex that occupies the entire free surface. In the interior there remains, however, a weak buoyancy-driven vortex with the opposite sense of rotation. For $Re = 0$ and increasing Rayleigh number there exists a sequence of successive pitchfork bifurcations from the conducting state. The respective neutral modes possess an increasing number of internal nodes. For the perturbed bifurcation ($Re \neq 0$), the solution branch (a) of order $n + 1$ is always connected to the branch (c) of order n. These higher nontrivial solutions, however, are unstable.

Owing to the invariance (7.26) of the Oberbeck–Boussinesq equations the two-dimensional vortex states are symmetrical with respect to $Re = 0$. The perturbed bifurcation for $Re < 0$ is indicated by the dashed lines in Fig. 7.40. On a continuous decrease of the Reynolds number beyond $Re = 0$ the weak state (b) transforms into the strong state (a), and vice versa. The boundary $Ra = Ra_*(Re)$ of the range of existence of the three basic states (a), (b), and (c) is indicated by a dotted line in Fig. 7.42. As expected, state (c) turns out always to be unstable, particularly with respect to two-dimensional perturbations. All unstable ranges indicated in Fig. 7.42 correspond to stationary modes.

For $Pr = 4$ and $\Gamma = 1$, the first instability is three-dimensional for $Re = 0$, with $m = 2$. The critical Rayleigh number is $Ra_c(m = 2) = 1616$ (see Fig. 7.7). It corresponds to the instability of the strong state. On an increase of the Rayleigh number (with $Re = 0$), the linear growth rate of the unstable mode increases until it reaches a maximum and then it finally vanishes again at $Ra'_c(m = 2) = 3586 > Ra_c(m = 0)$.[28] Immediately above $Ra = Ra'_c(m = 2)$ the basic state is linearly stable. In the (Re, Ra) plane the two bifurcation points at $Re = 0$ are connected to each other by the critical curve of the strong state for $m = 2$. Since the sign of the growth rate is conserved on both sides of

---

[27] The larger of the two thermocapillary corner vortices is located at the hot corner when the heating is from below.

[28] This behavior is in agreement with the form of the neutral curve (nose) discussed in Sect. 7.3.2 for $Pr = 1$ and 6.7.

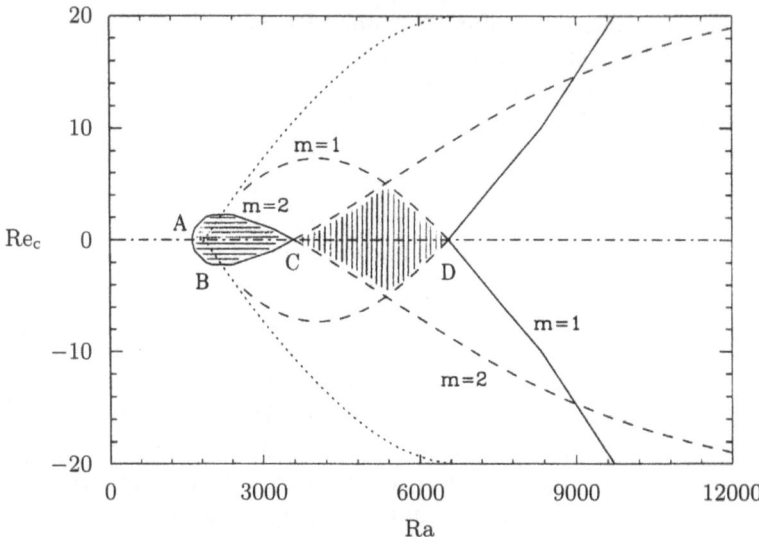

Fig. 7.42. Stability diagram for $\Gamma=1$, $Pr=4$, and $Bi=0$. ($\cdots$) Ra_*; (——) linear-stability boundary of the strong state (a); (- - -) linear-stability boundary of the weak state (b). Within the *horizontally hatched area* and to the *right* of the $m=1$ *full curve* the strong state is unstable. It is stable otherwise. Within the *vertically hatched area* the weak state is linearly stable. It is unstable otherwise. The letters have the following meaning: A, $Ra_c(m=2)=1616$; B, $Ra_c(m=0)=1825$; C, $Ra'_c(m=2)=3586$; D, $Ra_{c2}(m=1)=6557$

the critical curve, the strong state is linearly unstable with respect to $m=2$ perturbations in the closed horizontally hatched area (Fig. 7.42) only.

If, for $Ra_*(Re=0) < Ra < Ra'_c(m=2)$, the line $Re=0$ is crossed from negative to positive values of the Reynolds number, the strong state changes into the weak state. This weak state, therefore, is located on the side of the critical $m=2$ curve for $Re<0$ that is characterized by positive growth rates. Thus the weak state is unstable with respect to $m=2$ perturbations for $Re>0$. Owing to symmetry reasons, it is also unstable for $Re<0$. In the vertically hatched area (Fig. 7.42) the behavior is opposite. For $Re\approx0$ and $Ra'_c(m=2) < Ra < Ra_{c2}(m=1) = 6557$, both the strong and the weak state are linearly stable, the dashed line indicating the linearly stable range of the weak state. Apart from a critical $m=2$ curve, the area of stability of the weak two-dimensional state is also bounded by an $m=1$ stability curve. This curve can be explained as follows. For $Re=0$ both axisymmetric and, for $Ra > Ra'_c(m=2)$, the linearly stable states (a) and (b) become linearly unstable with respect to an $m=1$ mode at $Ra_{c2}(m=1)$. The calculated stability boundary for $Ra > Ra_{c2}(m=1)$ and $Re \neq 0$ belongs to the strong state. This is unstable for higher Rayleigh numbers. By the same argument as above, the weak state is also unstable there. For $Ra < Ra_{c2}(m=1)$, the

$m=1$ curve has been calculated for the weak state. The curve ends on the existence boundary of the weak state (dotted line).

From the stability diagram in Fig. 7.42 two interesting properties of the mixed convection can be deduced. First, the three-dimensional buoyant convection with basic wavenumber $m=2$ is easily suppressed for $\mathrm{Ra} > \mathrm{Ra_c}(m=2)$ by very weak thermocapillary effects ($|\mathrm{Re}| < 4$). Second, the axisymmetric convection is linearly stable within a closed supercritical parameter region. In the range $\mathrm{Ra} > \mathrm{Ra'_c}(m=2)$ two different axisymmetric basic states for $\mathrm{Re} \neq 0$, (a) and (b), should be realizable.

The continuation of the $m=1$ stability boundary on a larger scale is depicted in Fig. 7.43. Within the range of large Rayleigh numbers shown there

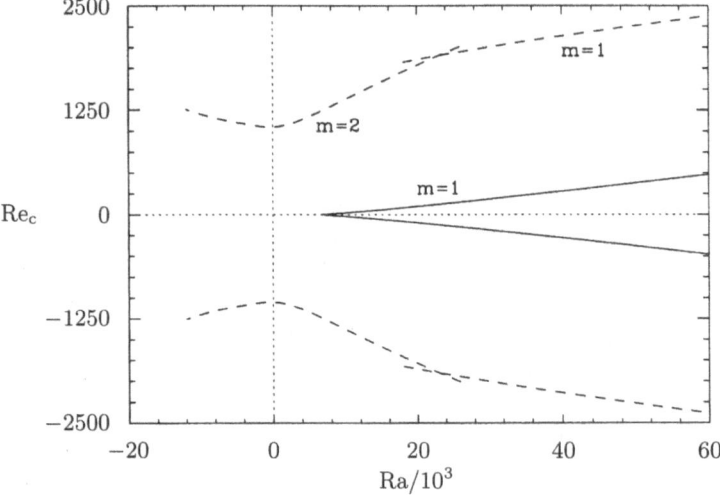

Fig. 7.43. Stability diagram for large Rayleigh and Reynolds numbers ($\Gamma=1$, $\mathrm{Pr}=4$, $\mathrm{Bi}=0$). For $\mathrm{Re} > 0$ the linearly stable range is bounded from below by a stationary $m=1$ instability (———), while it is bounded from above by an oscillatory instability (- - -) with either $m=2$ or $m=1$

always exists an interval of Reynolds numbers for which the two-dimensional combined convection (strong state) is linearly stable. On the scale of Fig. 7.43 the Reynolds numbers that restrict the linearly stable range from below and from above depend almost linearly on the Rayleigh number. If all linear-stability boundaries are supercritical, the onset of three-dimensional convection could be suppressed even for high Rayleigh numbers by quite moderate thermocapillary effects.[29]

---

[29] For any specific Boussinesq liquid and given geometry the (Ra, Re) plane is traversed along a ray originating from the origin (Re = Ra = 0) when the temperature difference ΔT is increased from zero.

An experimental verification of the calculated stability boundaries is still lacking. The parameter range for the complex bifurcation behavior around Ra ≈ 4000 for Pr = 4 seems, however, to be experimentally accessible if small-scale liquid bridges with weak thermocapillary effects are employed (see Velten et al. (1991)). Such experiments could also clarify the question of a possible hysteretic behavior, which cannot be answered in the context of a linear analysis.

7.6 Experimental Background

Most model experiments have been carried out using nearly cylindrical, small, nonisothermal liquid bridges with axial lengths of a few millimeters in gaseous atmospheres and under gravity conditions. Larger liquid bridges exhibit considerable surface deformations owing to the high Bond number Bo $= (\rho - \rho_a)gd/\sigma_0$ or are mechanically unstable (Myshkis et al. (1987), Slobozhanin & Perales (1993)). To reduce the Bond number a density-matched ambient liquid could be used. Such liquids typically are, however, much more viscous than air, and the momentum transport in the matrix fluid would not be negligible, possibly causing premature three-dimensional exterior flows. Moreover, the buoyancy forces would progressively increase with the linear dimension[30] (Bd $=$ Ra/Ma $= \mathcal{O}(d^2)$) and dominate the thermocapillary flow. These problems can be overcome by experiments under weightlessness (Schwabe & Scharmann (1984, 1985), Chun & Siekmann (1995), Carotenuto et al. (1996, 1998)) which allow the formation of large liquid bridges,[31] e.g. $d = \mathcal{O}(40$ mm$)$, (Chun & Siekmann (1995)). The characteristic timescales become, however, quite large ($d^2/\nu = \mathcal{O}(1$ mHz$)$ for silicone oil; cf. Kuhlmann (1995)).

In order to visualize the flow, transparent liquids of high Prandtl number (Pr > 1) in half-zone configurations have been frequently used (see e.g. Preisser et al. (1983), Velten et al. (1991), or Ostrach et al. (1985)). Here the experimental results are summarized on the basis of the comprehensive work of Velten et al. (1991).

These authors employed fluids with Pr $= 1, 7$, and 49. The onset Marangoni numbers and the critical frequencies were measured as functions of the aspect ratio, and different supercritical flow patterns were classified

---

[30] There are only a few liquids, for example water at 4°C, for which the thermal expansion coefficient vanishes ($\beta = 0$) and thus Ra ≈ 0. Since the temperature dependence of the surface tension is relatively weak, liquid bridges of lengths $\mathcal{O}(3$ mm$)$ require temperature differences of at least $\mathcal{O}(10$ K$)$ if the transition to three-dimensional flows is to be investigated. Under these conditions thermal expansion cannot be neglected.

[31] Under weightlessness the maximum aspect ratio for thermocapillary liquid bridges is mainly restricted by the Rayleigh limit (Chen et al. (1990), Rybicki & Floryan (1987b)).

by their dependence on Re and Γ. The measurements were based on signals from thermocouples positioned at different azimuthal angles close to the free surface but not touching it. The lengths and diameters of the liquid bridges were of the order of magnitude $\mathcal{O}(3\ \text{mm})$ and the relative free-surface deformations $\Delta\xi/R$ were less than 5% throughout. It can be anticipated, therefore, that the deformations had a negligible influence on the two-dimensional base flow and its stability.

The critical Marangoni numbers for $\Gamma=1$ ranged from $\text{Ma}_c(\text{Pr}=1)\approx 1500$, through $\text{Ma}_c(\text{Pr}=7)\approx 7500$, up to $\text{Ma}_c(\text{Pr}=49)\approx 15\,000$. The critical Marangoni number for $\text{Pr}=7$ agrees qualitatively with an extrapolation of the numerical data. If, in addition, the heat transfer at the free surface[32] and buoyancy effects are taken into account, a very good agreement between experiment and stability analysis is obtained. For heating from below, Wanschura (1996) finds $\text{Re}_c^{(\text{num})}=1356$, in comparison with $\text{Re}_c^{(\text{exp})}(\text{Pr}=7)=1215$. For heating from above, the results agree even better, $\text{Re}_c^{(\text{exp})}=970$ compared to $\text{Re}_c^{(\text{num})}=989$. The critical oscillation frequencies likewise agree very well. The experimental critical Reynolds numbers for $\text{Pr}=1$ are smaller than the theoretical values for $\text{Gr}=0$. This difference cannot be caused by buoyant convection (see Sect. 7.5). More likely, convective effects in the ambient gas phase combined with heat and mass transfer (evaporation) could be responsible for the differences.[33]

The order of magnitude of the critical Reynolds number for $1 \lesssim \text{Pr} \lesssim 100$ can be roughly estimated empirically as $\text{Re}_c(\text{Pr}, \Gamma=1)\approx 10^3$ (Kuhlmann (1995)). It decreases slightly with the Prandtl number ($\text{Re}_c(\text{Pr}=35, \Gamma=1)\approx 500$). Figures 7.44 and 7.45 display the dependence of the critical Reynolds numbers measured by Velten et al. (1991) for a fixed radius $R=3$ mm as a function of the aspect ratio for $\text{Pr}=7$ and $\text{Pr}=1$, respectively. From these data it can be concluded that the relation $\text{Re}_c(\Gamma)\approx \text{const.}$ is a crude but fair approximation for $\text{Pr}\gtrsim 1$ (constant) and $\Gamma=\mathcal{O}(1)$ not too small.

For all Prandtl numbers investigated by Velten et al. (1991) the measured oscillation periods ($f=\mathcal{O}(1\ \text{Hz})$) varied linearly with the length of the liquid bridge (for constant radius and liquid properties). The numerically determined periods have the same order of magnitude (Wanschura et al. (1995a)). The dependence of the oscillation frequency on Pr and Γ is not obvious. Since the perturbation temperature field is an essential feature of hydrothermal waves, one can anticipate, however, that the wave is synchronized with the

---

[32] From the the work of Preisser et al. (1983) a value of the Biot number $\text{Bi}\approx 6.4$ can be estimated.

[33] During the experiments of Velten et al. (1991) significant gas-phase convection as well as heat and mass transfer through the free surface took place, since the working temperatures T_0 of the liquid bridges ranged around 770 K ($\text{Pr}=1$) and 340 K ($\text{Pr}=7$), well above room temperature.

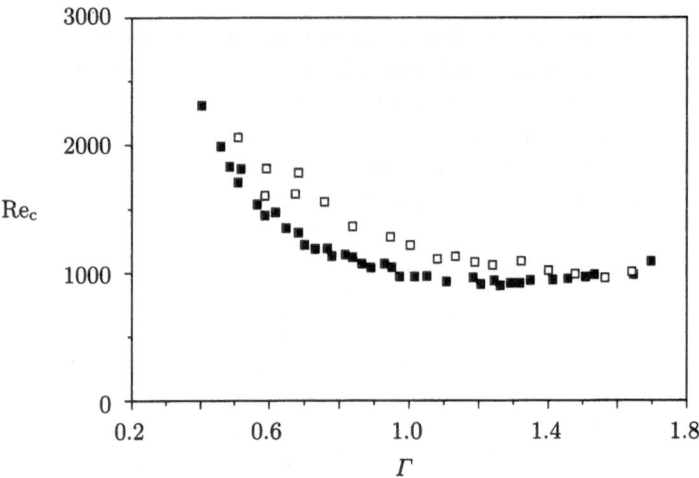

Fig. 7.44. Experimental critical Reynolds numbers as a function of the aspect ratio for $Pr = 7$ according to Velten et al. (1991). Heating from above (■); heating from below (□)

characteristic base-state eddy turnover time.[34] In this picture, the isosurfaces of the perturbation temperature field are spirally wrapped around the core of the toroidal base-state vortex with a winding number m. An example is shown in Fig. 7.46.

Since heat diffusion becomes less important as Pr becomes higher, this interpretation should hold even better as the Prandtl number becomes larger. In fact Muehlner et al. (1997) have measured the spiral character of a hydrothermal wave with $m = 1$ for $Pr = 35$ and $\Gamma = 1$ and found a nearly linear phase variation of $\approx 95°$ in the z direction for constant φ on the free surface. If the characteristic base-state velocity at threshold were constant, the eddy turnover time, and thus the oscillation period of the perturbation, would be proportional to the length d as is observed in the experiment. In order to correlate the measured frequencies, Carotenuto et al. (1996, 1998) have given the empirical formula for $Pr > 1$

$$\omega_c = \frac{\kappa Ma_c^{2/3}}{\sqrt{2Rd^3}} \tag{7.27}$$

and have found very good agreement with many experimental data.[35]

---

[34] Leypoldt et al. (1998) have shown by numerical simulations for $Pr = 4$ and 7 that the oscillation period at the onset is of the order of the axisymmetric basic eddy turnover time.

[35] The derivation of this formula given by Carotenuto (1998) remains, however, unsatisfactory. Equation (7.27) is based on a conductive inertial scaling of the base-state velocity (Table 3.1, $Re \to \infty$, $Pr \ll Re^{1/3}$) where $w \sim Re^{-1/3}$, rather than on the high-Prandtl-number scaling $w \sim Ma^{-1/7}$ which has been confirmed

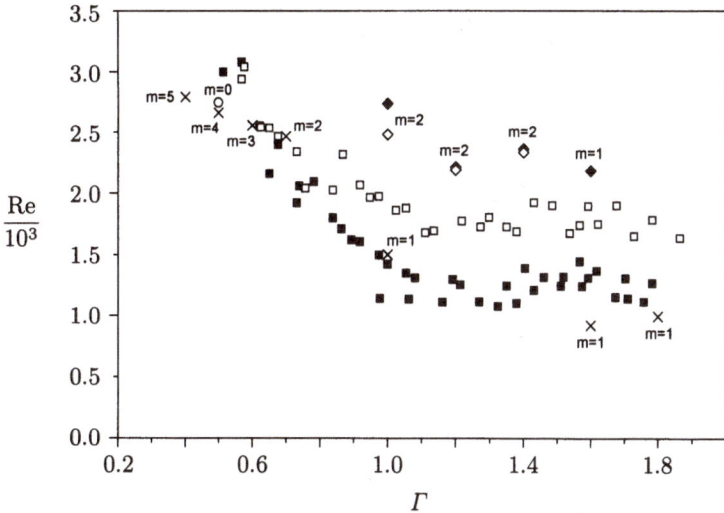

Fig. 7.45. Critical Reynolds numbers as a function of the aspect ratio for $Pr=1$. Experiments with heating from above (■) and heating from below (□) by Velten et al. (1991); linear-stability boundaries ($m=2$, $Bi=0$) by Wanschura et al. (1995a) (♦) and Neitzel et al. (1993) (◊); energy limits ($T_a=0.5$, $Bi=0.3$) by Shen et al. (1990) for $m=0$ (○) and for $m\neq0$ by Neitzel et al. (1991) (×)

Fig. 7.46. Isosurface of constant basic-state stream function $\psi_0 = -1.9\times10^{-3}$ (*green*) and isosurfaces of positive (*red*) and negative (*blue*) temperature perturbations ($\Theta=\pm10^{-2}$) of the neutral mode for $Pr=4$, $Re=1047$, and $Bi=Gr=0$. One can see the spiral character of the temperature perturbations in the center. They also extend, though much more weakly, to the outer side of the vortex core. The *lines* indicate a cube with height d and base $2R\times2R$

For the critical wavenumbers Preisser et al. (1983) found the empirical correlation

$$m_c \approx \frac{2.2}{\Gamma} \ . \tag{7.28}$$

This scaling of the azimuthal wavelength with the height d of the liquid bridge agrees well with the result $m_c \approx 2/\Gamma$ found numerically for *both* high and low Prandtl numbers (Sect. 7.4). Since the experimental data for the critical values and the modal structures agree qualitatively and in part quantitatively with the linear-stability results, it can be concluded in general that the phenomena observable close to the instability threshold can be well understood in the framework of the instability mechanisms discussed in Sect. 7.4.

Velten et al. (1991) confirmed that the amplitude of the temperature oscillations grow with the square root of the distance from the critical point $\propto (\mathrm{Re} - \mathrm{Re_c})^{1/2}$. Hysteresis was not observed.[36] Thus the instability is a supercritical Hopf bifurcation. Far above the threshold, and depending on Γ, a rich variety of different spatio-temporal patterns exist. Fig. 7.47 shows an example. The classification of the supercritical states is difficult partially because of the fact that clockwise- and counterclockwise-propagating hydrothermal waves may exist temporarily, whose amplitudes cannot be predicted by a linear analysis. A recent revised interpretation of Velten et al.'s (1991) data was given by Frank & Schwabe (1997), reporting on selected scenarios of the transition to chaos and observations of pulsating and traveling waves. Owing to the difficulties in getting comprehensive and precise experimental data for the temperature and velocity fields, full three-dimensional numerical simulations are becoming increasingly important for the prediction and understanding of these supercritical flows.

7.7 Three-Dimensional Simulations

Linear stability analysis (Sect. 7.4) is not able to predict the amplitude or the structure of the flow above the critical point. This can only be achieved by numerical calculations using the full time-dependent Oberbeck–Boussinesq equations (2.14–2.16). To date, a number of numerical simulations have been carried out. Rupp et al. (1989) performed the first three-dimensional simulations focusing on the onset of time-dependent flow in low- and high-Prandtl-number liquid bridges. Levenstam & Amberg (1995) simulated the flow in a low-Prandtl-number cylindrical liquid bridge, clarifying the bifurcation structure. With the aim of simulating the conditions of a past experiment under

---

by Kamotani & Ostrach (1998). Moreover, the length scale was selected to best fit the data.

[36] Earlier reports of a hysteretic behavior (Velten et al. (1989)) must be attributed to an impurity accumulation on the free surface in the course of the measurements.

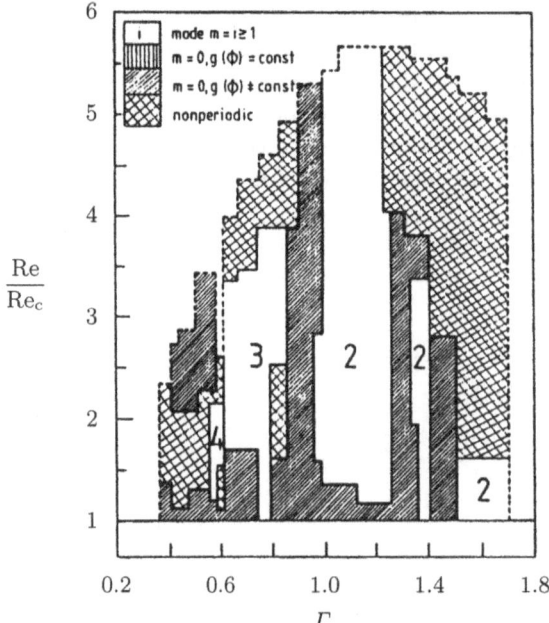

Fig. 7.47. Classification according to Velten et al. (1991) of the experimentally observed pattern formation of supercritical thermocapillary convection in a liquid bridge of NaNO₃ (Pr = 7) heated from above. The numbers within the *blank areas* denote the azimuthal wavenumbers of azimuthally traveling waves. Within the *cross-hatched area* aperiodic behavior was measured. Flows occuring in the remaining areas were denoted as *mixed modes* by the authors

weightlessness, Savino & Monti (1996a) investigated the influence of different ramping rates of the Marangoni number ($\partial_t \mathrm{Ma} > 0$) on the time lag after which unstable three-dimensional flows in high-Prandtl-number liquid bridges reach a given amplitude. Further low-Prandtl-number calculations were performed by Savino & Monti (1996b). The most comprehensive work is due to Leypoldt et al. (1998). The following summary is based on this work.

7.7.1 Low Prandtl Numbers

The existence of stationary three-dimensional convection in small-Prandtl-number liquid bridges was first demonstrated by Rupp et al. (1989). For $\Gamma = 1.2$ and $\mathrm{Bi} = \mathrm{Gr} = 0$ they found stationary patterns with symmetry $m = 2$ for Prandtl numbers in the range $0.007 \leq \mathrm{Pr} \leq 0.5$, in qualitative agreement with linear stability analysis.

Levenstam & Amberg (1995) simulated the flow for $\mathrm{Pr} = 0.01$, $\Gamma = 1$, and $\mathrm{Bi} = \mathrm{Gr} = 0$. In good agreement with linear stability analysis (Fig. 7.20, Table 7.1), they found a supercritical bifurcation out of the basic flow to a stationary mode with fundamental wavenumber $m = 2$ for $\mathrm{Re_c} = 1960$.

Beyond the threshold, the flow is essentially toroidal with a saddle-shaped, deformed vortex core. With increasing Reynolds number the core is displaced towards the cold wall and the free surface at $\varphi=0$ and π $(m=2)$, while it is shifted towards the hot wall and radially inward for $\varphi=\pi/2$ and $3\pi/2$ (up to a constant phase φ_0 depending on the initial conditions). The supercritical character of the bifurcation can be deduced from Fig. 7.48. The amplitudes

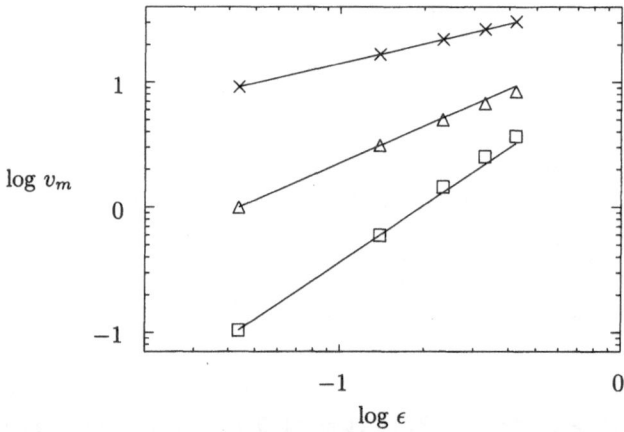

Fig. 7.48. Amplitudes of the harmonics v_m of the stationary three-dimensional supercritical azimuthal flow as a function of ϵ for $Pr=0.02$, $\Gamma=1$, and $Bi=Gr=0$ at $(r,z)=(1,0)$ $(Re_c^{(num)}=2108, m_c=2)$. $(\times)$ $m=2$; $(\triangle)$ $m=4$; $(\square)$ $m=6$

of the harmonics of the azimuthal velocity field $(m_c=2)$ grow like $v_2\sim\epsilon^{1/2}$, $v_4\sim\epsilon^1$, and $v_6\sim\epsilon^{3/2}$, where

$$\epsilon=\frac{Re-Re_c}{Re_c},\tag{7.29}$$

is the normalized distance from the critical point. The corresponding numerical values are given in Table 7.3.

Table 7.3. Amplitudes of the harmonics of the azimuthal velocity $v_m=\hat{v}_m\epsilon^{p_m}$ for saturated slightly nonlinear states with $m_c=2$ and $\Gamma'=1$, $Gr=Bi=0$. The critical Reynolds numbers are $Re_c(Pr=0.02)=2108$ and $Re_c(Pr=4)=1030$. The data are taken at $(r,z)=(1,0)$ for $Pr=0.02$ and at $(r,z)=(0.5,0)$ for $Pr=4$. After Leypoldt et al. (1998)

	Pr=0.02		Pr=4	
m	$\hat{v}_m$	p_m	$\hat{v}_m$	p_m
2	47.41	0.53	6.56	0.50
4	21.70	0.98	2.23	0.99
6	11.78	1.51	0.93	1.45

The three-dimensional stationary toroidal flow becomes unstable to a time-dependent flow at a secondary critical Reynolds number. For $Pr=0.02$ Leypoldt et al. (1998) found $Re_{c2}(\Gamma=1, Gr=Bi=0)=7160$. Slightly above critical, the oscillations are nearly harmonic in time with $\omega_c=92.8$. The character of the flow and the critical values do not differ much from the those at $Pr=0$. Hence, the secondary oscillatory instability must be inertial in nature, like the primary instability.[37] Since the basic flow for the secondary bifurcation is not homogeneous in the φ direction (not axisymmetric), the spatial structures of the oscillatory perturbations do not correspond to azimuthal normal modes. Instead, all fluctuating velocity components with odd azimuthal wavenumbers ($m=1, 3, 5, ...$) grow out of the basic flow with $m=2$, with the square root of the distance from the secondary critical point. The instability is thus a supercritical Hopf bifurcation. The oscillatory motion appears in the form of a spatially anharmonic standing wave. A closer examination reveals that the dominant feature of the oscillatory part of the flow consists of a pair of almost rectilinear vortices with opposite vorticity aligned parallel to the z axis. They are periodically created near $r=0.5$ and $\varphi=0, \pi$. With increasing intensity, both vortices travel to the center ($r=0$), where they annihilate while a new pair of vortices, now with opposite vorticity, is generated. The secondary oscillating vortices arise in a region where the basic flow is similar to a plane stagnation-point flow (see Fig. 7.49). The plane stagnation-point-like flow is caused by the deformation of the basic stationary vortex, whose core is stretched radially outwards along $\varphi=0, \pi$ while being pushed radially inwards at $\varphi=\pi/2, 3\pi/2$. The appearance of vortices with axes in the direction of the principal strain bears some similarity to the instability of linear plane stagnation-point flows (Lagnado et al. (1984)) and the existence of finite-amplitude vortices, although steady, in plane stagnation-point flows (Kerr & Dold (1994)).

The order of magnitude of the critical Reynolds number for the onset of oscillations in low-Prandtl-number half-zone models agrees qualitatively with that measured by Cröll et al. (1989) in a full floating-zone experiment. Since the full zone features two counterrotating toroidal vortices, owing to the temperature maximum on the free surface, the experimentally observed oscillations cannot be directly explained in terms of the mechanisms found in numerical half-zone simulations. Linear-stability calculations by Wanschura et al. (1997b) indicate, however, that the instability mechanisms and the transition Reynolds numbers are very similar in both half and full low-Prandtl-number zones, at least for the first stationary instability.

7.7.2 High Prandtl Numbers

When considering supercritical hydrothermal waves it must be taken into account that two waves propagating in opposite azimuthal directions become

---

[37] For $\Gamma=1$ and $Gr=Bi=0$, Levenstam & Amberg (1995) found $Re_{c2}(Pr=0.01)=6250$ and $Re_{c2}(Pr=0)=5962$.

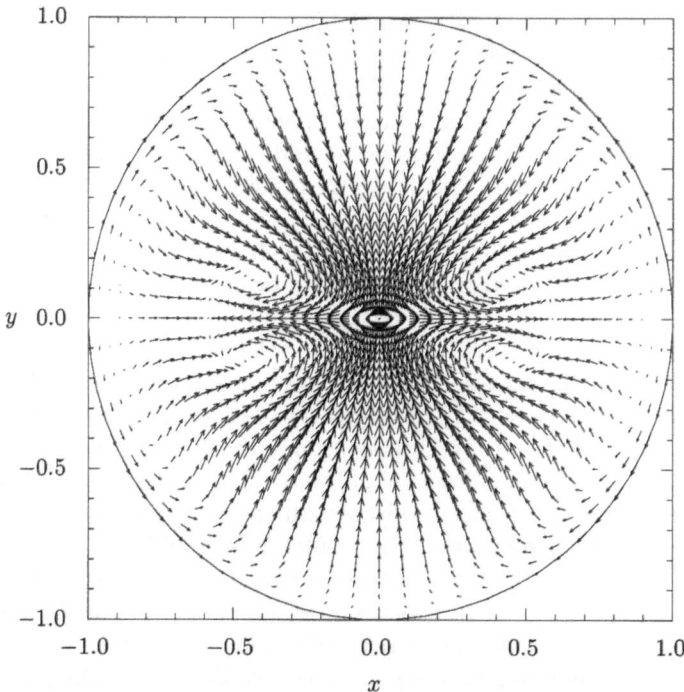

Fig. 7.49. Projection of the time-averaged slightly supercritical three-dimensional velocity field onto the plane $z=-0.3$ for $Pr=0$, $\Gamma=1$, $Bi=Gr=0$, and $Re=6000$. By continuity, fluid emerges in the positive z direction near the line $\varphi=0,\pi$. The critical values according to Leypoldt et al. (1998) are $Re_{c2}=5957\pm15$ and $\omega_c=80.4$

unstable, the relative amplitudes of which are not known a priori. A further complication arises owing to the wave fronts of the hydrothermal waves not being planes with $\varphi=\text{const}$. The local wave vector has components in the r and z directions, since the phase $G(r,z)$ is a function of both r and z (the amplitudes in (7.5) are complex). This is consistent with the observed spiral temperature field (Muehlner et al. (1997)). The neutrally stable hydrothermal waves traveling in the positive (F_+) and negative azimuthal directions (F_-) have the form

$$F_\pm = A(r,z)e^{i[\pm m\varphi-\omega t+G(r,z)]} + \text{c.c.}, \qquad G, A \in \mathbf{R}, \tag{7.30}$$

where F is any scalar perturbation field. A superposition of the two waves with relative amplitude $\vartheta=|F_-|/|F_+|\in[0,1]$ yields

$$F = A(r,z)\sqrt{(1+\vartheta)^2\cos^2(m\varphi-\varphi_0)+(1-\vartheta)^2\sin^2(m\varphi-\varphi_0)}$$
$$\times e^{i(\Phi-\omega t)} + \text{c.c.}, \tag{7.31}$$

with

$$\Phi = \arctan\left(\frac{1-\vartheta}{1+\vartheta}\tan(m\varphi - \varphi_0)\right) + G(r,z)\,. \tag{7.32}$$

For $\vartheta = 1$, the azimuthal nodal planes are determined by the arbitrary phase φ_0. Thus, for an analysis of thermocapillary convection by phase measurements, the structure (7.31) of the wave must be taken into account.

The stability boundaries found by Rupp et al. (1989) for high-Prandtl-number flows agree qualitatively with the linear-stability results (see Fig. 7.21). Owing to the small grid resolution and *critical slowing down*,[38] the simulation data are not as accurate as those given in Table 7.2.

More accurate three-dimensional simulations of the flow in cylindrical liquid bridges of high Prandtl number have been carried out by Leypoldt et al. (1998) using a finite-volume method in the r and z directions combined with a pseudospectral method in the azimuthal direction. The extrapolated critical Reynolds numbers for the onset of oscillations agree well (within less than $\pm 3\%$) with the linear-stability boundaries.

The supercritical character of the instability is demonstrated in Fig. 7.50 by plotting the saturated-flow-state amplitudes of the azimuthal velocity v for a pure traveling hydrothermal wave as a function of ϵ for $\mathrm{Pr}=4$.

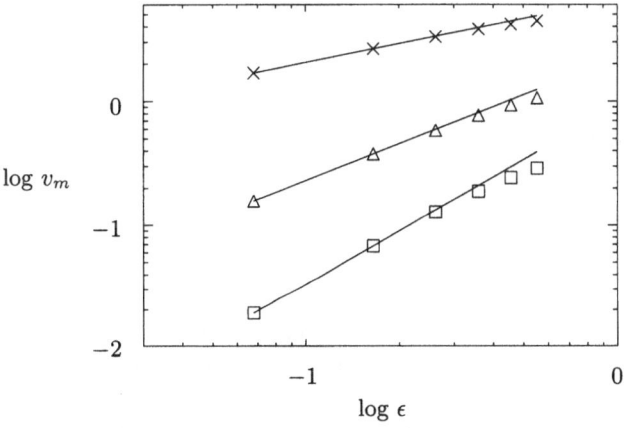

Fig. 7.50. Amplitudes of the harmonics v_m of the stationary three-dimensional supercritical azimuthal flow as a function of ϵ for $\mathrm{Pr}=4$, $\Gamma=1$, and $\mathrm{Bi}=\mathrm{Gr}=0$ at $(r,z)=(0.5,0)$ ($\mathrm{Re}_c^{(\mathrm{num})}=1030$, $m_c=2$). ($\times$) $m=2$; ($\triangle$) $m=4$; ($\square$) $m=6$

Depending on the symmetry of the initial conditions, traveling, standing, or mixed-type hydrothermal waves can grow out of the two-dimensional base state if the Reynolds number is above critical. When the initial perturbation has a magnitude $\mathcal{O}(1)$, either type of flow typically reaches a saturated amplitude within one thermal diffusion time d^2/κ. For all cases investigated

---

[38] The divergence of the timescale ($\sigma^{-1} \to \infty$) when the critical point is approached is called critical slowing down.

by Leypoldt et al. (1998) standing waves were found to be unstable. Small deviations from a pure standing wave grow in time and lead to a transition to a pure traveling wave. For infinitely small initial deviations from the standing-wave state it may take an exponentially long period of time before the transition to a traveling wave occurs. An example is shown in Fig. 7.51, from which the exponential growth and saturation of a standing wave (Fig. 7.51a) and a traveling wave (Fig. 7.51b) can be seen.

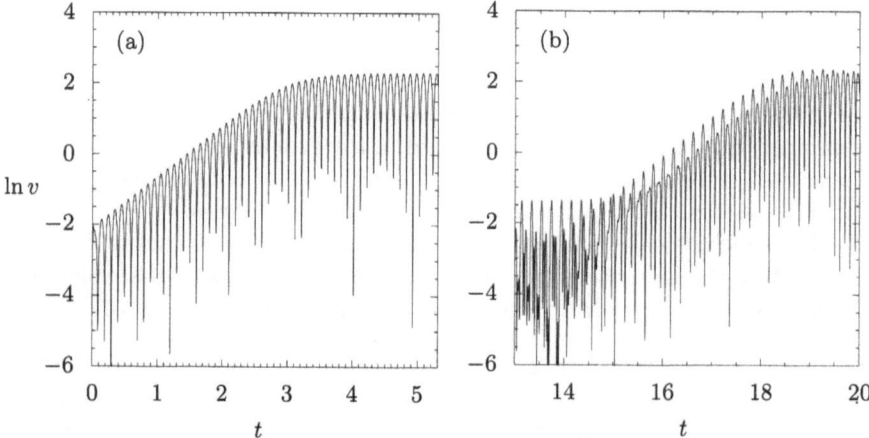

Fig. 7.51. Exponential growth and saturation of a standing wave (**a**) for $\mathrm{Pr}=4$, $\mathrm{Re} = 1200$, $\Gamma = 1$, and $\mathrm{Gr} = \mathrm{Bi} = 0$. The logarithm of the azimuthal velocity at $r = 0.95$ and $z = 0$ at an azimuthal antinode ($\varphi = 0$) is shown. (**b**) shows the exponential growth of azimuthal velocity perturbations of the standing wave leading to a traveling wave at the same (r, z) location but an azimuthal node of the standing wave ($\varphi=\pi/4$). Note the timescale. After Leypoldt et al. (1998)

Standing and traveling waves exibit a number of interesting nonlinear properties. The nonlinear interaction of the spatial harmonics of a hydrothermal wave gives rise to spectral components Θ_m of the temperature field with $m=0$. For a pure traveling wave, this axisymmetric component is steady. For standing waves, however, axisymmetric temperature components are generated that oscillate at twice the hydrothermal wave's frequency. Since the integral convective heat transfer to the the rigid walls at $z=\pm 1/2$ is only carried by axisymmetric components of the temperature field, the supercritical Nußelt number is time-independent for traveling waves, whereas it oscillates at twice the fundamental frequency for standing waves.

Leypoldt et al. (1998) found that the time-averaged Nußelt numbers decrease on passing from the two-dimensional supercritical flow to standing waves and to traveling waves: $\mathrm{Nu_{TW}} < \mathrm{Nu_{SW}} < \mathrm{Nu_{2D}}$. These heat transfer relations are different from those in many other systems, e.g. the Rayleigh–

Bénard problem, in which a convective state is selected that maximizes the heat transfer.

Another interesting property arises because of the breaking of the azimuthal-direction symmetry by traveling waves. Mode interactions generally lead to a nonzero axisymmetric ($m=0$) component of the azimuthal flow. As a result a hydrothermal wave causes an azimuthal mean flow depending on r and z. Leypoldt et al. (1998) found that the mean flow for $Pr=4$ is directed opposite to the direction of propagation of the hydrothermal wave in most areas of the (r, z) plane. Only close to the cold corner is the mean flow in the direction of the traveling wave. The integral (net) mean flow, nevertheless, is directed oppositely and increases linearly with ϵ. The order of magnitude of the numerically determined mean flow agrees well with the experimental observations of Frank & Schwabe (1997). Owing to the symmetry, standing waves cannot cause an azimuthal mean flow.

The dynamics of slightly supercritical hydrothermal waves can be described by the following set of amplitude equations (Cross (1988), Cross & Hohenberg (1993)):

$$\tau_0 \dot{R} = \epsilon R - g_s R^3 - g_c L^2 R, \tag{7.33}$$

$$\tau_0 \dot{\alpha}_R = c_0 \epsilon - c_2 g_s R^2 - c_3 g_c L^2, \tag{7.34}$$

$$\tau_0 \dot{L} = \epsilon L - g_s L^3 - g_c R^2 L, \tag{7.35}$$

$$\tau_0 \dot{\alpha}_L = c_0 \epsilon - c_2 g_s L^2 - c_3 g_c R^2, \tag{7.36}$$

where the right- and left-traveling hydrothermal waves are $f_+ = R(t)e^{i\alpha_R(t)}F_+(r, z, t)$ and $f_- = L(t)e^{i\alpha_L(t)}F_-(r, z, t)$, respectively. The real amplitudes (R, L) and phases (α_R, α_L) of the neutral modes $F_{\pm}$ vary on a slow timescale. Owing to the short length and the periodic boundary conditions in φ, slow spatial variations in φ do not occur.

For this system, there exist only two types of supercritical steady state solution; standing waves

$$R = L = \left(\frac{\epsilon}{g_s + g_c} \right)^{1/2}, \tag{7.37}$$

$$\alpha_R = \alpha_L = \left(c_0 - \frac{c_2 g_s + c_3 g_c}{g_s + g_c} \right) \frac{\epsilon t}{\tau_0} + \text{const.}, \tag{7.38}$$

and traveling waves, e.g. for $L \neq 0$,

$$L = \left(\frac{\epsilon}{g_s} \right)^{1/2}, \qquad R = 0, \tag{7.39}$$

$$\alpha_L = (c_0 - c_2) \frac{\epsilon t}{\tau_0} + \text{const.} \tag{7.40}$$

One of these is always linearly unstable. The other one is stable only if both solution branches bifurcate supercritically (Crawford & Knobloch (1991)). Standing waves are stable for $\epsilon > 0$ if $g_c/g_s < 1$ and $g_c + g_s > 0$, while traveling waves are stable if $g_c/g_s > 1$ and $g_s > 0$. Otherwise both states are unstable (see, e.g. Iooss (1987)).

By a series of numerical computations Leypoldt et al. (1998) determined the coefficients appearing in (7.33–7.36). The coefficients given in Table 7.4 yield $g_c/g_s = 2.76$ and $g_s > 0$, consistent with the numerically observed stability of traveling waves and the instability of standing waves.

Table 7.4. Numerically determined coefficients of the amplitude equations (7.33–7.36) after Leypoldt et al. (1998). The neutral mode is normalized such that $v(r = 0.5, z = 0) = 1$

τ_0	g_s	g_c	c_0	c_2	c_3
0.111	0.0136	0.0375	1.545	0.763	−0.345

Equations (7.38) and (7.40) contain the nonlinear frequency corrections to the critical value ω_c, which are different for the two types of wave. The frequency corrections $\dot{\alpha}_{R,L}$ are linear in ϵ. For the present example (Pr = 4, Table 7.4), the finite-amplitude standing wave's frequency does not differ much from that of the linear small-amplitude wave with frequency $\omega_c + c_0 \epsilon / \tau_0$. The frequency correction $\dot{\alpha}_L = (c_0 - c_2)\epsilon/\tau_0$ of the traveling wave, however, is smaller by a factor of ≈ 1.9 than that of the standing wave, hence $\omega_{TW} < \omega_{SW}$.

For a comparison with the flow patterns measured by Velten et al. (1991) (see Fig. 7.47), Leypoldt et al. (1998) performed numerical calculations in the range $Re_c < Re < 2Re_c$ for Pr = 7 and various aspect ratios, taking into account buoyancy (Bd = $0.13(d/\text{mm})^2$) and free-surface heat transfer (Bi = 6.4). In all cases investigated the asymptotic states were traveling waves with fundamental wavenumbers in agreement with the predictions of the linear stability theory (Sect. 7.4) and the experimental results[39] of Velten et al. (1991).

7.8 Energy Stability Analyses

Preisser et al. (1983) showed that the hydrothermal waves bifurcate supercritically out of the axisymmetric basic state. This was confirmed experimentally by Velten et al. (1991) for all cases investigated under gravity conditions (Pr = 1, 7, and 49) and, more recently, numerically by Leypoldt et al. (1998) for zero gravity. In spite of these results several authors (e.g. Chun (1980) and Velten et al. (1989)) have reported a hysteretic behavior. The latter measurements motivated energy stability analyses aimed at finding conditions for the guaranteed stability of axisymmetric thermocapillary convection. The results of the energy analyses performed are summarized in the following.

---

[39] Except for *mixed modes*, for which Velten et al. (1991) did not specify a wavenumber.

Shen (1989) and Shen et al. (1990) considered the relative energy growth rate (5.31). They calculated numerically the axisymmetric basic state that enters $\mathcal{F}$ parametrically. The functional $\mathcal{F}(\boldsymbol{x})$ to be maximized depends on the temperature and velocity fields of the kinematically admissible perturbations. For an extremum the first derivatives of $\mathcal{F}$ with respect to all unknown field entities x_i (discretized values of $\boldsymbol{u}$ and Θ) must vanish:

$$\frac{\partial \mathcal{F}(\boldsymbol{x})}{\partial x_i} = 0. \tag{7.41}$$

The resulting system of algebraic equations (7.41) has the form of a generalized nonlinear eigenvalue problem

$$\mathsf{A} \cdot \boldsymbol{x} = \mathrm{Re}_* \mathsf{B}(\mathrm{Re}_*) \cdot \boldsymbol{x}, \tag{7.42}$$

where A and B are Hermitian matrices. $\mathsf{B}(\mathrm{Re})$ depends on Re through the basic state. The system can be solved iteratively by first solving

$$\mathsf{A} \cdot \boldsymbol{x} = \rho \mathsf{B}(\mathrm{Re}) \cdot \boldsymbol{x} \tag{7.43}$$

for given Reynolds number Re. Then Re, and with it the base state, is varied until $\rho_* = \mathrm{Re}_* = \mathrm{Re}$, where ρ_* denotes the smallest positive eigenvalue of (7.43). Finally, the coupling parameter λ (cf. (5.30)) is selected in order to maximize Re_*. This procedure yields the energy stability boundary

$$\mathrm{Re}_E = \max_\lambda \mathrm{Re}_*. \tag{7.44}$$

For details of the solution method, the reader is referred to Mittelmann (1990) and Mittelmann et al. (1991).

For axisymmetric flows $(m=0)$ with $\mathrm{Bi} = \mathrm{Gr} = 0$ and $\mathrm{Pr} = 1$, the energy stability limit $\mathrm{Re}_E(m=0)$ obtained by Shen et al. (1990) is shown in Fig. 7.45 for $\Gamma = 0.5$ in comparison with other data. With increasing aspect ratio the axisymmetric energy limit grows monotonically up to $\mathrm{Re}_E(\Gamma = 1.6) = 11\,472$ (the data points are beyond the boundaries of the range shown in Fig. 7.45). The axisymmetric energy limits are significantly larger than the linear-stability boundaries for nonaxisymmetric perturbations. Thus the energy analysis for $m = 0$ merely shows that *all axisymmetric* perturbations decay for Re below $\mathrm{Re}_E(m=0)$, in agreement with the experimental results. Axisymmetric thermocapillary oscillations have not yet been observed in nearly upright cylindrical liquid bridges (see, however, Sect. 9.1.1).

Neitzel et al. (1991)[40] extended the analysis of Shen et al. (1990) to include nonaxisymmetric perturbations. These can be written as normal modes (7.5), as the Euler–Lagrange equations of the variational problem (5.33) are linear in the perturbations. To find the energy limit, $\mathrm{Re}_*(m)$ must be minimized with respect to m. The wavenumber of the least stable energy mode

---

[40] Shen et al. (1990), Neitzel et al. (1991), and Mittelmann (1994) used the ambient temperature $\theta_a = -0.5 = \mathrm{const}$. This temperature profile influences the basic states and the neutral modes when $\mathrm{Bi} \neq 0$.

scales with the aspect ratio similarly to (7.28). Since the nonaxisymmetric energy limits fall closely within the experimental stability boundaries of Velten et al. (1991),[41] the energy stability analysis does not yield excessively conservative limits here.[42] The wavenumbers of the least stable modes (see Fig. 7.45) and those of the most unstable modes in the experiments deviate from each other to some extent. Since the energy limits differ from the linear-threshold values, the coincidence of the energy data with the experiments of Velten et al. (1991) shown in Fig. 7.45 must be considered fortuitous.

Figure 7.45 also allows a comparison of the energy analyses of Neitzel et al. (1991) with the linear-stability calculations of Neitzel et al. (1993) and Wanschura et al. (1995a) for $Pr = 1$. The theoretical data are consistent in that the linear-stability boundaries are higher than the energy limits. The systematic deviations of the linear-stability boundaries (they are larger than the experimental transition points) may be associated with effects that are not included in the theoretical model, e.g. natural convection in the ambient gas or evaporation of liquid at the free surface.

The energy stability analyses of Neitzel et al. (1991) for $Pr = 1$ have been extended to small Prandtl numbers by Mittelmann (1994) $(Gr = Bi = 0)$. As a result the energy limits for $Pr = 0.0078$ and $\Gamma = \mathcal{O}(1)$ $(Re_E(\Gamma = 1, m = 1, Pr = 0.0078) = 654)$ are significantly smaller than the linear-stability boundaries for $Pr = 0.02$.[43] The limit for $Pr = 1$ is $Re_E(Pr = 1, \Gamma = 1, m = 2) = 1679 < Re_c = 2539$, consistent with the linear analysis. For $Pr > 1$ and $\Gamma = 1$, however, the energy limits exceed the linear-stability boundaries. This is in contradiction to the definition of Re_c and Re_E and thus renders the energy limits for $Pr > 1$ questionable.

---

[41] Neitzel et al. (1991) used $Bi = 0.3$. From the isotherms measured by Preisser et al. (1983) for $Pr = 7$ one can estimate the Biot number to be $Bi \approx 6.4$. It depends strongly on the experimental ambient conditions.

[42] In many other cases, e.g. for plane Couette flow (Joseph (1976)), the energy limits are very small $(Re_E = 82.6, Re_c \to \infty)$.

[43] The critical Reynolds number $Re_c(\Gamma = 1, Pr = 0.0078)$ is only slightly smaller than $Re_c(\Gamma = 1, Pr = 0.02)$ (see Fig. 7.20). It is anticipated that the linear-stability boundaries for $\Gamma \in [0.3, 2]$ follow the trend and are only slightly smaller than the values for $Pr = 0.02$ (Fig. 7.27).

8. Convection in Rectangular Geometry

Thermocapillary convection in rectangular containers is a fundamental model for the so-called *open-boat* crystal-growth method (Hurle (1994)). The structures of low-Reynolds-number two-dimensional flows in rectangular geometries are similar to those in cylindrical liquid bridges. There are differences, however, particularly at high Reynolds numbers, that arise because of the geometry of the boundaries, the presence of three-dimensional end effects, and the modified action of buoyancy forces, which are typically directed perpendicular to the interface.

The limit of pure buoyant convection in cavities, though of particular interest for combined thermocapillary–buoyant flows, cannot be covered here. Among pure buoyancy-driven cavity flows, those with rigid boundary conditions have received most attention. The interested reader may consult the collection of benchmark calculations by Roux (1990) and some more recent numerical (Afrid & Zebib (1990), Janssen & Henkes (1995), Henkes & Le Quéré (1996)) and experimental (Braunsfurth & Mullin (1996)) work. For the stability of infinitely extended systems, see Hart (1983) and Kuo & Korpela (1988).

For the mathematical description we consider a rectangular volume of liquid with lengths L_x, L_y, and L_z in the x, y, and z directions. The side walls at $x=0$ and L_x are heated and kept at constant temperatures $T_0+\Delta T$ and T_0, respectively. Let the free surface be at $y = L_y$ (Fig. 8.1) and let us take the depth of the liquid layer L_y as the length scale of the problem. Two aspect ratios arise, defined by $\Gamma = L_x/L_y$ and $\Gamma_z = L_z/L_y$. Employing the scalings given in Table 8.1 and defining the Reynolds and Grashof numbers

$$\text{Re} = \frac{\gamma \Delta T L_y}{\rho \nu^2} \frac{1}{\Gamma}, \tag{8.1}$$

$$\text{Gr} = \frac{g \beta \Delta T L_y^3}{\nu^2} \frac{1}{\Gamma}, \tag{8.2}$$

the volume equations (2.14–2.16) remain unchanged. Here the acceleration due to gravity is $\boldsymbol{g} = -g\boldsymbol{e}_y$. On the rigid walls at $x = (0, \Gamma)$, $y = 0$, and $z = (0, \Gamma_z)$ the velocity field must vanish ($\boldsymbol{u} = 0$). Neglecting surface deformations[1]

---

[1] As shown by Hsieh (1992) for $\Gamma = \Gamma_z = 1$, $\text{Pr} = 100$, $\text{Gr} = 0$, and $\text{Re} = 400$, the surface deformations are of the relative order of magnitude $\mathcal{O}(10^{-3} - 10^{-2})$. If,

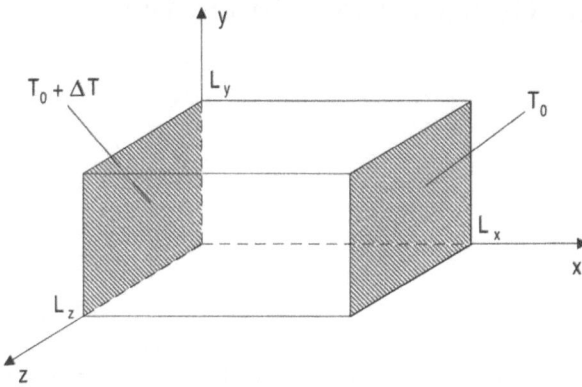

Fig. 8.1. Geometry and coordinate system for the rectangular open-boat model

Table 8.1. Scales for the nondimensionalization of the Navier–Stokes equations in the case of rectangular containers

Variable	x, y, z	t	$\boldsymbol{u} = (u, v, w)$	p	T
Scale	L_y	L_y^2/ν	$\gamma \Delta T L_y / \rho \nu L_x$	$\gamma \Delta T / L_x$	$\Delta T L_y / L_x$

$(\mathrm{Ca} \to 0)$, the boundary conditions for the velocity field on the free surface are

$$\left.\begin{array}{c} v = 0 \\ \partial_y u + \partial_x \theta = 0 \\ \partial_y w + \partial_z \theta = 0 \end{array}\right\} \quad \text{on} \quad y = 1 \,. \tag{8.3}$$

Using the reduced temperature $\theta = \Gamma(T - T_0)/\Delta T$, the thermal boundary conditions are given by

$$\theta = \Gamma \quad \text{on} \quad x = 0 \,, \tag{8.4}$$

$$\theta = 0 \quad \text{on} \quad x = \Gamma \,. \tag{8.5}$$

Finally, the ambient temperature field is assumed to be linear, with $\theta_\mathrm{a} = \Gamma - x$. If not noted otherwise, all other boundaries are taken as adiabatic $(\mathrm{Bi} = 0)$:

$$\partial_y \theta = 0 \quad \text{on} \quad y = 0, 1 \,, \tag{8.6}$$

$$\partial_z \theta = 0 \quad \text{on} \quad z = 0, \Gamma_z \,. \tag{8.7}$$

however, both the capillary number and the extent of the liquid layer are large, significant flow-induced $(\mathrm{Gr} = 0)$ surface deflections may arise.

8.1 Two-Dimensional Flows

Many investigations have concentrated on two-dimensional flows. Similarly to the mechanically-driven-cavity flow problem,[2] thermocapillary convection in a square container has evolved to a benchmark problem. In fact, some analogies exist between the two systems (see Carpenter & Homsy (1990)). The stability of two-dimensional flow states is of particular interest. A number of publications on the onset of two-dimensional oscillations have appeared (see e.g. Ben Hadid & Roux (1990), Ohnishi et al. (1992), and Mundrane & Zebib (1994)). Only a few results, however, are available for three-dimensional thermocapillary convection (but cf. Daviaud & Vince (1993) and, for annular systems with rectangular cross section, Schneider (1995)).

8.1.1 Short Containers ($\Gamma \leq 1$)

In the limit of small Prandtl and Reynolds numbers $((\mathrm{Re}, \mathrm{Pr}) \to 0)$ the flow and temperature fields can be written in terms of sums over even Papkovich–Fadle functions as in Sect. 4.2. All flow properties derived there apply at least qualitatively to the rectangular geometry: the fields $\psi = \mathcal{O}(\mathrm{Re})$ and $\Theta = \theta - (\Gamma - x) = \mathcal{O}(\mathrm{RePr})$ are symmetric with respect to $x = \Gamma/2$ (compare (4.58) and (4.59)). For $\Gamma \lesssim 1/1.4$ the creeping flow consists of a sequence of counterrotating Moffatt eddies (Moffatt (1964a), Joseph & Sturges (1975)). The extension in the y direction of the corresponding vortices is $\approx 1.396 \times \Gamma$ as in liquid bridges, they have a structure similar to the one shown in Fig. 4.9, and their strength decays exponentially in the negative y direction $\sim e^{\approx 4.21y/\Gamma}$.

The stationary nonlinear $(\mathrm{Re} \gg 1)$ two-dimensional thermocapillary convection in a square container $(\Gamma = 1)$ has been calculated numerically by Zebib et al. (1985) for different Prandtl and Reynolds numbers up to $\mathrm{Re} = 5 \times 10^4$. The flow consists of a main vortex and two small corner vortices at $y = 0$ and $x = (0, 1)$. A representative case is shown in Fig. 8.2. For high Reynolds numbers, boundary layers develop along all boundaries. On the assumption that the free-surface temperature gradient remains $\mathcal{O}(1)$ for all Reynolds and Prandtl numbers Zebib et al. (1985) showed that the thermocapillary boundary layer thickness δ scales like $\mathrm{Re}^{-1/3}$ (see also Sect. 3.2).

Extending the work of Zebib et al. (1985), Carpenter & Homsy (1990) showed numerically that a third secondary vortex separates on the rigid wall close to the hot corner $(x = 0, y = 1)$ if $\mathrm{Re} > 4 \times 10^4$ for $\mathrm{Pr} = 10^{-3}$ or $\mathrm{Re} > 1.3 \times 10^5$ for $\mathrm{Pr} = 1$. They found that the assumption $\partial_x \theta|_{y=1} = \mathcal{O}(1)$ does not always hold for high Prandtl numbers, since the free surface becomes

---

[2] In the so-called *lid-driven cavity* the flow is induced by moving one bounding wall tangentially to itself. As a simple model system with Dirichlet boundary conditions, it is often employed to test numerical methods. See e.g. Burggraf (1966), Schreiber & Keller (1983), Goodrich et al. (1990), or Deville et al. (1992).

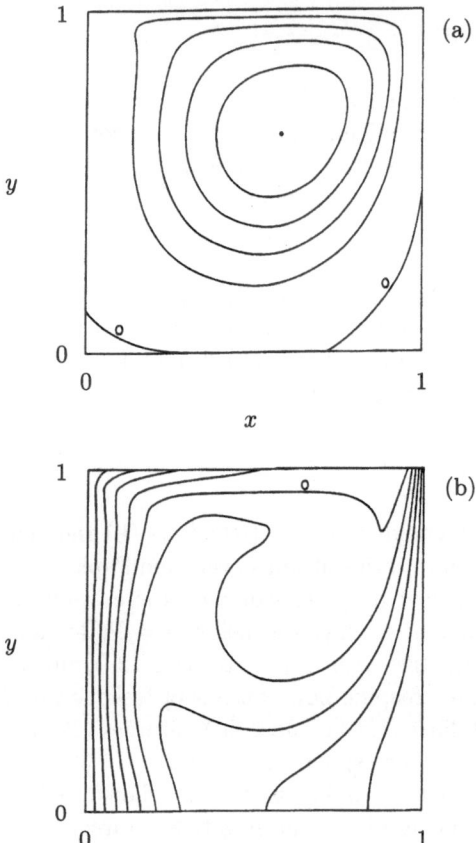

Fig. 8.2. Two-dimensional thermocapillary flow in a square container ($\Gamma = 1$), after Zebib et al. (1985), for $Re = 10^4$, $Pr = 1$, and $Bi = Gr = 0$. (**a**) streamlines, (**b**) isotherms

essentially isothermal, apart from small corner regions. This observation led to the conclusion that the asymptotic regime with $\delta \sim Re^{-1/3}$ is only reached for Reynolds numbers which become increasingly large the higher the Prandtl number is (compare Sect. 3.2.3). Further, Carpenter & Homsy (1990) found that the stationary two-dimensional flow for $\Gamma = 1$ and $Bi = 0$ is linearly stable with respect to two-dimensional perturbations at least up to $Ma = 1.3 \times 10^5$ for $Pr = 10$ and up to $Ma - 1.2 \times 10^5$ for $Pr = 30$.

Carpenter & Homsy (1989) investigated the additional influence of buoyancy[3] for $Pr = 1$ and aspect ratio $\Gamma = 1$. Their numerical calculations for fixed dynamic Bond number ($Bd = Gr/Re = const.$) showed a change of the scaling of the free-surface boundary layer thickness from $\delta \sim Re^{-1/2}$ to $\delta \sim Re^{-1/3}$ within a narrow range of Reynolds numbers. Supported by a scaling analysis, the numerically observed behavior led to the hypothesis that the combined

---

[3] Combined buoyant–thermocapillary convection in rectangular containers has also been investigated by Bergman & Ramadhyani (1986), Bergman & Keller (1988), and Ramaswamy & Jue (1992).

flow is thermocapillary-dominated for any fixed value of the dynamic Bond number Bd, if only the Reynolds number Re is sufficiently high. It was further shown that the third separated vortex close to the hot corner is suppressed by buoyancy.

The two-dimensional streamline pattern may be considerably affected by buoyancy. While the streamlines are nearly circular in the center of the vortex for pure thermocapillary-driven flows, the hot and thus lighter fluid transported along the free surface to the cold wall will not penetrate deep into the bulk but rather return directly below the surface flow. This buoyancy-induced density stratification leads to a separation of a comparatively hot and mainly thermocapillary-driven circulation in a layer below the free surface from colder buoyancy-driven flow structures in the bulk (Metzger & Schwabe (1988), Schwabe & Metzger (1989), Schwabe & Scharmann (1989)). As a consequence of the stratification the thermocapillary free-surface eddy is flattened and subject to high strain. Eventually, the streamlines take the shape of cat's eyes separated by a hyperbolic stagnation point in between. An example is shown in Fig. 8.3. As discussed in Sect. 8.2.3 the appearance of a free stagnation point may indicate a three-dimensional instability of these flows at higher driving forces.

Fig. 8.3. The cat's-eye flow near the free surface in a deep cavity heated from the right-hand side (the lower part of the cavity is not shown). The flow is driven by combined buoyant and thermocapillary forces. After Schwabe & Metzger (1989)

8.1.2 Stationary Flow in Long Containers ($\Gamma \gg 1$)

For shallow liquid layers with $\Gamma \gg 1$ and $\mathrm{Gr} = 0$ the main vortex discussed in Sect. 8.1.1 becomes laterally elongated. In the limit $\Gamma \to \infty$ and for small Reynolds numbers $\mathrm{Re} = \mathcal{O}(\Gamma^{-1})$ (cf. (8.1)) one obtains the asymptotic surface temperature $\theta = \Gamma - x$. The velocity field is then given by the plane Poiseuille–Couette profile (Birikh (1966), Sen & Davis (1982), Strani et al. (1983))

$$u = \left(\frac{3}{4} y^2 - \frac{1}{2} y \right) , \tag{8.8}$$

(cf. (6.3) with $z \to y - 1/2$). The temperature field in the bulk of the liquid depends on the thermal boundary conditions. For an ideal heat-conducting boundary at $y = 0$ one obtains, differently from (6.4),

$$\theta = \Gamma - x - \frac{1}{48} \mathrm{RePr} \left(3y^4 - 4y^3 + y \right) . \tag{8.9}$$

For $\Gamma \approx 4$ the streamlines in the middle of the container are already nearly horizontal (Strani et al. (1983)) and for $\Gamma > 4$ (8.8) and (8.9) represent good approximations to u and θ in that region.

Two-dimensional thermocapillary flows in long containers up to $\Gamma = 25$ have been calculated by Ben Hadid & Roux (1990) for Gr $= 0$ and Pr $= 0.015$. For heat-conducting boundary conditions at the bottom ($y = 0$) and for Reynolds numbers not too large, the temperature field depends on the velocity field in the vicinity of $x = (0, \Gamma)$ only. For Prandtl number Pr $= 0.015$ and up to Re $= \mathcal{O}(10^4)$ it is nearly conductive in the interior and almost decouples from the flow.

When the Reynolds number is small the flow consists of the asymptotic core flow (8.8, 8.9) with symmetric turning zones at $x = 0$ and Γ (Fig. 8.4a). On an increase of the Reynolds number three distinct zones develop. Close to the cold wall ($x = \Gamma$) a number of corotating vortices are created (for $\Gamma = 12.5$ two vortices appear for moderate Reynolds numbers, Fig. 8.4). The flow close to the hot wall ($x = 0$) behaves like the entrance flow of a channel (Schlichting (1968)). Ben Hadid & Roux (1990) showed that the interior flow ($x \approx \Gamma/2$) can be well approximated by the Poiseuille–Couette profile (8.8) as long as Re $< 20\Gamma$. For higher Reynolds numbers the end zones grow into the interior and the Poiseuille–Couette-like flow is destroyed.

For Re $\gtrsim 200\Gamma$ the classical boundary-layer scaling for shear-induced flows holds in the entrance region ($0 < x < \Gamma/2$). The dynamic equation ($\nabla \times (2.14)$) in two dimensions yields

$$\mathrm{Re}\, \boldsymbol{u} \cdot \nabla \omega \sim \nabla^2 \omega . \tag{8.10}$$

Making the approximation $\partial_y \sim \delta^{-1}$ in (8.10), where δ is the boundary-layer thickness, one obtains the order-of-magnitude relation

$$\mathrm{Re} \frac{u}{x} \sim \frac{1}{\delta^2} . \tag{8.11}$$

Together with the thermocapillary boundary condition, $1 = \partial_y u \sim u/\delta$, the scaling of the horizontal velocity within the entrance region is obtained as

$$u \sim \left(\frac{x}{\mathrm{Re}} \right)^{1/3} , \tag{8.12}$$

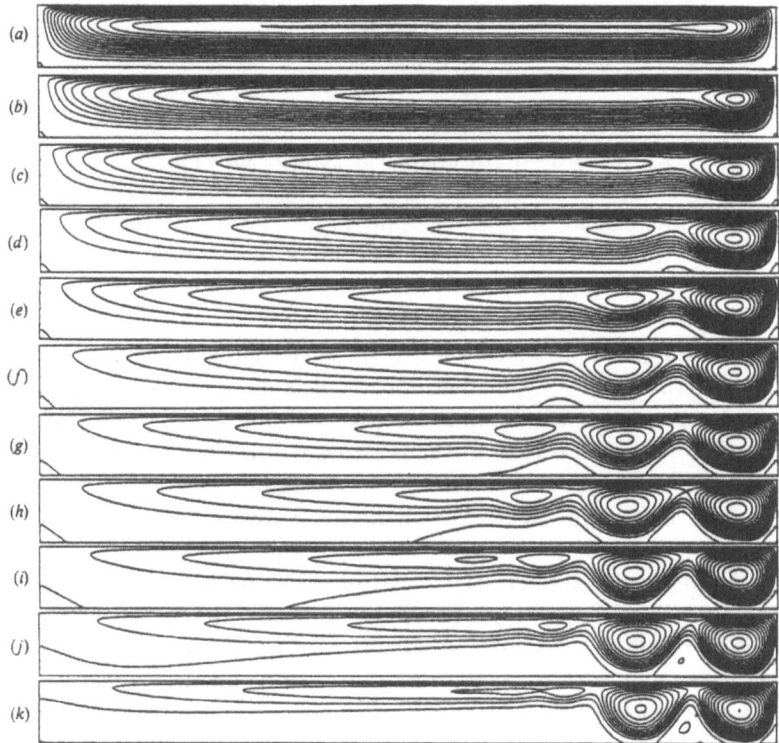

Fig. 8.4. Streamlines in a long, open, rectangular container with $\Gamma = 12.5$ and $\mathrm{Pr} = 0.015$, $\mathrm{Gr} = 0$, and $\mathrm{Bi} = \infty$. The conductive temperature profile is imposed on the free surface. The Reynolds numbers are: **(a)** $\mathrm{Re} = 66.7$, **(b)** $\mathrm{Re} = 333$, **(c)** $\mathrm{Re} = 667$, **(d)** $\mathrm{Re} = 1330$, **(e)** $\mathrm{Re} = 2000$, **(f)** $\mathrm{Re} = 3330$, **(g)** $\mathrm{Re} = 5000$, **(h)** $\mathrm{Re} = 6670$, **(i)** $\mathrm{Re} = 10^4$, **(j)** $\mathrm{Re} = 1.33 \times 10^4$, **(k)** $\mathrm{Re} = 2 \times 10^4$. After Ben Hadid & Roux (1990)

where x denotes the distance from the hot side ($x = 0$) of the container.

The flow structures found by Ben Hadid & Roux (1990) close to the end walls have been explained by Laure et al. (1990). They considered the limit $\Gamma \to \infty$ with $\mathrm{Pr} \to 0$ and investigated the spatial stability (see Sect. 5.1.1) of the Poiseuille–Couette flow (return flow) driven by a constant shear on the free surface. In this limit the flow decouples from the temperature field. Using the stream function ψ_0 for the basic state, the linearized equation governing small two-dimensional stationary perturbations ψ is

$$\nabla^4 \psi = \mathrm{Re} \left[(\partial_x \psi) \, \partial_{yyy} \psi_0 - (\partial_y \psi_0) \, \nabla^2 \partial_x \psi \right] . \tag{8.13}$$

This equation can be written as an evolution equation in the x direction,

$$\frac{\partial}{\partial x} \mathbf{Z}(x, y) = \mathsf{L}(\mathrm{Re}) \cdot \mathbf{Z}(x, y) , \tag{8.14}$$

where L is a linear differential operator and $\boldsymbol{Z} = (\psi, \psi_x, \psi_{xx}, \psi_{xxx})^{\mathrm{T}}$. Using the separation ansatz $\psi = f(y)e^{\gamma x}$ one obtains

$$\gamma^4 f + 2\gamma^2 f'' + f'''' = -\gamma \mathrm{Re}\left(\psi_0' f'' + \gamma^2 \psi_0' f - \psi_0''' f\right) , \tag{8.15}$$

where the prime denotes differentiation with respect to y. For $\mathrm{Re} \to 0$ it follows that $f'''' + 2\gamma^2 f'' + \gamma^4 f = 0$. Inserting the general solution (4.12) into the boundary conditions yields the eigenvalue equation $2\gamma = \sin 2\gamma$. The solutions are just the characteristic roots of the odd Papkovich–Fadle functions (cf. (4.14)). Because the eigenvalue equation is symmetric in γ, the streamlines in the end zones are symmetric with respect to $x = \Gamma/2$ for small Reynolds numbers.

The eigenvalue problem for $\mathrm{Re} \neq 0$ can be solved numerically. The set of eigenvalues describes the x dependence of the eigenmodes. Since all perturbations must decay from the boundaries, the relevant modes near the hot wall $(x=0)$ have $\Re(\gamma) < 0$ (decay in the positive x direction) and those near the hot wall $(x=\Gamma)$ have $\Re(\gamma) > 0$ (decay in the negative x direction). The most dangerous modes are the ones that decay slowest, i.e. those with the smallest absolute value of γ.

For small Reynolds numbers the behavior near the boundaries is quasisymmetrical as long as $\mathrm{Re} < 15.5$. For higher values of Re one of the eigenvalues turns real and the behavior is asymmetric. It can be shown (Laure et al. (1990)) that the mode with the real eigenvalue growing upstream against the surface flow towards the hot end wall becomes saturated by nonlinear terms for $\mathrm{Re} > 15.5$. This mode corresponds to the entrance flow. For $\mathrm{Re} > 400$ the downstream-growing mode with a complex eigenvalue also becomes saturated. Here the finite-amplitude nonlinear pattern corresponds to corotating vortices with a wavelength $\lambda \approx 2$ (for $100 < \mathrm{Re} < 1000$). This value is in good agreement with the numerical simulation of Ben Hadid & Roux (1990) (see Fig. 8.4).

The corotating vortices in Fig. 8.4 appear as a result of a spatial instability. They were also analyzed by Priede & Gerbeth (1997b). Priede & Gerbeth (1997c) showed that the above stationary and spatially growing modes remain spatially attenuated for $\mathrm{Pr} = 0$ no matter how large the Reynolds number is. For higher Prandtl numbers, however, the stationary transverse rolls[4] having a wavelength comparable to the depth of the layer may become convectively unstable and even be the most dangerous modes if the Bond number is sufficiently high (Bo $\gtrsim 0.3$, Priede & Gerbeth (1997b)).[5] The to-

---

[4] The axes of the counterrotating rolls are aligned perpendicular to the base flow.

[5] In low-Prandtl-number pure buoyancy-driven Hadley circulation stationary transverse rolls are convectively unstable, as was shown by Hart (1983) for insulating top and bottom boundaries (see also Laure & Roux (1989)). Nonlinear two-dimensional pure buoyant low-Prandtl-number cat's-eye flows were calculated by Drummond & Korpela (1987), who found that the instability is imperfect in finite-length containers and that the parallel flow is stabilized with respect to the cat's-eye flow when the Prandtl number is increased beyond $\mathrm{Pr} \gtrsim 0.12$.

tal, slightly supercritical two-dimensional flow then appears in the form of corotating vortices embedded in the basic shear flow and it resembles Kelvin's cat's-eye flow (Kelvin (1880)). Corresponding steady two-dimensional flows were observed experimentally in high-Prandtl-number liquids by Schwabe et al. (1992), Villers & Platten (1992), Schneider (1995), Saedeleer et al. (1996), Riley & Neitzel (1998), and others. The side walls present in experiments generally cause the bifurcation to be imperfect. Moreover, the cat's-eye flow exhibits a certain degree of asymmetry with respect to $\Gamma/2$ owing to the different character of the turning zones near the heated walls.

8.1.3 Two-Dimensional Oscillatory Flows

Pr $\ll$ 1. Smith & Davis (1983a) considered the linear stability of the Poiseuille–Couette flow (6.3, 6.4). For small Prandtl numbers the most dangerous mode propagates nearly in the z direction, perpendicular to the applied temperature gradient. The critical wavelength is large compared to the layer depth $\lambda \gg L_y$ and the critical Reynolds number diverges like $\text{Re}_c \sim \text{Pr}^{-1/2}$, i.e. the critical Marangoni number tends to zero like $\text{Ma}_c \sim \text{Pr}^{1/2} \to 0$ (Sect. 6.2). Hence, the first instability in infinite, nondeformable thermocapillary liquid layers with $\text{Gr} = 0$ occurs in the form of *three-dimensional* hydrothermal waves. This is consistent with the results of Carpenter & Homsy (1990) for a square cavity and those of Ben Hadid & Roux (1990) for finite extended layers, since these authors did not find any two-dimensional instabilities.

In their two-dimensional numerical simulation for $\Gamma = 4$ and $\text{Pr} = 0.02$, Ohnishi et al. (1992), in contrast to Ben Hadid & Roux (1990), found oscillatory flows when the Marangoni number exceeded the critical value $\text{Ma}_c \approx 400$. It was demonstrated that the onset of oscillatory thermocapillary convection depends strongly on the grid resolution,[6] which could explain the contradictory result of Ben Hadid & Roux (1990) for a coarser mesh. The critical Marangoni number for the finite two-dimensional system with $\Gamma = 4$ is considerably higher than that for three-dimensional perturbations in infinite layers $(\text{Ma}_c^{3D}(\text{Pr} = 0.02) = \mathcal{O}(10))$. Since the wavelength of the three-dimensional hydrothermal waves for $\text{Pr} = 0.02$ is large $(\lambda = \mathcal{O}(20))$, it is anticipated that the hydrothermal waves of Smith & Davis (1983a) appear in finite systems only when the extension perpendicular to the temperature gradient is sufficiently large, e.g. $L_z \gtrsim 20 L_y$.

Buoyancy forces can have a destabilizing effect on the otherwise stable two-dimensional thermocapillary flow. Ben Hadid & Roux (1992) considered

---

Kuo & Korpela (1988) corrected the linear-stability results of Hart and showed that the two-dimensional transverse rolls are indeed the most dangerous perturbations for $\text{Pr} < 0.033$ (insulating boundary conditions).

[6] The onset Rayleigh number for two-dimensional pure buoyant oscillatory flow in side-heated cavities with aspect ratio $\Gamma = 4$ likewise depends sensitively on the grid resolution; see Roux (1990).

mixed convection in long containers for $Pr = 0.015$. The critical Rayleigh number for the onset of two-dimensional oscillations was found to increase (stabilization), if the Reynolds number was increased from zero. In this case both driving forces augment each other and support the same sense of rotation of the vortex flow. For opposing driving ($Re < 0$), a destabilization occurs. The corresponding stability diagram for $\Gamma = 4$ and $Pr = 0.015$ has been completed by Mundrane & Zebib (1994). It is shown in Fig. 8.5. For

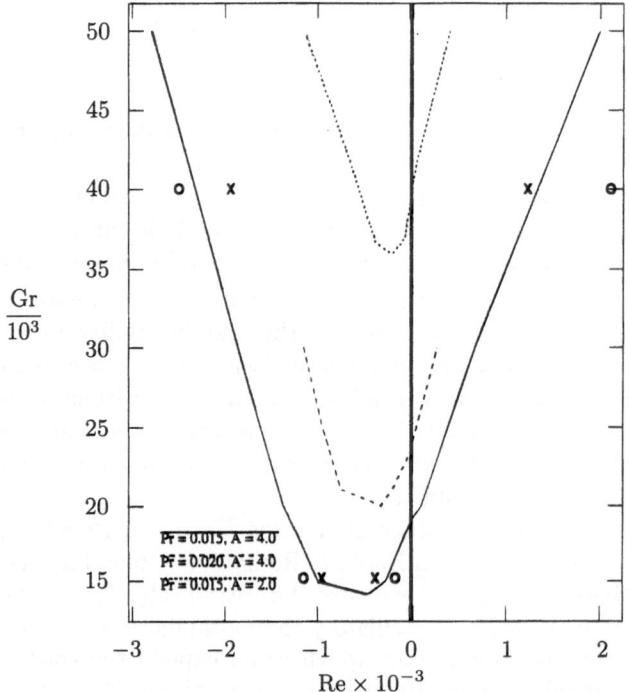

Fig. 8.5. Critical Grashof number Gr_c as a function of the Reynolds number Re for the onset of two-dimensional oscillations in a rectangular system with adiabatic boundary conditions, after Mundrane & Zebib (1994). (——) $Pr = 0.015$, $\Gamma = 4$; (- - -) $Pr = 0.02$, $\Gamma = 4$; (· · · ·) $Pr = 0.015$, $\Gamma = 2$. The points marked as ($\times$) and ($\circ$) denote oscillatory and stationary flows, respectively, calculated by Ben Hadid & Roux (1992)

$Re < 0$ a maximum destabilization exists. On a further reduction of the Reynolds number beyond the minimum in Fig. 8.5, the critical Rayleigh number increases again. In this parameter range, the flow is thermocapillary-dominated. For a discussion of the instability mechanisms the reader is referred to Mundrane & Zebib (1994).

Pr > 1. Unlike the case for small Prandtl numbers, the hydrothermal-wave instability is nearly two-dimensional for high Prandtl numbers (Chap. 6).

Peltier & Biringen (1993) simulated the two-dimensional flow for $Pr = 6.8$ and $Gr = Bi = 0$. They found a supercritical oscillatory bifurcation when the aspect ratio is larger than $\Gamma > 2.3$. These results were confirmed by Xu & Zebib (1998) and extended to $Pr = 1$, 4.4, 6.78, 10, and 13.9. Multiple regions of instability were even found at higher values of Γ. A stability diagram is shown in Fig. 8.6.

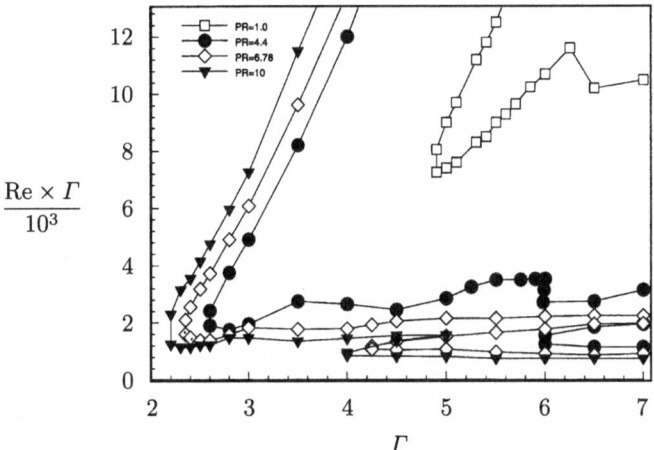

Fig. 8.6. Critical values $Re \times \Gamma$ for the onset of two-dimensional thermocapillary flow oscillations in rectangular containers. ($\square$) $Pr = 1$; ($\bullet$) $Pr = 4.4$; ($\Diamond$) $Pr = 6.78$; ($\triangle$) $Pr = 10$. After Xu & Zebib (1998)

A comparison with hydrothermal waves is in order. Typical critical Marangoni numbers for small aspect ratios are $Ma_c(\Gamma = 2.5) \approx 3200$ and $Ma_c(\Gamma = 4) \approx 2500$, both of which are much larger than the corresponding critical Marangoni number for infinite layers[7] ($Ma_c(Pr = 6.8) = \mathcal{O}(200)$ at $\lambda \approx 2.5$). When the aspect ratio is increased, however, the critical threshold for two-dimensional hydrothermal waves is approached (Xu & Zebib (1998)). It must be concluded that the lateral boundaries have a strongly stabilizing effect for low-aspect-ratio flow oscillations and even completely suppress the instability for $\Gamma < 2.3$. Moreover, the instability mechanism is altered by close side walls and it differs from that of hydrothermal waves. For $\Gamma = 2.6$ Peltier & Biringen (1993) explained the oscillations in terms of a vortex that periodically expands and contracts in the y direction. This process is associated with a periodic appearance of a cold spot in the vicinity of the hot wall. The cold spot is caused by convective transport of cold fluid from the cold wall along the adiabatic bottom boundary. This behavior was confirmed by Xu & Zebib (1998). Clearly, a finite geometry (L_y) is essential for

---

[7] It must be taken into account, however, that the slope of the surface temperature in the bulk is weaker than suggested by the Marangoni number, because of significant temperature variations near the end walls.

this process. It is interesting to note that the minimum of the critical curve in Fig. 8.6 occurs at $\Gamma \approx 2.5$, which is about the wavelength of the most dangerous hydrothermal wave in plane layers. Experimental results on thermocapillary convection in finite-size containers ($\text{Pr} = 4$) have been reported by Villers & Platten (1992).

8.2 Three-Dimensional Flows

8.2.1 Large Aspect Ratios

Three-dimensional numerical simulations have been carried out by Xu & Zebib (1998) for selected high Prandtl numbers in the range $\text{Pr} = 1$–13.9. When the aspect ratio Γ_z is $\mathcal{O}(1)$ the oscillations are nearly two-dimensional, except for a small region near the end walls $z = (0, \Gamma_z)$. On an increase of the length Γ_z the hydrothermal waves still propagate upstream against the basic surface flow. In addition, lengthwise oscillations (in the z direction) in the form of standing waves arise. This behavior can be explained in terms of two plane hydrothermal waves of equal amplitude propagating at angles $\pm \alpha$ with respect to the applied temperature gradient. For increasing Prandtl number, an ever-higher aspect ratio Γ_z is required for the appearance of lengthwise standing waves. This may be due to the decrease of the critical angle of propagation of the hydrothermal waves (see Fig. 6.3c) with Pr. As an example, the wave in a cavity of aspect ratio $\Gamma = \Gamma_z = 15$ is shown in Fig. 8.7. For this shallow cavity the critical Reynolds number obtained from the slope of the basic-state surface temperature in the bulk and the angle of propagation α are already in good agreement with the respective data for plane layers.

8.2.2 Side-wall Effects

When the aspect ratios are of $\mathcal{O}(1)$ side-wall effects become significant. To elucidate the three-dimensional flow modifications caused by the presence of lateral walls we consider the stationary three-dimensional thermocapillary convection ($\text{Gr}=0$) in a cube ($\Gamma=\Gamma_z=1$) with adiabatic boundary conditions (Saß et al. (1996)).

As in two dimensions, secondary eddies driven by the primary thermocapillary vortex exist in the corners at $y=0$ and $x = (0, 1)$. The secondary eddies develop out of viscous Moffatt eddies (Moffatt (1964a)) and they are most pronounced at midplane, $z \approx 0.5$. As for plane thermocapillary flows their size increases with increasing Reynolds number. However, they are much smaller in the cube than in the two-dimensional square cavity.

In addition to these structures, two symmetric secondary vortices with vorticity components primarily in the x direction (parallel to the surface flow) appear directly below the free surface near the corners at $y=1$ and $z=(0, 1)$

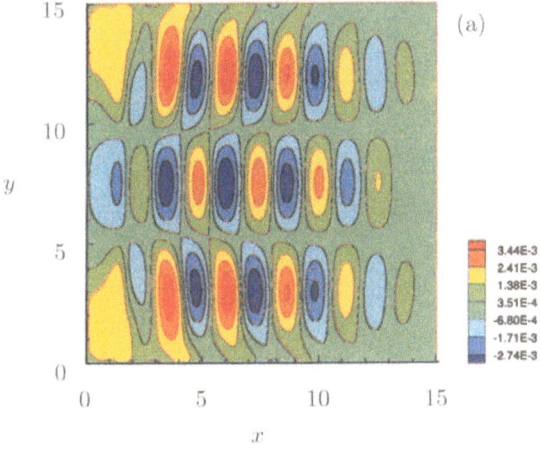

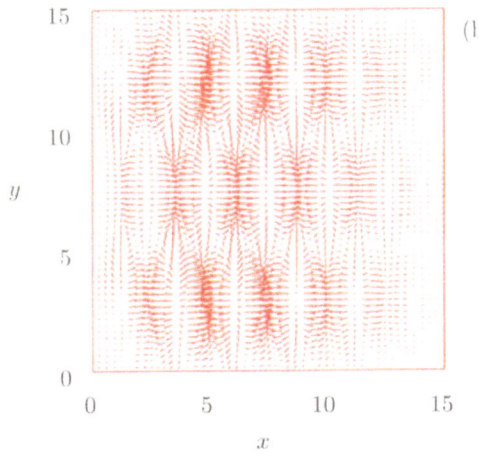

Fig. 8.7. Hydrothermal wave in a cavity with aspect ratio $\Gamma = \Gamma_z = 15$ for $Pr = 13.9$ and $Bi = Gr = 0$ at $Re = 710/\Gamma = 47.3$. Instantaneous surface values of the fluctuations of the temperature (**a**) and the velocity field (**b**) are shown. From Xu & Zebib (1998). Compare Fig. 6.4

(see Fig. 8.8). These vortices are caused by thermocapillary effects and can be explained in terms of lengthwise surface temperature gradients (see Fig. 8.9). Since the primary vortex flow is weaker close to the rigid side walls than in the center of the cavity, the surface temperature is nearly conducting close to the side walls. By transporting hot fluid from the heated wall, a moderate thermocapillary convection in the center of the cavity leads to an increase of the temperature in the center of the free surface. The resulting lengthwise temperature gradients (Fig. 8.9a) drive the observed secondary surface flow towards the side walls (Fig. 8.8). When the Marangoni number is high, thermal boundary layers appear and the temperature in the center of the free surface will be lower than for moderate Marangoni numbers (Fig. 8.9b). Now the main effect is a secondary lengthwise thermocapillary flow directed towards the centerline $z = 1/2$ (see Fig. 8.10). Only in the very vicinity of

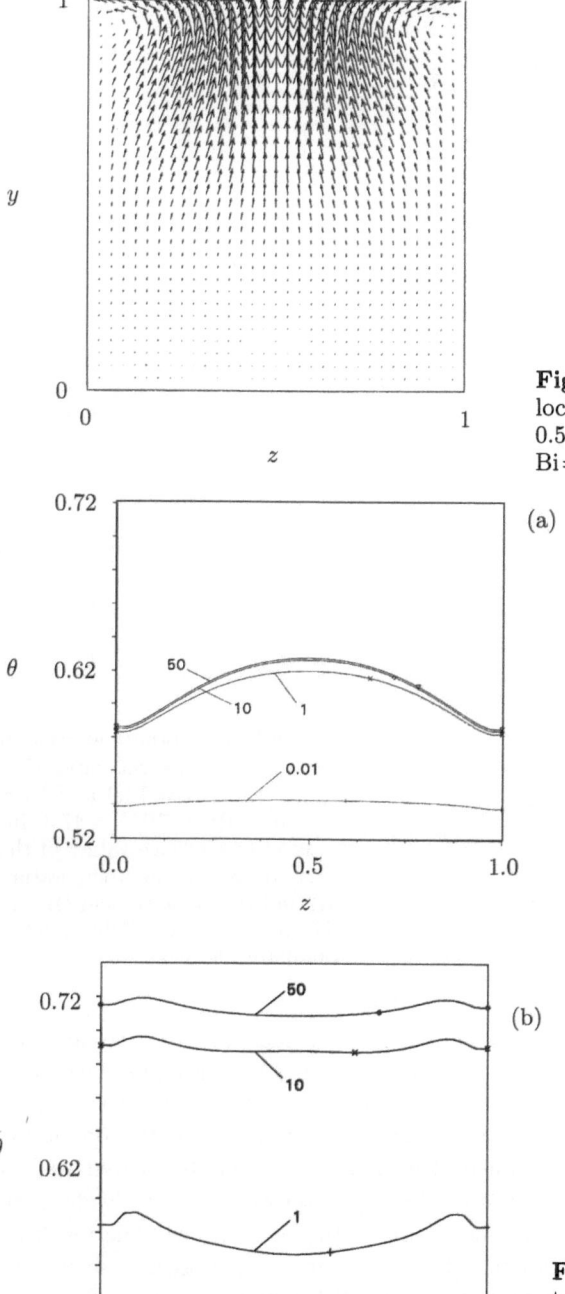

Fig. 8.8. Projection of the velocity field onto the plane $x = 0.5$ for $\mathrm{Pr} = 1$, $\mathrm{Re} = 100$, and $\mathrm{Bi} = \mathrm{Gr} = 0$

(a)

(b)

Fig. 8.9. Surface temperature distributions θ as functions of z at $(x = 0.5,\ y = 1)$ for $\mathrm{Ma} = 100$ (**a**), $\mathrm{Ma} = 10^4$ (**b**), for various Prandtl numbers given as labels

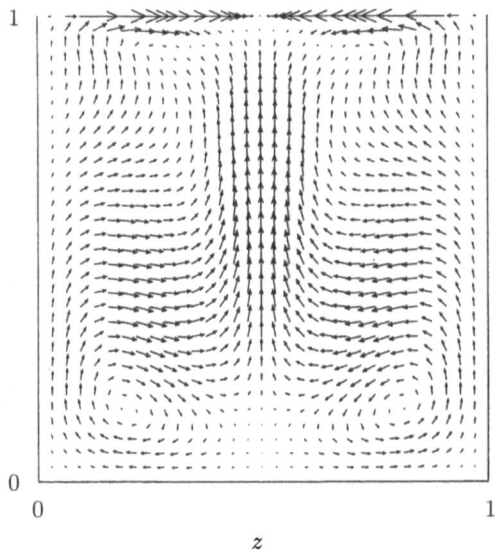

Fig. 8.10. Projection of the velocity field onto the plane $x = 0.5$ for $\mathrm{Pr} = 1$ and $\mathrm{Ma} = 5 \times 10^4$

each side wall does a weak flow remain directed towards the wall (not visible on the scale of Fig. 8.10). For high Marangoni numbers the secondary flows may reach sizable amplitudes. The ratio $|w_{\max}|/|u_{\max}|$, for example, is 25% for $\mathrm{Ma} = 5 \times 10^4$ and $\mathrm{Pr} = 1$.

Since the axis of the primary vortex is perpendicular to the side walls, secondary Bödewadt vortices (Bödewadt (1940)) are induced at high Reynolds numbers. They extend in a curved fashion along each side wall and can be well distinguished in the lower part of the container (the flow adjacent to the side walls at $z = (0, 1)$ is directed towards the centerline $x = y = 1/2$ in the lower half of Fig. 8.10). In the upper part of the cavity they are not visible, since thermocapillary forces oppose the Bödewadt flow. A cross section through these vortices at $y = 0.625$ is shown in Fig. 8.11. Owing to the superposed primary flow field in the positive and negative x directions at $y \approx 1$ and $y \approx 0$, respectively, the secondary vortices' centers in the (x, z) plane close to the cold wall would appear in the *corners* of the cold wall for $y \approx 1$, whereas they would appear in the middle ($x = 0.5$, $z = 0.5$) for $y \approx 0$. The apparent positions of the Bödewadt vortices in the vicinity of the hot wall behave in just the opposite manner.

Within the range of parameters investigated by Saß et al. (1996) no multiple two- or three-dimensional solutions were found. Since the numerical calculations were based on the assumption of steady flow, no conclusion can be drawn regarding the stability of the calculated flows with respect to time-dependent perturbations.

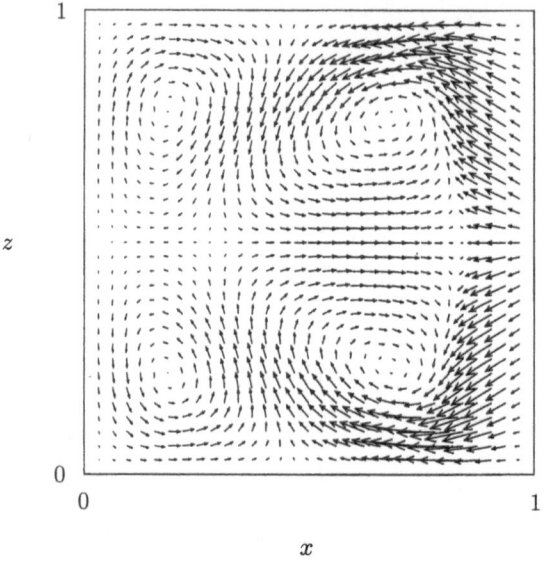

Fig. 8.11. Projection of the velocity field onto the plane $y = 0.625$ for $\mathrm{Pr} = 1$ and $\mathrm{Ma} = 5 \times 10^4$

8.2.3 Steady and Spatially Periodic Three-Dimensional Flows

Pure thermocapillary flows seem to be much more stable with respect to three-dimensional perturbations for aspect ratios $\Gamma \lesssim 2.3$ than for larger aspect ratios, at least for high Prandtl numbers (cf. Fig. 8.6). Apart from the secondary three-dimensional flows caused by rigid side walls, the two-dimensional base state for aspect ratios $\Gamma = \mathcal{O}(1)$ can become three-dimensional through a stationary symmetry-breaking bifurcation when gravity is present.

To investigate the mechanisms and the conditions under which these instabilities appear it is useful to consider volumes that are infinitely extended in lengthwise direction ($\Gamma_z \to \infty$). Then side-wall effects are absent.[8]

One of the first investigations along these lines was due to Gillon & Homsy (1996).[9] They experimentally investigated the combined buoyant–thermocapillary convection in a liquid volume with $\Gamma = 1.47$, $\Gamma_z - 3.8$, $\mathrm{Pr} = 8.4$, and dynamic Bond number $\mathrm{Bd} = \mathrm{Gr}/\mathrm{Re} = \rho g \beta L_y^2 / \gamma = 2.5$. It was observed that the basic flow, which was two-dimensional except for minor side-wall effects, became unstable for high Marangoni numbers $\mathrm{Ma} = \mathcal{O}(10^5)$, giving way to stationary three-dimensional convection cells with a lengthwise wavelength of $\mathcal{O}(1)$. Similar experimental results have been obtained by

---

[8] Schwabe et al. (1992) avoided side-wall effects by selecting an annular system for their experiments.

[9] The aspect ratio $\Gamma_z = 1$ used by Schwabe & Metzger (1989) was obviously too small to observe the patterns.

Daviaud & Vince (1993) for a certain range of aspect ratios $\Gamma = \mathcal{O}(1)$ and, for an annular system heated from the center, by Favre et al. (1997).

In order to simulate numerically the experiment of Gillon & Homsy (1996), Mundrane & Zebib (1993) calculated the three-dimensional flow for the same parameters (but Bd = 5.76) and Bi = 0. They employed periodic boundary conditions in the z direction corresponding to the experimentally determined value $\lambda = 1.4$. For Ma $= 2.93 \times 10^3$ the flow was found to be two-dimensional and stable. For Ma $= 1.95 \times 10^5$ the stationary calculation yielded a three-dimensional flow pattern. A projection of this flow onto the (y, z) plane at $x = 1.10$ is shown in Fig. 8.12. The calculated flow structure

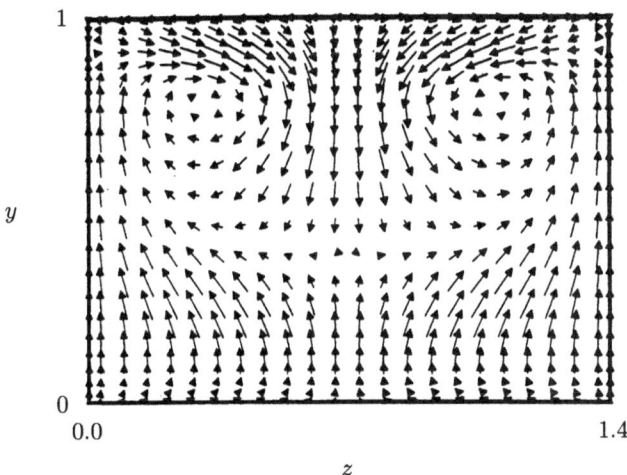

Fig. 8.12. Velocity field in a thermocapillary cavity with $\Gamma = 1.47$ projected onto the plane $x = 1.10$, after Mundrane & Zebib (1993). The parameters are Re $= 1.95 \times 10^5$, Bd $= 5.76$, Pr $= 8.4$, and Bi $= 0$. The wavelength in the z direction is $\lambda = 1.4$ in units of L_y

agrees qualitatively with the projection of the streamlines observed by Gillon & Homsy (1996).

Similarly, Schneider (1995) also observed stationary convection cells in a radially heated annular system with an upper free surface, filled with ethanol (radii $R_i = 20$ mm, $R_o = 40$ mm; $\Gamma = (R_o - R_i)/L_y < 4.5$). The observed wavelength in the azimuthal direction normalized by the layer's height L_y agrees well with the one found by Gillon & Homsy (1996) in the rectangular container.

A satisfactory physical explanation of the observed and calculated stationary instability has not yet been given. The analysis is complicated by the presence of both thermocapillary and buoyancy forces. Since the vertical velocity at the cell boundaries in Fig. 8.12 transports relatively cold fluid from the bulk to the free surface, the Marangoni effect suppresses the secondary flow in the (y, z) plane. The instability should, therefore, be caused by another mechanism. It is important to note that the basic thermocapillary vortex is squeezed vertically owing to the buoyancy-induced density stratification and is thus subject to high strain before the instability sets in (cf. Sect. 8.1.1). In fact, the base flow in the annular-gap system was observed to consist of an elongated stretched vortex just before the onset of instability (Schneider (1995)). Similarly, Gillon & Homsy (1996) noticed a stretched thermocapillary-driven vortex immediately below the free surface with two smaller corotating eddies embedded in the main circulation. An example of such a flow is shown in Fig. 8.3.

Since the order of magnitude of the critical Reynolds number ($\mathcal{O}(10^4)$) for the stationary cellular instability is not low, an inertial mechanism could be responsible for the pattern formation. In this context it seems worth mentioning that a two-dimensional flow with a strikingly similar cat's-eye structure, shown in Fig. 8.13, was recently found to be unstable to a steady cellular pattern at rather low Reynolds numbers (Kuhlmann et al. (1996)). The flow was

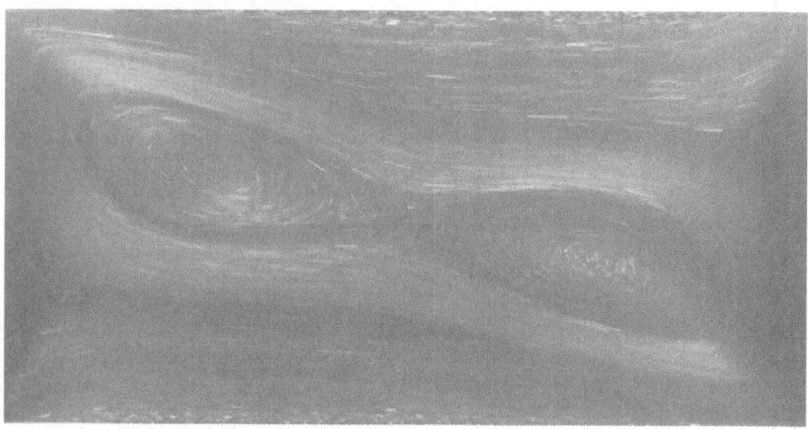

Fig. 8.13. Basic cat's-eye flow just before the onset of three-dimensional instability in a two-sided lid-driven cavity. The walls on the *left* and *right* sides move with equal absolute values of the velocity up and down, respectively

observed in an isothermal two-sided lid-driven cavity. Kuhlmann et al. (1996) have shown that the energy driving the instability originates from the straining motion near the central stagnation point. The type of instability should, therefore, be insensitive to the details of the boundary conditions (thermocapillary or mechanical driving). It was shown, moreover, that the generic mechanism can be traced back to the elliptic instability of vortices (Bayly (1986), Pierrehumbert (1986), Waleffe (1990)). The instability relies on a strong straining motion in the center of the vortex. The strain typically manifests itself in the form of elliptical streamlines or, for high strain rates, as a free hyperbolic stagnation point.[10] In view of the topological equivalence of both types of base flow and of supercritical cellular flow,[11] it is very likely that the combined buoyant–thermocapillary instability is also due to this mechanism.

The steady three-dimensional cellular flow can become time-dependent. Braunsfurth & Homsy (1997) considered the cavity flow for $Pr = 4.4$ and equal aspect ratios $\Gamma = \Gamma_z$ in the range 1–8. It was observed that in the fully developed steady cellular flow state the surface velocity is directed opposite to both the buoyant and the thermocapillary forces in a region adjacent to the cold wall, the flow reversal being most pronounced at a section ($z = $ const.) between two neighboring convection cells. On an increase of the Reynolds number, and with it also the Rayleigh number, the strength of the cellular flow increases and the region of flow reversal grows from the cold wall towards the hot side. As the stagnation line approaches the hot wall, the surface temperature gradient in the vicinity of the hot wall increases and drives a small but intense eddy,[12] while it is low along the remainder of the free surface. For Marangoni numbers above a second critical value[13] the hot-corner eddy was observed to periodically oscillate in strength and diameter. The instability mechanism has been discussed in terms of a positive feedback due to a temperature fluctuation at the stagnation line: if the temperature decreases at the stagnation line because of a perturbation, the temperature gradient near the hot corner is increased and the corner eddy becomes stronger. Simultaneously, the steady three-dimensional cellular flow from the cold wall towards the stagnation line feels less opposition because of a reduced surface stress on the cold side of the cavity. This effect also contributes to the amplification of the strength of the hot-corner vortex, yielding a positive feedback. Note that a strong three-dimensional steady cellular flow must be present to provide the surface flow reversal in order that this type of instability can occur.

---

[10] It is known that linear hyperbolic stagnation-point flow is unstable because of vortex stretching (Lagnado et al. (1984)). It can be argued (Kuhlmann et al. (1998a)), however, that the present hyperbolic stagnation-point flow embedded in a noncircular closed-streamline flow can support the same instability mechanism as elliptical flow. Both flows are periodic and should thus likewise enable a resonant amplification of inertial waves.

[11] The wavelength $\lambda = \mathcal{O}(2.5)$ of the cellular pattern in the lid-driven cavity flow is nearly twice that observed in the thermocapillary systems of Gillon & Homsy (1996) and Schneider (1995). The *effective* layer thickness in the thermocapillary systems is, however, smaller than L_y, since the strained thermocapillary vortex occupies approximately the upper half of the cavity only.

[12] In the experiment the eddy was induced within the meniscus, since the contact line was not fixed. Hence, the meniscus shape may have a significant influence on the onset of oscillations.

[13] For $\Gamma = 3.56$ and $Ra = 4.3 \times 10^4$ the critical Marangoni number is $Ma_{c2} \approx 5.5 \times 10^4$.

9. Extensions to Simple Models

Up to this point we have mainly considered minimum models of thermocapillary convection, neglecting secondary effects. On the basis of the fundamental behavior, more realistic and hence more complicated models can be investigated. For instance, some aspects of the influence of natural convection have already been discussed in Sects. 7.5 and 8.1.3. To date, the influence of many other factors has not been fully understood or studied systematically. Some of these factors will be summarized in Sect. 9.1. Among them are thermocapillary flow instabilities in noncylindrical liquid bridges, dynamic surface deformations, and the influence of solutocapillary convection in binary mixtures. Section 9.2 treats flow control by external forces.

9.1 Intrinsic Properties of Thermocapillary Systems

9.1.1 Static Free-Surface Shape

The particular case of a straight cylindrical liquid bridge considered in Chap. 7 can only be realized by fixing the liquid volume at $V_0 = \pi R^2 d$ and in the limit Bo$\rightarrow 0$ (e.g. under the condition of weightlessness). If $V \neq V_0$ the free surface will no longer be cylindrical. In addition, the hydrostatic pressure difference may cause significant free-surface deflections. This applies particularly to industrial realizations of the floating-zone process (Hurle (1994)).

Under zero gravity and in the limit Ca$\rightarrow 0$ dynamic surface deformations cannot occur (see Sect. 2.3). Then the static shape of the liquid bridge will be mirror-symmetric with respect to $z = 0$ and axisymmetric, and the shape can be parametrized by either the fluid volume or the contact angle.[1] When the free surface is thermally insulating, heat conduction alone leads to reduced surface temperature gradients close to the contact lines if the liquid volume is decreased ($V < V_0$) (Shevtsova et al. (1996)). For this reason the streamlines do not reach deep into the corners for slender liquid bridges (Fig. 9.1a). Conversely, the surface temperature gradients become large in the corners if V is increased above V_0, and the streamlines penetrate deep into the corners

---

[1] For very large volumes the liquid bridge may undergo a transition to a three-dimensional state (Slobozhanin & Perales (1993), Slobozhanin et al. (1997)).

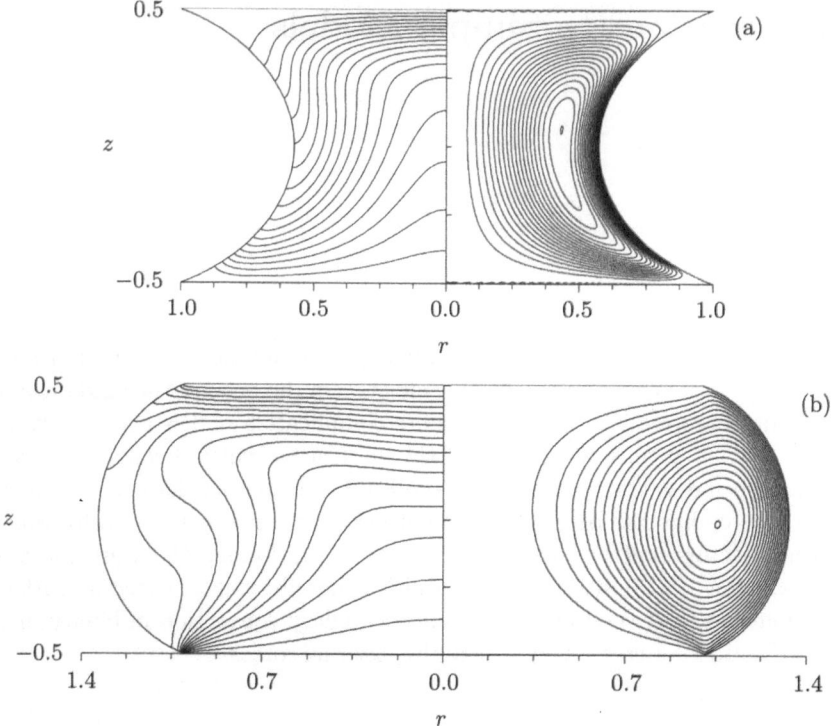

Fig. 9.1. Isotherms (*left*) and streamlines (*right*) in a thermocapillary liquid bridge with $\Gamma = 1$, $\mathrm{Pr} = 1$, $\mathrm{Re} = 10^3$, and $\mathrm{Gr} = \mathrm{Bi} = \mathrm{Ca} = 0$ for contact angles $\alpha = 20°$ (**a**) and $160°$ (**b**), calculated on a 101×101 graded mesh

(Fig. 9.1b). In the cold corner this thermally induced effect is assisted by a kinematic effect owing to the large contact angle (compare Sect. 3.1 and Kuhlmann et al. (1998b)) and by the convective compression of the isotherms (see also Kamotani & Platt (1992), Kozhoukharova & Slavchev (1986), Sasmal & Hochstein (1994), Tao et al. (1995)).

Associated with a shape change is a modified two-dimensional base flow whose stability generally differs from that of the cylindrical case. Cao et al. (1992), Hu et al. (1994), and Masud et al. (1997) have shown experimentally that the critical Marangoni number for the onset of oscillatory flow in half-zones depends strongly on the liquid volume V. Representative measurement results (critical temperature differences) are shown in Fig. 9.2 for a high-Prandtl-number silicone oil and moderate Bond numbers. The critical oscillation frequencies depend on r_m/R in a similar way. The radius r_m is defined as $r_m = \min(\xi)$ if $\min(\xi) < R$ (slender liquid bridges) and $r_m = \max(\xi)$ if $\max(\xi) > R$ (barrel-shaped liquid bridges), where $\xi(z)$ is the radial po-

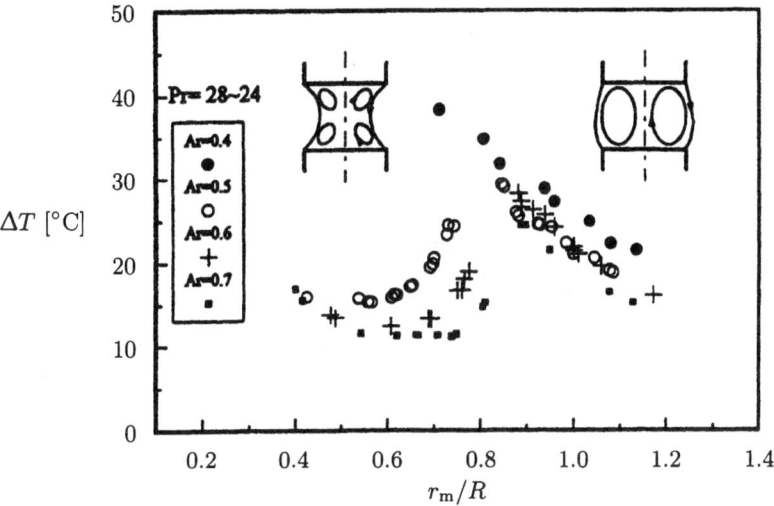

Fig. 9.2. Experimental critical temperature differences according to Masud et al. (1997) for Dow Corning silicone oil liquid bridges with Prandtl number Pr=24–28 (evaluated at the mean temperature) as a function of r_m/R. The bridge is heated from above and the critical wavenumber is $m_c = 1$ in all cases. Data are given for aspect ratios $\Gamma = 0.8$ (●), $\Gamma = 1.0$ (○), $\Gamma = 1.2$ (+), and $\Gamma = 1.4$ (□). The Bond number for $\Gamma = 1.4$ was Bd=0.26. The Marangoni number can be calculated from the temperature difference using $\nu = 2$ cSt, $\rho = 0.87$ g/cm^3, and $R = 1$ mm (Y. Kamotani (private communication, 1998))

sition of the free surface.[2] The stability boundary shown consists of two different branches. The threshold Marangoni number increases rapidly close to $r_m/R \approx 0.8$. For barrel-shaped liquid bridges with larger values of r_m/R another instability branch exists.

Numerical investigations of the volume dependence of the critical Marangoni numbers are sparse. For Pr = 1, 10, and 50 and long bridges with Γ in the range between 0.8 and 2.8, Chen & Hu (1998) obtained qualitatively similar results ($m_c = 1$ throughout) to those shown in Fig. 9.2 by means of a linear-stability calculation using the regularization (7.24).[3] More accurate results have been obtained by Nienhüser et al. (1998) for Pr=0.02. As expected from the low-Prandtl-number results for $V/V_0=1$, the most dangerous modes are stationary. Neutral curves for $\Gamma=1$ are shown in Fig. 9.3.

---

[2] Note that the control parameter r_m/R defined here is neither a unique nor a continuous function of the volume V if Bo $\neq 0$, since it is possible that min $(\xi) < R <$ max (ξ) (compare Fig. 9.4). For sufficiently small Bo the range of discontinuity of r_m/R near $V=V_0$ is narrow.

[3] The results of Chen & Hu (1998) must be considered qualitative, since neither the truncation order of the double Chebyshev expansion nor any validation was provided. Moreover, the regularization (7.24) cannot be used for large Prandtl numbers (Pr $\gtrsim$ 5), because the driving forces become increasingly localized at the hot corner when the Marangoni number is high (cf. Sect. 3.2.3).

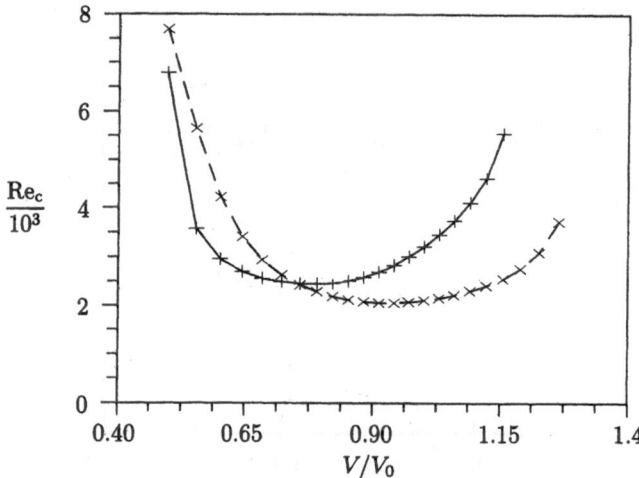

Fig. 9.3. Critical Reynolds number Re_c as a function of the normalized volume V/V_0 for $\Gamma = 1$, $Pr = 0.02$, and $Gr = Bi = Ca = 0$. (——) $m = 1$; (- - -) $m = 2$. The finite-difference grid was 61×121 (radial × axial) with a geometrical stretching factor 0.90 towards the boundaries. After Nienhüser et al. (1998)

The critical Reynolds number has a minimum of $Re_c \approx 2000$ at $V/V_0 \approx 0.95$ and the base flow is stabilized for high and low normalized volumes V/V_0. At low volumes ($V/V_0 \lesssim 0.75$) the critical mode changes from $m_c = 2$ to $m_c = 1$. A physical explanation of the dependence of the critical curves on the volume of fluid is still lacking. Moreover, the influence of the Bond number has not yet been investigated systematically.[4]

Cylindrical liquid bridges are remarkably stable with respect to two-dimensional oscillatory perturbations (Sect. 7.8, Shen et al. (1990)). Similarly, the onset of oscillations in the noncylindrical high-Prandtl-number thermocapillary liquid bridges investigated to date is three-dimensional. Recently, however, two-dimensional oscillatory flows at rather low Marangoni numbers, obtained by means of a two-dimensional time-dependent numerical simulation for a liquid bridge with a noncylindrical static shape in a homogeneous axial gravity field, were reported (V. M. Shevtsova (private communication, 1996), see also Shevtsova (1998)). Snapshots of the instantaneous isotherms and streamlines within one period of oscillation are shown in Fig. 9.4. It remains to be shown that the calculated two-dimensional oscillations are observable in a three-dimensional simulation and in a corresponding experiment.

---

[4] Two-dimensional thermocapillary convection in rectangular cavities filled with fluids having a deformed free surface has been investigated numerically by Keller & Bergmann (1990) for a range of contact angles around 90°.

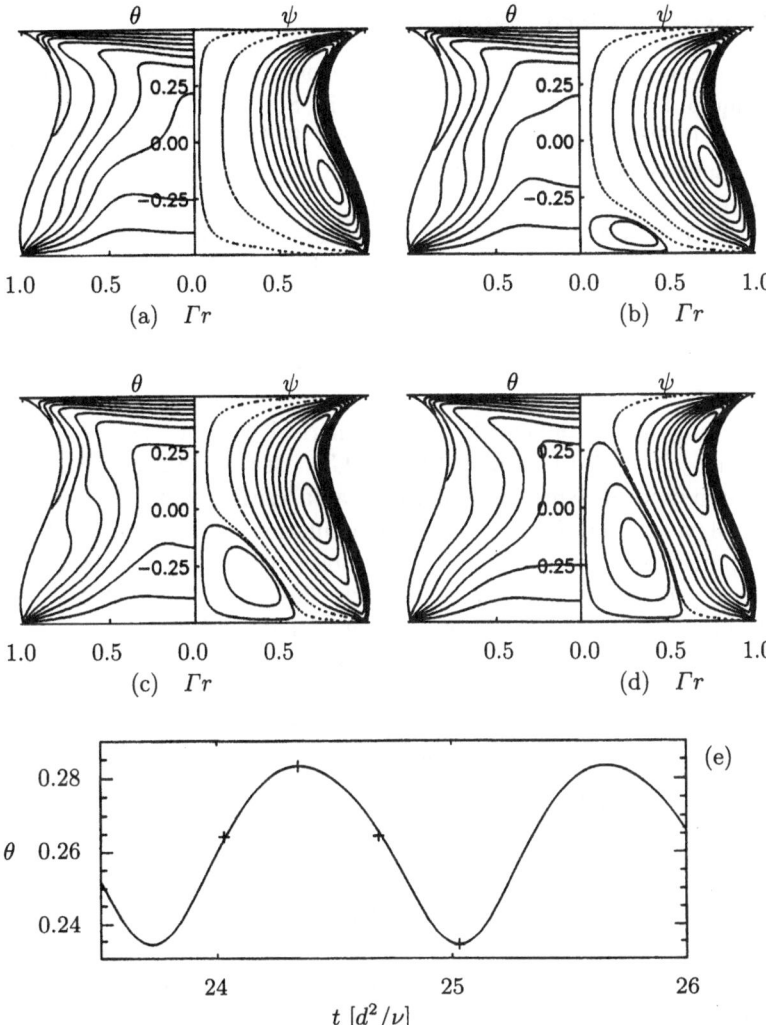

Fig. 9.4. Two-dimensional oscillatory thermocapillary convection in a liquid bridge with a statically deformed free surface (Ca → 0) according to V. M. Shevtsova (private communication, 1996). (**a–d**) Snapshots of isotherms and streamlines (the levels are not equidistant, for a better visualization of the separated vortex) taken at instants indicated by plus signs (+) in (**e**), which shows the temperature signal at a fixed location. The parameters are $\Gamma = 1.33$, $V/V_0 = 0.81$, $Bo = 7.2$, $Pr = 105$, $Re = 229$, $Gr = -534$ (heating from above), and $Bi = 0$

9.1.2 Dynamic Surface Deformations

The influence of thermocapillary convection on the shape of the free surface of liquid bridges has been discussed in Sect. 4.2.4 for Stokes flow. Hyer et al. (1991) and Zhang & Alexander (1992) calculated numerically the two-dimensional convection in half- and full-zone models for higher Marangoni numbers. Flows in two-dimensional rectangular containers were investigated by Cuvelier & Driessen (1986) and Sasmal & Hochstein (1994). For an overview of numerical techniques for dynamic free-surface flows the reader is referred to Floryan & Rasmussen (1989) and Kothe (1998).

The influence of the free-surface deformability (Ca ≠ 0) on the onset of oscillatory flows in thermocapillary liquid bridges has been controversial for a long time (see Sect. 7.4.3). Apparently contradictory results exist for the flow in rectangular containers. In their numerical calculations for an open cavity with $L_x/L_y = 2$, Pr = Ca = 0.01, and Gr = Bi = 0, Chen & Hwu (1993) obtained dynamic free-surface oscillations at a critical value of Ma ≈ 2.2. The critical value Ma_c for the onset of oscillations was found to depend nearly exponentially on Ca^{-1}. From this observation they concluded that the oscillatory flow completely vanishes in the limit Ca → 0 and that the surface deformability is an essential requirement for the calculated oscillations. Mundrane et al. (1995) carried out numerical calculations for the same parameters.[5] For Ca = 0.01 and for Reynolds numbers up to Re = 2500 they found, however, only *stationary* flows, with dynamic surface deformations less than 10^{-2} times the layer's thickness. Their calculations are in good agreement with those of Liakopoulos & Brown (1993). Moreover, by comparing calculations using different methods (finite volumes and boundary elements, respectively), it was demonstrated that the stationary flow-induced surface deformation can be well approximated by a first-order perturbation expansion according to (4.38–4.41). The free-surface deformations found by Mundrane et al. (1995) are of the same order of magnitude as those calculated by Hsieh (1992) for thermocapillary flow with Pr = 100 in a differentially heated open cube. The magnitudes of these surface deformations are smaller by several orders than the values given by Chen & Hwu (1993). Thus the majority of the numerical calculations indicate that dynamic surface deformations are not the primary cause of the onset of time-dependent flow in rectangular thermocapillary cavities. This hypothesis is in agreement with the results for liquid bridges (Sect. 7.4) and it is consistent with the onset of two-dimensional oscillatory flow in rectangular cavities with a flat free surface as calculated by Peltier & Biringen (1993) and Xu and Zebib (1998) (Sect. 8.1.3). There remains the possibility that the oscillatory flow states are not unique and depend on the initial conditions.

In contrast to the foregoing, surface deformations in extended thermocapillary liquid layers with $\Gamma = L_x/L_y \gg 1$ definitely cannot be neglected, even for small capillary numbers Ca. Under the condition Ca/k ≪ 1,

---

[5] And additionally for $L_x/L_y = 1$ and 4.

Smith & Davis (1983b) have shown by an asymptotic stability analysis for
long-wave perturbations (wavenumber $k \to 0$) that a return flow (6.3–6.4)
becomes unstable with respect to two-dimensional surface waves at a critical
Reynolds number

$$\mathrm{Re_c} = \frac{k^2}{3\mathrm{Ca}\left(\frac{1}{10} + \frac{1}{4}\frac{\mathrm{Pr}}{\mathrm{Bi}}\right)}, \tag{9.1}$$

provided that Pr, Bi, and $k^2\mathrm{Re}/\mathrm{Ca}$ are $\mathcal{O}(1)$. For $\mathrm{Pr} = \mathrm{Bi} = 0$ they found

$$\mathrm{Re_c} = \frac{32k^2}{41\mathrm{Ca}}. \tag{9.2}$$

In the limit $k \to 0$, the capillary number must tend to zero like $\mathrm{Ca} \sim k^2 \to 0$
in order that the critical Reynolds number remains $\mathcal{O}(1)$. Hence for finite
capillary numbers the return flow is unstable with respect to long-wave surface perturbations. Only in the limit of long waves is the critical mode two-
dimensional, because no Squire theorem exists for $k \neq 0$. According to (9.2),
the critical Marangoni number for the onset of surface waves in low-Prandtl-
number liquids satisfies $\mathrm{Ma_c} \sim \mathrm{Pr}$, while the critical threshold for hydrothermal waves scales like $\mathrm{Ma_c} \sim \mathrm{Pr}^{1/2}$. Thus surface waves may appear as the
first instability when the Prandtl number is sufficiently low.

   For thin thermocapillary liquid layers with a given temperature distribution in the gas phase Floryan & Chen (1994) have formulated conditions that
must be satisfied by the ambient temperature field in order that the surface
deformations remain small. Tan et al. (1990) and Burelbach et al. (1990) investigated theoretically (in the viscous approximation) and experimentally
the stationary two-dimensional thermocapillary flow with the large dynamic
surface deformations which are caused by a long-wave horizontal temperature variation of the heated bottom plate. The deformations can become so
large that the liquid film dries out locally. An overview of the evolution of
thin liquid films subject to various bulk and surface forces has been given by
Oron et al. (1997) on the basis of the long-wave theory.

9.1.3 Non-Monotonic Surface Temperature Profiles

In the floating-zone method the temperature distribution on the free surface of the melt takes a maximum value between the feed and the seed
crystal, the phase boundaries beeing practically at the melting temperature.
When buoyancy is absent this global temperature distribution leads to convection structures consisting mainly of two counterrotating thermocapillary-
driven vortices. The influence of different temperature profiles T_a on the vortex flow structure has been investigated by Rybicki & Floryan (1987a) and
Davis (1989) for creeping flows in liquid bridges (see also Sect. 4.2.5).

Menke (1994) numerically simulated the flow in two-dimensional liquid bridges with deformed boundaries[6] for small Prandtl numbers. The temperature distribution in the ambient gas T_a was assumed Gaussian with a maximum at $z = 0$. When the boundary conditions are symmetric with respect to $z = 0$ and the initial conditions are selected as $(u = 0, T = T_0)$ a time-dependent oscillatory flow is found to persist for $t \to \infty$, preserving the reflection symmetry, if the Marangoni number is above critical. If, however, the boundary conditions are perturbed asymmetrically for a limited period of time, a different oscillatory flow is obtained for $t \to \infty$ with broken mirror symmetry. The critical Marangoni number for asymmetric oscillations is considerably smaller than that for symmetric oscillations.[7]

Further two-dimensional calculations of the stationary convection in symmetrically heated liquid bridges and deformable surfaces have been carried out by Lan & Kou (1990), Zhang & Alexander (1992), and others.

One can anticipate that the first instability in symmetrically heated liquid bridges is three-dimensional, similar to that in half-zones. For an ambient temperature T_a which is symmetric with respect to $z=0$ the stationary two-dimensional base state is mirror-symmetric if $\mathrm{Gr}=0$. It is then easy to show that the normal modes of the linear stability problem (5.2, 5.3) must be either symmetric or antisymmetric with respect to $z=0$. Thus for a linear stability analysis only half of the full zone neeeds to be considered, with appropriate symmetry conditions at $z=0$. Wanschura et al. (1997b) calculated the linear stability of the flow in full zones with $\Gamma = 2$ for the case of a Gaussian temperature profile Θ_a with a full width at half maximum temperature of 0.41 times the zone length.[8] The most unstable modes were symmetric with $u=v=\partial_z w=\Theta=0$ at midplane, $z=0$. The critical wavenumber $(m=2)$ and the magnitudes of the critical Reynolds numbers were found to be comparable with those for the half-zone. The type of instability, however, depends on the Biot number. For $\mathrm{Bi} \to \infty$ the temperature field is passive, since the surface temperature is fixed and equal to the ambient temperature Θ_a. In this case the instability corresponds to that for zero Prandtl number. The mechanisms are the same as for the half-zone discussed in Sect. 7.4.1. On a decrease of the Biot number azimuthal temperature variations occur in the neutral mode and the associated thermocapillary forces have a stabilizing effect. Near $\mathrm{Bi} \lesssim 55$ a crossover to a hydrothermal-wave instability takes place. The respective

---

[6] The deformations of both the free surface and the phase boundaries were very small.

[7] Similar two-dimensional simulations for symmetrically heated full zones have been carried out by Kazarinoff & Wilkowski (1989, 1990). Above a certain threshold of the Marangoni number asymmetrical oscillations were found. These did not occur, however, for zero capillary number. Since the numerical code failed to converge for $\mathrm{Ca}=0$ and because of the low grid resolution, these results should be considered with care. Validations are as yet lacking.

[8] The full width at half maximum temperature was incorrectly specified as 0.5 times the zone length by Wanschura et al. (1997b). The correct value is 0.41. In addition, the results shown in their Fig. 3 are for $\Gamma=2$ and not for $\Gamma=1$.

critical Reynolds number reaches a minimum near Bi $\approx$ 18. On a further reduction of the Biot number the critical Reynolds number increases again and diverges for Bi$\rightarrow$0, because all temperature gradients vanish in the limit of an adiabatic liquid bridge.

Neglecting surface deformations, the full three-dimensional nonlinear flow problem has been solved by Baumgartl et al. (1990) for the aspect ratio $\Gamma = 2.4$ and small Prandtl numbers. On an increase of the Reynolds number the two counterrotating toroidal base-flow vortices are replaced by a stationary three-dimensional flow which consists of a large convection roll with its horizontal axis in the interior of the liquid volume. This is a symmetric mode in the above sense and qualitatively consistent with the finding of Wanschura et al. (1997b). On a further increase of Re the flow becomes time-dependent and the axis of the single roll starts to rotate around the (vertical) z axis, resulting in an azimuthally traveling wave. This scenario is different from that for the half-zone, for which the low-Prandtl-number oscillatory flow is an azimuthally standing wave (see Sect. 7.7.1).

A convection pattern similar to the case of symmmetrical heating arises when the free surface is subject to a monotonic temperature gradient but the surface tension has an extremum for some intermediate temperature T^* between the wall temperatures. The thermocapillary stress must then change its sign on the free surface where $T = T^*$. Corresponding two-dimensional simulations have been carried out by Kanouff & Greif (1994) for a differentially heated square container. The liquid, with Pr = 0.1, was assumed to have a surface-tension extremum at the mean value $T^* = T_0 + \Delta T/2$. If the extremum is a minimum the fluid flows along the free surface from the center towards the boundaries. In this case only mirror-symmetric stationary flow states were obtained. If, however, the surface flow is directed from the boundaries towards the center of the cavity (a maximum of the surface tension), asymmetric two-dimensional flow oscillations occurred above a critical Marangoni number. In the course of the oscillations both vortices alternately grew and shrank in size such that one of them occupied almost the whole cavity during part of the cycle.

9.1.4 Moving Liquid–Solid Interfaces

Besides the deformability of the free surface between liquid and ambient gas, also the liquid–solid phase boundary is dynamically determined during the crystal-growth process. Since the local advancement of the crystal-growth front is controlled by the transport of latent heat, the temperature field must be calculated both in the liquid and in the solid phase if the location of the phase boundary is to be determined. Let the position of the phase boundary of the melt zone be given by $z = \zeta(r, \varphi)$ and the coordinate system fixed to the solid phase. Then the boundary conditions are (e.g. Davis (1990), Davis (1993))

$$u \cdot t = 0, \tag{9.3}$$

$$[\rho_l u + (\rho_l - \rho_s) \, \partial_t \zeta \, e_z] \cdot n = 0, \tag{9.4}$$

$$T_l = T_s = T_m \left(1 + \frac{\tilde{\gamma}}{L} \nabla \cdot n\right), \tag{9.5}$$

$$n \cdot \left[k \, \nabla T\right]_l^s = L \, \partial_t \zeta \, e_z \cdot n. \tag{9.6}$$

Here T_m is the melting-point temperature of a plane phase front, $n(r, \varphi)$ the unit normal vector directed out of the liquid phase and into the solid phase, and $t(r, \varphi)$ either of the two independent tangent vectors. The densities of the solid and liquid phases are denoted by ρ_s and ρ_l, respectively, k is the heat conductivity, L is the latent heat (per volume of solid), $\tilde{\gamma}$ is the specific surface energy, and $[...]_l^s$ denotes the difference at the phase boundary between the values in the solid and the liquid phase of any quantity.

The *no slip* condition is provided by (9.3). Equation (9.4) is the normal mass balance and describes the phase-change convection due to the different densities of the liquid and solid phases. Capillary undercooling due to the local curvature of the phase boundary (Gibbs–Thomson effect) is taken into account in (9.5). It may, however, be neglected in may cases, because it is important only on very small scales. For binary liquids the melting point is also shifted by the constitutional undercooling mc, where m is the slope of the liquidus and c the concentration of the solute in the melt. In this case an additional equation for the conservation of the dissolved species must be provided. Finally, (9.6) specifies the discontinuity of the normal heat flux caused by the latent heat.

Two-dimensional stationary calculations of the temperature and velocity fields in symmetrically heated liquid zones with deformable liquid–gas and liquid–solid boundaries have been carried out by Lan & Kou (1991a) for NaNO$_3$ (Pr $\approx$ 9) and Si (Pr $\approx$ 0.01). In all cases investigated the phase boundary was found to be convex towards the liquid side. The dynamic deformations of the liquid–gas interface were very small. Further work on this problem can be found in Bergman & Webb (1990), Lan & Kou (1990,1991b), and Mühlbauer et al. (1995).

Davis (1993) gave an overview of the interaction between convection and the dynamic phase boundary. In single-component systems, hydrodynamic instabilities are favored by the presence of a phase change at the solid–liquid interface. Bifurcations which are steady for rigid boundary conditions typically become hysteretic for dynamic phase-change boundaries. In binary systems the Mullins–Sekerka instability may occur (Mullins & Sekerka (1964)). In this case the phase front propagating into a constitutionally undercooled melt becomes unstable with respect to surface corrugations, whose wavelenghts are usually very small compared to the macroscopic dimensions and the hydrodynamic length scales. Therefore, hydrodynamic and morphological instabilities are very often considered separately and their coupling is neglected. The various morphological instabilities as well as the typical wave and hexagonal patterns of the phase boundary have been discussed by Coriell & McFadden (1993). Further reviews have been given by Langer (1980) and Glicksman et al. (1986).

9.1.5 Temperature-Dependent Material Parameters

In many applications and several model experiments the applied temperature differences are very large. In these cases not only the density and the surface tension, which are approximated as linear functions of T within the Boussinesq approximation, but also other material parameters, such as the viscosity, must be considered temperature-dependent.

Bückle & Perić (1992) performed numerical calculations using a quadratic approximation for the temperature dependence of μ, ρ, Pr, and c_p. Results obtained for two-dimensional combined buoyant–thermocapillary convection in an open cavity with Pr = 17 (ethanol) were compared with the quasi-two-dimensional experimental data of Schwabe & Metzger (1989). It was shown that the Oberbeck–Boussinesq equations represent a good approximation if the temperature differences remain below $\Delta T < 10$ K. While for $\Delta T = 40$ K the computed field quantities deviated qualitatively from the experimental data when using the Boussinesq approximation, the numerical results with temperature-dependent material parameters were in good agreement with the experimental results. Hence, for complex flows with large temperature differences the validity of the Boussinesq approximation must always be examined. Corresponding criteria have been given by Gray & Giorgini (1976) and were confirmed by Bückle & Perić (1992) for the cases investigated.

The influence of a temperature-dependent viscosity on the onset of hydrothermal waves in a cylindrical liquid bridge with Pr = 4 and $\Gamma = 1$ was investigated by Kozhoukharova et al. (1998). Using an extension of the linear stability analysis described in Sect. 7.2 and assuming a linear variation of the kinematic viscosity with temperature, they found that the critical Reynolds number decreases with decreasing viscosity group, which is defined as $R_\nu = \zeta \Delta T / \nu_0$, where ν_0 is the reference kinematic viscosity and ζ the linear Taylor coefficient of an expansion of $\nu(T)$. For $R_\nu \approx -0.5$, which is typical for model experiments with silicone oil, the critical Reynolds number is reduced by $\approx 10\%$. This destabilization was explained in terms of the reduced viscosity of the hot free-surface layer of fluid characteristic of the basic toroidal flow state (cf. Fig. 7.2). Associated with this is an increased radial basic-state temperature gradient, which is known to provide the energy for hydrothermal waves (Sect. 7.4.2). Moreover, the azimuthal return flow of the perturbation mode near the free surface is facilitated by the reduced viscosity of the free-surface layer. Since the threshold shift is a function of the azimuthal wavenumber m, the critical wavenumber, which remains $m_c = 2$ for $\Gamma = 1$ and $R_\nu > -0.5$, may change for certain aspect ratios and sufficiently high absolute values of R_ν.

9.1.6 Binary Mixtures and Adsorption

When modeling the melt flow of multicomponent or highly doped semi-conductor crystals, at least two different species must be taken into account. Since the surface tension can depend strongly on the concentration of the second component (see e.g. Tison et al. (1992) for bismuth in tin), it is approximated by

$$\sigma(T, c) = \sigma_0(T_0, c_0) - \gamma(T - T_0) - \gamma_S(c - c_0),   \tag{9.7}$$

where γ_S is the solutal coefficient of the surface tension.

For binary mixtures, the volume equations (2.14–2.16) must be supplemented by an equation for the concentration $c = \rho_S/\rho$ of the minority component. Here ρ_S is the mass density of the dissolved species and ρ the total density. If, in addition, the Soret effect (Landau & Lifshitz (1959)) is operative the scaled volume equations[9] (cf. Wanschura et al. (1995b), Wanschura (1996)) read

$$\partial_t \boldsymbol{u} + \boldsymbol{u} \cdot \nabla \boldsymbol{u} = -\nabla p + \Delta \boldsymbol{u} + \mathrm{Gr}\,(\theta - \psi_S c)\,\boldsymbol{e}_z,   \tag{9.8}$$

$$\nabla \cdot \boldsymbol{u} = 0,   \tag{9.9}$$

$$\partial_t \theta + \boldsymbol{u} \cdot \nabla \theta = \frac{1}{\mathrm{Pr}}\,\Delta \theta,   \tag{9.10}$$

$$\partial_t c + \boldsymbol{u} \cdot \nabla c = \frac{1}{\mathrm{Sc}}\,(\Delta c + \Delta \theta),   \tag{9.11}$$

where the concentration c is scaled with $\Delta T\,k_T/T_0$. The thermal diffusion ratio is denoted by k_T and the Schmidt number is defined as $\mathrm{Sc} = \nu/D$, where D is the diffusion. The Soret separation ratio is $\psi_S = -(\beta_S/\beta)(k_T/T_0)$, where $\beta_S = -\rho^{-1}(\partial \rho/\partial c)$ is the solutal expansion coefficient.

If the phase boundaries are assumed stationary and mass transport through the boundaries is neglected, the diffusion current density $\boldsymbol{i}$ (mass per unit area and time) perpendicular to the phase boundary must vanish. In nondimensional form this condition reads

$$\boldsymbol{i} \cdot \boldsymbol{n} = -(\nabla c + \nabla \theta) \cdot \boldsymbol{n} = 0.   \tag{9.12}$$

In addition to the usual boundary conditions, tangential stresses caused by concentration variations must also be taken into account in (2.29). This leads to

$$\boldsymbol{t} \cdot \left[\nabla \boldsymbol{u} + (\nabla \boldsymbol{u})^{\mathrm{T}}\right] \cdot \boldsymbol{n} + \boldsymbol{t} \cdot \nabla(\mathrm{Re}\,\theta + \mathrm{Re}_S\,c) = 0,   \tag{9.13}$$

where the solutal Reynolds number is defined as

$$\mathrm{Re}_S = \frac{\gamma_S \Delta T d}{\rho \nu^2}\,\frac{k_T}{T_0}.   \tag{9.14}$$

---

[9] In the present case it is more convenient to scale the velocities and the pressure with ν/d and $\rho \nu^2/d^2$, respectively.

Using the above model, Wanschura (1996) investigated the stability of the two-dimensional thermo- and solutocapillary convection in liquid bridges for Gr = 0. Owing to the high Schmidt numbers of real systems the solutal boundary layers are very thin for strongly convecting flows and make the numerical stability analysis a difficult task. The results obtained for relatively small Schmidt numbers demonstrate, however, the relevant instability mechanisms. Within the model (9.8–9.11), concentration gradients are exclusively generated by Soret separation.[10] For the above nondimensionalization of c the Soret separation causes concentration gradients, which oppose the temperature gradients. For this reason thermo- and solutocapillary surface forces compensate each other if $\mathrm{Re} = \mathrm{Re}_\mathrm{S}$ and $\boldsymbol{u} = 0$ (the diffusion current density vanishes everywhere in the volume: $\nabla c = -\nabla \theta = -\boldsymbol{e}_z$).

In the limit of vanishing solutocapillary Reynolds number ($\mathrm{Re}_\mathrm{S} = 0$) the pure thermocapillary case is recovered (Sect. 7.4). Then the concentration field is passive and determined by $\boldsymbol{u}$ and θ. Formally, the same volume equations are obtained in the pure solutocapillary limit of vanishing thermocapillary Reynolds number ($\mathrm{Re} = 0$) and $\mathrm{Pr} \to 0$. Only the boundary conditions for the scalar fields differ in the two limits: $\theta(z = \pm 1/2) = \pm 1/2$ in comparison to $\partial_z c(z = \pm 1/2) = -1$. The mechanisms of the pure solutocapillary instabilities as a function of the Schmidt number are completely analoguous to the thermocapillary case.[11] In particular, both the stationary (Sc $\lesssim$ 1) and the oscillatory (Sc $\gtrsim$ 1) instabilities exist. The critical values $(\mathrm{Re}_\mathrm{S}^\mathrm{c}, \omega_\mathrm{S}^\mathrm{c})$ merely differ slightly from those in the thermocapillary case ($\mathrm{Re}_\mathrm{c}, \omega_\mathrm{c}$; see Wanschura (1996)).

The stability boundaries for $(\mathrm{Re}, \mathrm{Re}_\mathrm{S}) \neq 0$ and $\mathrm{Pr} = 0.02$ are shown in Fig. 9.5 for different Schmidt numbers. Owing to (9.11) and (9.12) the concentration field for Sc $= 0$ is given, up to a constant, by the temperature field ($c = -\theta + \mathrm{const.}$). Then the effective Reynolds number is $\mathrm{Re} - \mathrm{Re}_\mathrm{S}$ and the stability boundary is a straight line through the critical value for pure thermocapillary flow with unit slope: $\mathrm{Re}_\mathrm{c} = \mathrm{Re}_\mathrm{c}(\mathrm{Re}_\mathrm{S} = 0) + \mathrm{Re}_\mathrm{S} = 2062 + \mathrm{Re}_\mathrm{S}$ (see Fig. 9.5). The instability mechanisms for nonzero Schmidt numbers have been discussed in detail by Wanschura (1996). On an increase of the Schmidt number strong radial concentration gradients appear. The concentration transported from the bulk to the surface by the stationary three-dimensional neutral mode (Sect. 7.4.1) leads to strong azimuthal concentration gradients and results in significant thermocapillary forces in the azimuthal direction. For $\mathrm{Re}_\mathrm{S} > 0$ these thermocapillary stresses augment the flow that produces them and the stability boundary decreases rapidly. For $\mathrm{Re}_\mathrm{S} < 0$, however, the azimuthal

---

[10] In crystal growth the dissolved component usually accumulates in front of the phase boundary of the growing crystal, since the equilibrium partition coefficient is $k = c_\mathrm{solid}/c_\mathrm{liquid} < 1$. Owing to the low diffusivity, melt convection then leads to solutal boundary layers with high concentration gradients perpendicular to the growth front. See also Garandet et al. (1994) and Müller & Ostrogorsky (1994).

[11] The pure solutocapillary system is equivalent to a thermocapillary system in which an axial heat current at $z = \pm 1/2$ is imposed.

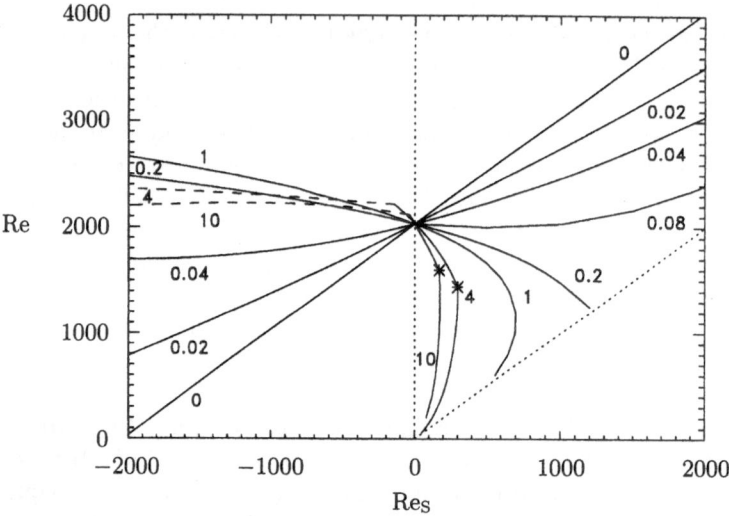

Fig. 9.5. Stability boundaries of two-dimensional thermo- and solutocapillary convection in a liquid bridge for $\Gamma = 1$ and $Pr = 0.02$ plotted in the (Re, Re_S) plane for different Schmidt numbers (shown as labels). For all cases the critical mode has $m = 2$. Only for small Reynolds numbers Re does the critical azimuthal wavenumber change to $m = 1$, at the locations indicated by $(*)$. (———) Stationary instabilities; (- - -) oscillatory instabilities

surface stresses induced by the neutral mode oppose the flow that generates them and Re_c is increased (stabilization, see Fig. 9.5). When the forces opposing the surface flow of the neutral mode become too strong a frustration results. The frustration is resolved by a phase shift between the concentration and the flow field and leads to an oscillatory instability (- - - in Fig. 9.5) similar to that in Sect. 7.4.2. Further continuations of the stability boundaries are shown in Fig. 9.6.

Tison et al. (1992) carried out experiments using a tin–bismuth mixture. The Schmidt number was of order $\mathcal{O}(10^3)$. Since $Re_S > 0$ for this system, the linear-stability boundary Re_c is much lower than in the pure thermocapillary case and can be calculated reliably despite the high Schmidt number (Wanschura (1996)). Flows in liquid bridges with similar material parameters should, therefore, become three-dimensional at very small temperature differences $(\Delta T = \mathcal{O}(mK))$.

Guérin et al. (1991) investigated the three-dimensional convective instability of the conductive state in a cylindrical melt of a binary mixture in front of a plane crystal-growth front. Owing to the segregation effect a concentration gradient is created ahead of the propagating front, which, in a gravity field, leads to an unstable density stratification. By an appropriate scaling of the equations an analogy was established to the Rayleigh–Bénard problem.

Using a simplified thermo-solutocapillary model for the flow in a rectangular open container, Polezhaev & Ermakov (1992) numerically found two-

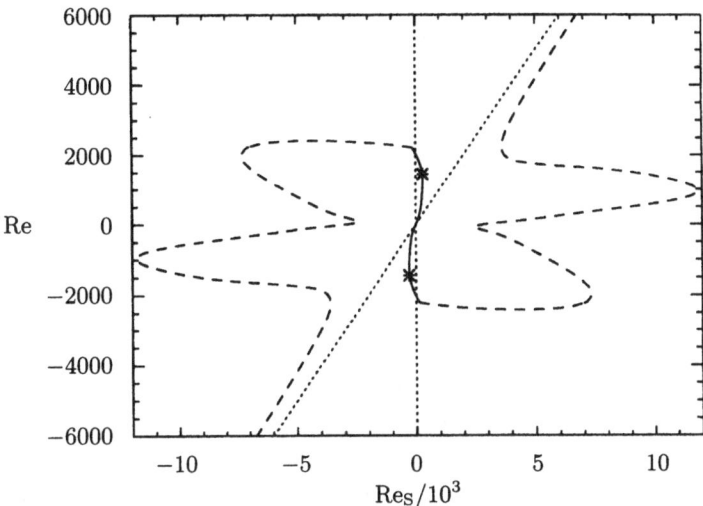

Fig. 9.6. Typical stability boundaries of the two-dimensional combined thermo- and solutocapillary convection in a liquid bridge for $\Gamma = 1$, $m = 2$, $Pr = 0.02$, and $Sc = 4$ shown in the (Re, Re$_S$) plane. The critical wavenumber changes from $m = 2$ to $m = 1$ on a decrease of the absolute value of Re at the locations indicated by ($*$). (——) Stationary, (- - -) oscillatory

dimensional oscillatory solutions of the governing equations for $Pr = 0$ and $Sc = 10$. For two-dimensional calculations closely related to the floating-zone method the reader is referred to Lan & Kou (1993a).

The possibility of instabilities involving solutocapillary effects has already been pointed out by Schwabe et al. (1978) and was investigated in more detail by Schwabe et al. (1996). The latter authors employed a model float zone of NaNO$_3$ with an organic surface-tension-reducing agent having a segregation coefficient $k < 1$, such that the impurity accumulates in front of the crystallizing interface. This accumulation reduces the total surface stresses that drive the flow in the melt and transport hot fluid along the free surface to the crystallization front. Associated with this reduction of the capillary flow is a reduction of the heat transfer to the crystal, giving rise to an even higher growth speed. This self-amplifying advancement of the crystallization front comes to an end at some distance from the hot, rigid wall present in the experiment. Thereafter, the solute boundary layer is rapidly removed by the thermocapillary flow and a back-melting occurs which is also self-amplifying, since the melting material is now deficient in the solute. This process repeats itself in an irregular fashion with distinct spikes of the flow intensity.

If the dissolved substances are surface-active they accumulate in a thin surface, layer which is usually a monolayer (Gibbs adsorption). Within a macroscopic theory this layer can be considered two-dimensional. Instead of

(9.12), the conservation of the surface concentration C (mass per unit area) for an isothermal fluid then requires the (dimensional) boundary condition[12]

$$i \cdot n = \partial_t C + \nabla_\| \cdot (u_\| C) + C (\nabla \cdot n)(n \cdot u) - D_s \nabla_\|^2 C \qquad (9.15)$$

(Stone (1990)), where n is the unit outward normal vector (of the liquid), $\nabla_\| = (I - nn) \cdot \nabla$ the tangential nabla operator, and D_s the surface diffusivity. In this formulation the time derivative must be taken keeping the surface coordinates fixed (Wong et al. (1996)). Eventually, the mass transport through the interface must also be taken into account. If the adsorption and desorption processes are very fast the diffusion current in the bulk of the liquid from and to the free surface is the limiting process. Then $i = -\rho D \nabla c$, with corresponding modifications if the Soret effect is to be included. The first term on the right-hand side of (9.15) describes the local accumulation, the second term is the convective transport in the surface layer, the third term represents concentration changes through an expansion or compression of the surface area, and the fourth term stands for the surface diffusion. The Gibbs equation (Davis (1990)) establishes a relation between the surface concentration C and the volume concentration c. For small concentrations the following linear relation holds:

$$\gamma_S c = \frac{k_B T_0}{m_0} C, \qquad (9.16)$$

where k_B is the Boltzmann constant and m_0 the mass of a molecule of the dissolved species. The Gibbs depth is defined as $(m_0 \gamma_S)/(\rho k_B T_0)$. A solution of the coupled equations (9.8–9.11) and (9.15) together with the other remaining boundary conditions is difficult. Therefore, little is known about the influence of adsorption at liquid–gas or liquid–liquid interfaces on three-dimensional pattern formation in the liquid volume.[13]

In some cases of adsorbed species a surface viscosity must be taken into account in addition to the solutal effects (see e.g. Mansell et al. (1994)). The required boundary condition is complicated and has been derived in a general form by Scriven (1960), who assumed that the surface layer can be modeled as a two-dimensional Newtonian fluid.

9.1.7 External Convection: Double Layers

In most cases of thermocapillary convection the free surface of the liquid meets an ambient gas whose dynamic viscosity is much smaller than that of the liquid ($\rho_a \nu_a \ll \rho \nu$). Thus the shear stresses exerted on the interface by a weak ambient-gas flow can usually be neglected. This simplification is not

---

[12] A strict derivation of the boundary condition using a microscopic model of Brownian particles that move in a surface potential and a discussion of the range of validity have been given by Brenner & Leal (1982).

[13] The influence of adsorption on weak thermocapillary convection in shallow rectangular containers was investigated by Homsy & Meiburg (1984).

allowed, however, if the flow in the ambient atmosphere is strong, for example if it is driven by buoyancy forces (Velten et al. (1991)).

In the encapsulated crystal-growth process a second liquid (subscript 2), for instance boron oxide (B_2O_3), is layered over the melt (subscript 1) to protect it from contamination. Now thermocapillary forces can act on the outer liquid–gas interface as well as on the interior liquid–liquid interface. For $\Gamma = 2$ Liu et al. (1993) calculated the two-dimensional thermocapillary convection in plane double layers bounded both from below and from above either by perfectly conducting rigid walls or by a free surface on top and by a heat-conducting rigid wall at the bottom. The latter case has also been considered by Doi & Koster (1993) and Wang et al. (1994) for different heat transfer coefficients at the free surface (see also Liu & Roux (1994)). In addition to stationary two-dimensional numerical calculations, analytical solutions have been given for the parallel flow far away from the lateral boundaries (similar to that for the plane flow considered in Chap. 6). For $Gr = 0$ the strength of the flow in the melt (1) depends on the parameter Ma^*:

$$Ma^* = Ma_1 - \frac{1}{2}\frac{\rho_2 \nu_2 \kappa_2}{\rho_1 \nu_1 \kappa_1}Ma_2, \quad \text{with} \quad Ma_i = \frac{\gamma_i \Delta T L_y}{\rho_i \nu_i \kappa_i}, \quad (9.17)$$

where L_y is the sum of the layer depths. For $Ma^* = 0$, or $\gamma_1/\gamma_2 = 0.5$, the thermocapillary convection in layer (1) is completely suppressed. In this case the thermocapillary shear stress acting on the interior (liquid–liquid) interface is just compensated by the shear force that is caused by the return flow in the outer layer (2). Since end walls modify the flow in finite-length containers it cannot be completely suppressed. Nevertheless, the numerically calculated strength of the thermocapillary convection in the inner layer (2) is minimized when $\gamma_1/\gamma_2 \approx 0.5$ and the Reynolds number is not too large. This behavior is consistent with the result of Villers & Platten (1990) who, by comparison with experimental data (Villers & Platten (1988)), showed that the infinite-layer model yields a good approximation to the flow in the middle of a finite-length container, provided the aspect ratio is $\Gamma \gtrsim 3$. For higher Reynolds numbers, lower values of γ_1/γ_2 may minimize the circulation in layer (2) (Doi & Koster (1993)).

The stationary creeping flow in plane double layers with a deformable free surface and a plane nondeformable internal interface has been calculated by Rao & Biswal (1995) for large aspect ratios Γ, using an asymptotic theory. Viviani & Cioffi (1992) computed the low-Reynolds-number flow in a cylindrical double layer by means of a biorthogonal series expansion (cf. Sect. 4.1). The two-dimensional combined buoyant–thermocapillary convection in a cylindrical geometry (inner liquid GaAs, outer liquid B_2O_3) has been studied numerically by Li et al. (1993).

To date little is known about the hydrodynamic stability of one- and two-dimensional thermocapillary flows in double layers.[14] A first step in this

---

[14] For an overview of oscillatory instabilities in liquid double layers in which the temperature gradient is perpendicular to the interior interface (Marangoni con-

direction was made by Wang & Kahawita (1998), who performed a few two-dimensional numerical simulations of the onset of time-dependent flow in a low-Prandtl-number liquid covered by another layer of fluid, taking into account thermocapillary and buoyancy effects.

9.2 External Body Forces

The most important external force field for crystal growth from the melt is the homogeneous gravity field g. It couples to the density ρ. If the density difference $\rho - \rho_a$ is nonzero, hydrostatic pressure differences across the free surface, measured by the Bond number Bo, may lead to large surface deformations. In addition to this isothermal effect, gravity couples to the temperature-induced density variations and may lead to natural convection. The combined natural and thermocapillary convection has been discussed in Sect. 7.5 for liquid bridges and in Sect. 8.1 for open cavities.

Besides gravity, other volume forces are of interest. For example, the feed and seed crystals in the float-zone process are often rotated independently around their axis to influence the melt flow and to smooth out deviations from axisymmetry. Moreover, magnetic fields are used to prevent the onset of oscillatory flow in semiconductor melts.

9.2.1 Uniform Rotation

When the rigid boundaries at $z = \pm 1/2$ of an isothermal cylindrical liquid bridge rotate with different but constant angular velocities Ω_i about the z axis, secondary axial and radial flows are induced by centrifugal forces in addition to the primary azimuthal flow. For differential rotation ($\Omega_1 \neq \Omega_2$) with equal signs ($\Omega_1 \Omega_2 \geq 0$), the secondary flow consists of a single vortex, whereas two stacked toroidal vortices with different senses of rotation appear when the end walls rotate in opposite directions ($\Omega_1 \Omega_2 < 0$). Since the secondary vortices are driven by inertia forces, their strength relative to the primary flow vanishes in the limit of low rotation-rate differences. Several flow patterns and dynamic free-surface shapes for low rotation rates have been calculated in the limit of low capillary number by Harriott & Brown (1983). For counterrotation ($\Omega_1 \Omega_2 < 0$) the secondary rotation-induced vortices have an opposite sense of rotation to that in the thermocapillary flows in full zones. Hence, the former flows have the potential to suppress thermocapillary convection to a certain extent.

Chun & Wuest (1982) carried out experiments with thermocapillary half-zones ($\Gamma = 1.6$) of high Prandtl number ($C_{18}H_{38}$) heated from above. The Marangoni number was kept at a nearly constant, supercritical value Ma $\approx$

---

vection) and which is bounded from above and from below by rigid walls, see Nepomnyashchy & Simanovskii (1995).

3.4×10^4. ON increasing the rotation rate of the hot wall (augmenting case) a decrease of the flow oscillation frequency was observed. For a sufficiently high rotation rate the flow oscillations were suppressed completely, yielding a steady flow. When the cold wall was rotated (opposing case), or if both walls rotated in the same direction, the flow oscillation frequency was found to increase with the rotation rate. Even in this case, the amplitude of the oscillatory flow was suppressed for high rotation rates.

Kamotani & Kim (1988) investigated the influence of a constant rotation of the cold wall only on the onset of thermocapillary flow oscillations in high-Prandtl-number liquid bridges ($\mathrm{Pr} = 27$ and $\mathrm{Pr} = 67$). The experimentally determined critical Marangoni numbers attained a minimum value (destabilization of the basic state) when the rotation rate Ω coincided with the angular phase velocity ω/m (where m is the fundamental wavenumber) of the azimuthally travelling hydrothermal wave. The maximum relative reduction of $\mathrm{Ma_c}$ was found to be 30% and occurred for aspect ratios $\Gamma < 1.4$. For higher rotation rates the critical Marangoni number was found to increase approximately linearly with Ω. In agreement with the results of Chun & Wuest (1982) the frequencies ω of the hydrothermal waves were increasing functions of the rotation rate Ω.

The influence of a rapid rotation in the limit of small Ekman numbers $\mathrm{Ek} = \nu/\Omega R^2$ has been studied by Smith (1986b), using an asymptotic expansion for the flow in a half-zone. Neglecting surface deformations ($\mathrm{Ca} \to 0$) and nonlinear terms (weak thermocapillary flow), convection takes place in a boundary layer of thickness $\mathcal{O}(\mathrm{Ek}^{1/3})$ adjacent to the free surface. The flow has a cellular pattern and decays exponentially in the radial direction away from the free surface. In the interior of the liquid bridge the fluid is in solid-body rotation. The confinement of the thermocapillary-driven flow to a narrow radial layer is a consequence of the Taylor–Proudman theorem (Greenspan (1969)), according to which no axial velocity variations for rapidly rotating ideal fluids may occur. In addition, Smith (1986b) calculated the flow induced by centrifugal buoyancy and gave the first-order corrections to the interface shape.

The two-dimensional stationary thermocapillary convection in a differentially heated liquid bridge with $\mathrm{Pr} = 1$ and $\Gamma = 1$ was investigated numerically by Nataraj (1989) with end-wall rotation in equal and opposite directions. For equal rotation (in direction and absolute value) of both end walls the asymptotic behavior predicted by Smith (1986b) was confirmed. Even for Reynolds numbers as large as $\mathcal{O}(2000)$ the thermocapillary vortex is confined to a layer near the free surface when the rotation rate is increased, and a second weak recirculation cell of opposite sense of rotation is formed. The interior of the fluid remains approximately in solid-body rotation. If both end walls are rotated with equal rate but opposite sense, the rotation-induced secondary flow near the hot boundary is compatible with the sense of rotation of the thermocapillary vortex. This leads to a large-diameter vortex adjacent to the hot wall, extending far towards the cold wall. For the parameters investigated, the other secondary vortex with opposite sign of vorticity, which would appear symmetrically to the former vortex if thermocapillary forces were absent, is pushed into the interior of the liquid bridge.

Lan & Kou (1991b) computed the stationary two-dimensional thermocapillary and natural convection in a symmetrically heated liquid bridge with $Pr = 9$, taking into account surface deformations of both the free surface and the melting and solidifying interfaces. The calculated interface shapes agree well with experimental data. For rapid and opposite end-wall rotation the forced secondary flow suppresses the opposing thermocapillary flow and the interface between liquid and solid becomes flatter. The model has been extended to binary mixtures with a segregation coefficient $k < 1$ by Lan & Kou (1993b) in order to model convective dopant transport. Several cases of equal and opposite sense of rotation of the seed and feed crystals were considered, aiming at a homogeneous concentration distribution along the growing interface. The stationary radial concentration profile in the growing crystal depends strongly on the rotation rates. For equal but opposite rotation the concentration profile was found to be nearly constant. The absolute values of the rotation rates must be increased in proportion to the thermocapillary convection in order to keep the radial concentration profile almost constant.

9.2.2 Vibrations

Vibration-induced flows are of particular interest under weightless conditions, since they may cause undesired perturbations in several materials-processing systems. Although the mean value of the acceleration is zero, high-amplitude vibrational perturbations can give rise to flows with nonvanishing mean. This property, on the other hand, could be utilized to compensate thermocapillary-driven flows.

A homogeneous time-dependent acceleration field of the form $g = g_0 \cos \varpi t$ with $g_0 = $ const. and frequency ϖ can induce fluid motion if the density is not constant. If the vibration frequency ϖ is large compared to the characteristic inverse timescales $\varpi \gg \max(\nu/d^2, \kappa/d^2)$, but still sufficiently small that the fluid can be considered incompressible, $\varpi \ll c/d$ (c is the speed of sound), the hydrodynamic equations can be decomposed by means of a suitable averaging technique into a part varying rapidly on a timescale ϖ^{-1} and a slowly varying part. When the displacement amplitude $b = g_0/\varpi^2$ is small, the induced velocity field will be small and the Boussinesq equations can be linearized with respect to the oscillatory components of u and θ. In this case the response of the velocity field is harmonic. For larger values of b, a mean flow varying on a slow timescale is generated by nonlinear interactions. This mean flow can be due to various mechanisms.

One mechanism of mean-flow generation is due to the temperature-induced density variations in the bulk. The importance of this volumetric effect is measured by the vibrational Rayleigh number

$$\text{Ra}_v = \frac{(\beta \Delta T d\varpi b)^2}{\nu \kappa} . \qquad (9.18)$$

The corresponding averaged equations have been derived by Gershuni & Zhukhovitsky (1979) and a number of different systems have been investigated (Gershuni & Zhukhovitsky (1986)).

Another classical mechanism is the generation of a mean flow by an oscillatory boundary layer of thickness $\mathcal{O}(\nu/\varpi)^{1/2}$, which appears when a rigid wall oscillates tangentially or when the flow above a stationary rigid wall oscillates rapidly (Schlichting (1968)). The induced mean flow, called *acoustic streaming*, is directed parallel to the wall away from the points of highest oscillation amplitude and extends into the bulk of the liquid. Its strength is governed by the vibrational Reynolds number $\text{Re}_v = \varpi b^2 / \nu$.

In the presence of free surfaces, the vibrations typically excite surface capillary waves. When the frequency of vibration is high and the viscosity of one fluid is large, the surface waves are strongly damped (Levich (1962)). The damping of the capillary waves is accompanied by a loss of wave momentum which is balanced by a mean viscous stress driving a mean flow. This third mechanism was studied first by Longuet-Higgins (1953).

Owing to the different driving mechanisms of mean flow, the resulting flow structures can be quite complicated, depending on the governing parameters. Unlike the case of isothermal oscillations of liquid bridges[15] and vibration-induced flows in nonisothermal systems without free surfaces,[16] only a few investigations have been concerned with combined thermocapillary–vibration flows.

Neglecting surface waves, Chen et al. (1995a) numerically investigated the temporal response of the axisymmetric flow in a symmetrically heated thermocapillary liquid bridge with $\text{Pr} = 26$ subject to axial harmonic vibrations. For all cases considered, the response was found to be essentially harmonic in time. Owing to the periodic buoyancy forces, both of the toroidal vortices (total flow) periodically grow and shrink alternately. In an extreme case the large vortex occupies almost the whole volume during part of the cycle. For high vibration amplitudes the response becomes strongly anharmonic.[17]

Gershuni et al. (1992) calculated the time-averaged flow in a symmetrically heated thermocapillary liquid bridge with $\text{Pr} = 0.02$ subject to axial high-frequency vibrations, taking into account the volumetric streaming effect. In the absence of thermocapillary effects the streaming flow consists of two toroidal vortices symmetrical about the temperature maximum at mid-

---

[15] See Sanz (1985), Meseguer & Sanz (1985), Bauer (1992), Mollot et al. (1993), Chen & Tsamopoulos (1993), Lee et al. (1996) and, for annular systems, Ehmann & Siekmann (1995).

[16] See e.g. Farooq & Homsy (1994).

[17] Experimental investigations of thermocapillary liquid bridges under harmonic axial acceleration have also been carried out by Tang et al. (1995); see also Grugel et al. (1994).

plane. The free-surface flow is directed from the cold walls towards the temperature maximum and is thus opposite to the conventional thermocapillary free-surface flow. In the combined thermocapillary–vibrational case the mean flow is composed of a pair of thermocapillary vortices near the free surface and another pair of counterrotating vibration-induced vortices in the bulk. On an increase of the vibration amplitude the vibrational flow dominates and the thermocapillary vortices are completely suppressed. The kinetic energy of the time-averaged flow field has a minimum value at a certain vibrational Rayleigh number (9.18). It was thus demonstrated that axial vibrations can effectively suppress thermocapillary convection.

A similar result has been obtained by Anilkumar et al. (1993). They experimentally investigated the thermocapillary convection in a differentially heated liquid bridge of silicone oil where only the cold wall was vibrated axially. Under isothermal conditions the vibration-induced average velocity was found to be proportional to the acceleration amplitude $b\varpi^2$. In the presence of thermocapillary flow, there exists an acceleration amplitude for which both types of vortex flow nearly compensate each other.

Lyubimov et al. (1997) considered the combined action of all three of the above mechanisms of streaming on the thermocapillary flow in a full zone on the basis of on a generalized Boussinesq approximation for the mean flow. Under weightlessness the fast-oscillating part of the flow field is modeled as a potential flow in their approach, driven only by the damped capillary waves taken into account by suitable free-surface boundary conditions. For a low-Prandtl-number system and $We = 7500$, $Ra_v/Re_v = 46.5$, and $Ma = 400$, the volumetric mean-flow contribution is typically dominant. Since it has a similar structure to the pure thermocapillary flow, an effective suppression of the latter is possible and the total kinetic energy of the mean flow reaches a minimum at a vibrational Rayleigh number which depends on the Marangoni number.

The independent vibration of both end walls of a liquid bridge with frequencies near the resonance frequency for surface oscillations has been investigated theoretically by Nicolás & Vega (1996). They considered the nonlinear flow of a nearly inviscid fluid. Density variations were not taken into account. In contrast to the cases considered by Anilkumar et al. (1993) and Gershuni & Zhukhovitzky (1986), the mean flow is induced by surface oscillations in this case. For a symmetrically heated liquid bridge and small vibration amplitude, the vibrational mean flow augments the thermocapillary flow. For high vibration amplitude, however, the sense of rotation of the vibration-induced vortices changes. They now oppose the thermocapillary flow, leading to a reduction of the flow velocities in the bulk. It should be noted that the velocity extrema on the free surface close to the cold cylinders do not depend on the Reynolds number. Even for high Reynolds numbers, a small but intense vibration-induced vortex remains in each cold corner, with a sense of rotation opposing that of the thermocapillary vortex. The authors conclude that vibrations with a frequency close to the mechanical resonance are not suited to suppress thermocapillary convection. They suggest that vibration frequencies far away from resonance, together with a properly chosen phase difference between the two oscillating cylinders, are more appropriate for a compensation of the thermocapillary flow.

The stability of two-dimensional flows in which the thermocapillary motion is partly compensated by vibrational mean flows has not yet been investigated.

9.2.3 Magnetic Fields

Silicon, an important material for industrial applications, has a relatively high electrical conductivity. This property is utilized in the electromagnetic heating of melt zones. In this technique an induction coil generates an electromagnetic field in the megahertz range. The field induces electrical currents in a thin skin layer under the free surface of the zone, thereby heating the melt by Joule heating (Bohm et al. (1994)).

Electromagnetic fields can also be used to influence the fluid flow in liquid metals. Magnetic fields are particularly well suited for this purpose (Moreau (1990)). The effect of a magnetic field on a conducting, moving fluid is based on the following mechanism. The component of the velocity field perpendicular to the magnetic field induces an electrical current perpendicular to both fields. The induced current, in turn, together with the same magnetic field leads to a Lorentz force which is opposite to the direction of the velocity field that generates the current. Therefore, all elements of the liquid with velocity components perpendicular to the direction of the magnetic field experience a retarding force. This mechanism is employed in crystal growth in order to prevent the onset of oscillatory convection (see e.g. Cröll et al. (1994)).

For a calculation of magnetohydrodynamic flows Maxwell's equations must be considered, in addition to the Navier–Stokes equations:

$$\nabla \cdot \boldsymbol{D} = \rho \,, \tag{9.19}$$

$$\nabla \cdot \boldsymbol{B} = 0 \,, \tag{9.20}$$

$$\nabla \times \boldsymbol{E} = -\partial_t \boldsymbol{B} \,, \tag{9.21}$$

$$\nabla \times \boldsymbol{H} = \boldsymbol{j} + \partial_t \boldsymbol{D} \,. \tag{9.22}$$

The importance of magnetic fields for flow control in crystal growth has been discussed by Hurle & Series (1994) together with a mathematical model for semiconductor melts (see also Roberts (1967) and Landau & Lifshitz (1985)). Different approximations to the full magnetohydrodynamic equations have been tested and employed numerically by Baumgartl (1992) for calculation of the convection in rigid cylinders.

Since the convective velocities and the corresponding rates of change of the electromagnetic fields are small compared to the speed of light, electromagnetic waves can be neglected. In this case the displacement-current density vanishes $(\partial_t \boldsymbol{D} = 0)$. The electrical current density $\boldsymbol{j}$ is given by Ohm's law

$$\boldsymbol{j} = \sigma \left(\boldsymbol{E} + \boldsymbol{u} \times \boldsymbol{B} \right) , \tag{9.23}$$

where σ is the electrical conductivity, $\boldsymbol{E}$ denotes the electric field and $\boldsymbol{B}$ the magnetic induction. To obtain the dynamic equation for the magnetic field, $\boldsymbol{E}$ is eliminated between (9.21) and (9.23) using (9.22):

$$\partial_t \boldsymbol{B} - \nabla \times (\boldsymbol{u} \times \boldsymbol{B}) + \frac{1}{\sigma \mu} \nabla \times (\nabla \times \boldsymbol{B}) = 0 . \tag{9.24}$$

From Maxwell's equations we obtain for the Lorentz force density $\boldsymbol{L}$

$$\boldsymbol{L} = \boldsymbol{j} \times \boldsymbol{B} = \frac{1}{\mu} \left(\nabla \times \boldsymbol{B} \right) \times \boldsymbol{B} , \tag{9.25}$$

where μ is the magnetic susceptibility. The Lorentz force density enters the momentum balance. Owing to the high electrical conductivity, the forces exerted on free charge carriers can be neglected. This leads to the (unscaled) model equations

$$\partial_t \boldsymbol{u} + \boldsymbol{u} \cdot \nabla \boldsymbol{u} = -\frac{1}{\rho} \nabla p + \nu \Delta \boldsymbol{u} + g \beta T \boldsymbol{e}_z + \frac{1}{\rho \mu} \left(\nabla \times \boldsymbol{B} \right) \times \boldsymbol{B} , \tag{9.26}$$

$$\nabla \cdot \boldsymbol{u} = 0 , \tag{9.27}$$

$$\partial_t T + \boldsymbol{u} \cdot \nabla T = \kappa \Delta T , \tag{9.28}$$

$$\partial_t \boldsymbol{B} - \nabla \times (\boldsymbol{u} \times \boldsymbol{B}) = -\frac{1}{\sigma \mu} \nabla \times (\nabla \times \boldsymbol{B}) , \tag{9.29}$$

$$\nabla \cdot \boldsymbol{B} = 0 . \tag{9.30}$$

In addition to the usual boundary conditions (see Sect. 2.2) the electrical current density perpendicular to dielectric boundaries must vanish, i.e. $\boldsymbol{n} \cdot \boldsymbol{j} = 0$.

If $\boldsymbol{u}$ and t are scaled with U and d/U, respectively, it is natural to define the magnetic Reynolds number

$$\mathrm{Re_m} = \sigma \mu U d . \tag{9.31}$$

Since the magnetic Reynolds numbers are small in most applications (Hurle (1994)), the magnetic field in (9.29) is essentially diffusion-dominated and it can be expanded for small deviations caused by convection. For a magnetic field parallel to the z direction, $\boldsymbol{B}$ is constant to lowest approximation and given by $B_0 \boldsymbol{e}_z$. Then $\nabla \times \boldsymbol{E} = 0$ and the electric field $\boldsymbol{E} = -\nabla \Phi$ can be expressed through a potential. Hence, the model simplifies to

$$\partial_t \boldsymbol{u} + \boldsymbol{u} \cdot \nabla \boldsymbol{u} = -\nabla p + \Delta \boldsymbol{u} + \mathrm{Gr}\, \theta \boldsymbol{e}_z + \mathrm{Ha}^2 \left(\boldsymbol{u} \times \boldsymbol{e}_z - \nabla \Phi \right) \times \boldsymbol{e}_z , \tag{9.32}$$

$$\nabla \cdot \boldsymbol{u} = 0 , \tag{9.33}$$

$$\partial_t \theta + \boldsymbol{u} \cdot \nabla \theta = \frac{1}{\mathrm{Pr}} \Delta \theta , \tag{9.34}$$

$$\Delta \Phi = \boldsymbol{e}_z \cdot \nabla \times \boldsymbol{u} , \tag{9.35}$$

where the viscous scales, as for (9.8–9.10), have been used. The square of the Hartmann number

$$\mathrm{Ha} = B_0 d \left(\frac{\sigma}{\rho\nu}\right)^{\frac{1}{2}} \qquad (9.36)$$

is a measure of the ratio of magnetic to viscous forces.

For strong magnetic fields, Hartmann boundary layers appear. To derive the boundary-layer thickness it is useful to consider the two-dimensional flow in a plane spanned by x and z with $\boldsymbol{B} = B_0\boldsymbol{e}_z$. From (9.35) with electrically insulating boundary conditions, we get $\nabla\Phi = 0$ and the momentum equation can be written as

$$(\partial_z\psi\partial_x - \partial_x\psi\partial_z)\,\Delta\psi = \Delta^2\psi - \mathrm{Gr}\,\partial_x\theta - \mathrm{Ha}^2\partial_z^2\psi, \qquad (9.37)$$

where $u = \partial_z\psi$ and $w = -\partial_x\psi$. The leading-order balance near the boundaries is given by

$$\Delta^2\psi \sim \mathrm{Ha}^2\partial_z^2\psi. \qquad (9.38)$$

With the boundary-layer thickness for a horizontal wall ($z = $ const.) denoted by $\delta_\perp$, this yields $\delta_\perp^{-4} \sim \mathrm{Ha}^2\delta_\perp^{-2}$, giving the scaling $\delta_\perp \sim \mathrm{Ha}^{-1}$. For the layer thickness $\delta_\parallel$ on a vertical wall with $x = $ const., one gets from (9.38) $\delta_\parallel^{-4} \sim \mathrm{Ha}^2$ and thus $\delta_\parallel \sim \mathrm{Ha}^{-1/2}$.

The model (9.32–9.35) has been employed by Priede et al. (1994) to investigate the onset of hydrothermal waves in plane thermocapillary layers without lateral through flow, under weightless conditions ($\mathrm{Gr} = 0$), for the case when the magnetic field is aligned perpendicular to the layer. By restricting the considerations to hydrothermal waves that propagate perpendicular to the applied temperature gradient, i.e. in the y direction,[18] the linear stability problem could be solved analytically. By order-of-magnitude considerations and numerical calculations it was shown that the critical Reynolds number depends quadratically on the Hartmann number, while the critical wavenumber scales linearly with Ha. This behavior is crucially determined by the basic flow, which is essentially confined to thin Hartmann boundary layers of thickness Ha^{-1}.[19]

The convective stability of plane thermocapillary layers in a coplanar magnetic field has been considered by Priede & Gerbeth (1997c) (see also Priede & Gerbeth (1995)). In this case the basic flow does not depend on Ha if end effects are neglected, and it is given by (6.3) and (6.4). It can be shown from the perturbation equations that the effect of the magnetic

---

[18] Coordinates as in Chap. 6.

[19] The effect of a constant magnetic field perpendicular to the upper surface of a cavity with $\Gamma = 4$ on the two-dimensional flow has been investigated by Ben Hadid et al. (1997) for different bounday conditions, with emphasis on buoyant convection. Corresponding three-dimensional simulations for $\Gamma = 4$ and $\Gamma_z = 1$ can be found in Ben Hadid & Henry (1997).

field on the stability of the flow is then $\sim e_k \cdot e_B$, where e_k and e_B denote the unit vectors in the directions of k and B, respectively. Therefore, the magnetic field has no effect on that neutral mode for which $k \perp B$. In particular, its neutral Reynolds number remains unchanged. Modes with wave-vector components parallel to the magnetic field are damped and the corresponding neutral Reynolds numbers increase with B. Hence, stabilization is only possible to a maximum in general, and the wave vector of the critical mode will align with $k \perp B$. If, however, no neutral mode with $k \perp B$ exists at finite Reynolds numbers, a stabilization of the base flow without bounds is possible by increasing the magnetic field strength. This occurs for small-Prandtl-number liquid metallic layers, for which it is known that the transverse mode ($k \parallel e_x$) has the highest neutral Reynolds number. This even diverges as Pr $\downarrow 0.018$. Thus, an unlimited stabilization with respect to hydrothermal waves is possible with a transverse magnetic field ($B \parallel e_y$) for sufficiently low-Prandtl-number fluids.

The opportunities for the suppression of the oscillatory thermocapillary convection in melt zones have been explored by several authors. In experiments with silicon Cröll et al. (1994) found that strong axial magnetic fields (up to 0.5 T) completely eliminate flow oscillations from the bulk of melt zones[20] (see Fig. 1.3). A number of numerical studies (Baumgartl et al. (1990), Baumgartl & Müller (1992), Herrmann et al. (1992)), some of which compared their results with experimental data, arrived at the same conclusion. Neglecting inertia terms and dynamic surface deformations, Morthland & Walker (1996) calculated the two-dimensional thermocapillary flow in melt zones for several configurations in an axisymmetric magnetic field. When the free surface does not deviate too much from cylindrical, the thermocapillary convection is expelled from the bulk and becomes confined to a layer of thickness $\mathcal{O}(\text{Ha}^{-1/2})$ near the free surface. These results explain the experimental finding of Cröll et al. (1994), who did not observe any micro-segregation (caused by oscillatory convection) in the bulk of the grown crystal, although a layer close to the edge (free surface) exhibited strong dopant striations (due to time-dependent thermocapillary flow).[21]

The first stability analyses of the two-dimensional thermocapillary flow in asymmetrically heated liquid bridges under the influence of a homogeneous magnetic field in the z direction were carried out by Prange (1997) and Prange et al. (1998). They employed the model equations (9.32–9.35) and solved the stability problem using the numerical method described in Sect. 7.2, extended to accomodate the additional magnetic field. In qualitative agreement with Morthland & Walker (1996), the center of the two-dimensional toroidal basic-state vortex is shifted to the free surface with

---

[20] See also Cröll et al. (1998) and Dold et al. (1998).

[21] Morthland & Walker (1997) have shown that in a strong axial magnetic field a nonaxisymmetric ambient temperature distribution can drive an azimuthal thermocapillary surface flow, leading to a significant return flow in the bulk. This may lead to a radial segregation despite the presence of the magnetic field.

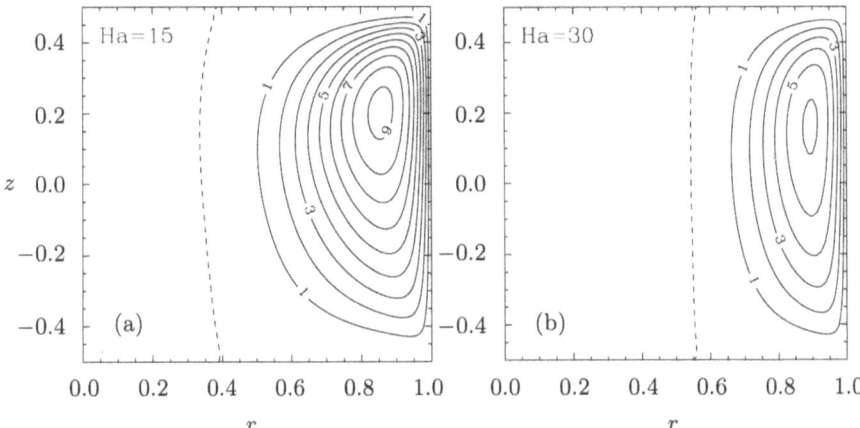

Fig. 9.7. Radial compression and suppression of the basic-state vortex by an axial magnetic field in a cylindrical liquid bridge with $\Gamma=1$, $Pr=0.02$, $Gr=Bi=0$, and $Re=2061$, the critical Reynolds number for $Ha=0$. The stream function is shown for $Ha=15$ (**a**) and $Ha=30$ (**b**) (cf. Fig. 7.1a for $Ha=0$). The heating is from the lower side. The *dashed line* indicates $\psi=0$. After Prange et al. (1998)

increasing Hartmann number (Fig. 9.7). The increase of the stationary stability boundary of the basic state is shown in Fig. 9.8 for $\Gamma=1$, $Pr=0.02$, and $Gr=Bi=0$. In the range shown $m=2$ is the critical wavenumber. The critical Reynolds number increases strongly with the Hartmann number and readily

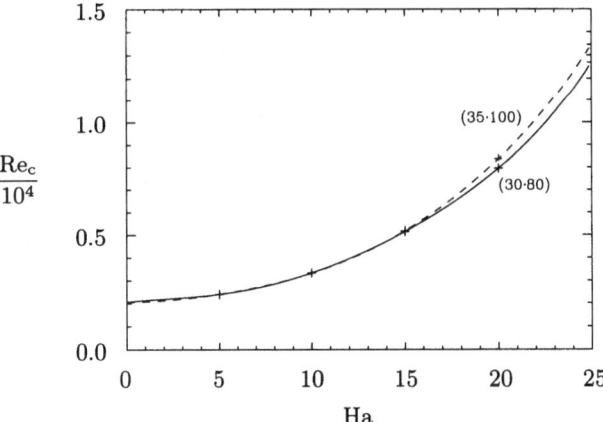

Fig. 9.8. Critial Reynolds number for the onset of stationary three-dimensional convection in a differentially heated cylindrical liquid bridge as a function of the Hartmann number in a homogeneous axial magnetic field. The parameters are $\Gamma=1$, $m=2$, $Pr=0.02$, and $Gr=Bi=0$. Data points (+) are given for different grids $N\times M$, where N and M are the number of radial Chebyshev-collocation and axial finite-difference points, respectively. After Prange et al. (1998)

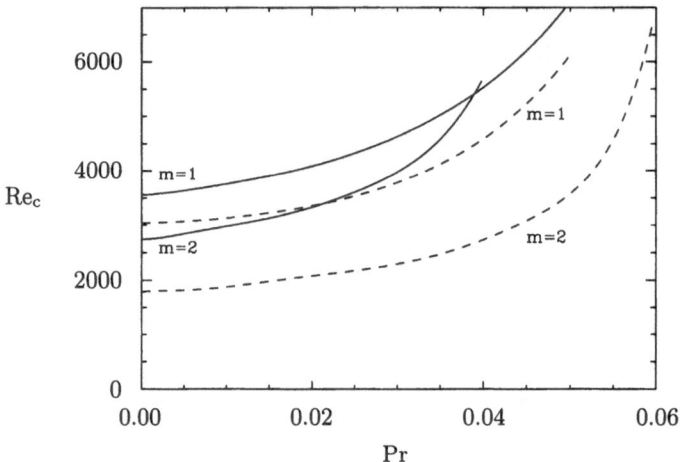

Fig. 9.9. Neutral Reynolds numbers for the onset of three-dimensional convection as functions of the Prandtl number in an asymmetrically heated cylindrical liquid bridge subject to a homogeneous axial magnetic field, according to Prange (1997). (- - -) Ha$=0$, (———) Ha$=10$. The parameters are $\Gamma=1$ and Gr$=$Bi$=0$

reaches Re$_c=2.2\times10^4$ for Ha$=30$. The influence of the Prandtl and Hartmann numbers on the the stability boundary and on the critical wavenumber is shown in Fig. 9.9 for $\Gamma=1$ and Gr$=$Bi$=0$. As for Ha$=0$ (compare Fig. 7.20), the critical Reynolds number also grows rapidly with Prandtl number when an axial magnetic field is present. For Ha$\neq0$ the critical mode may change in the range considered. The crossover Prandtl numbers for which these changes occur seem to depend strongly on the Hartmann number. For Ha$=10$, for example, the critical wavenumber changes from $m=2$ to $m=1$ near Pr≈0.04.

These examples demonstrate the sensitive dependence on the external magnetic field of the critical Reynolds number Re$_c$ and the associated wavenumber m_c for the onset of stationary three-dimensional convection. For an investigation of the onset of time-dependent flows and for a quantitative comparison with the experiments of Cröll et al. (1994), even higher Reynolds numbers must be considered.

A. Appendix

A.1 Some Material Parameters

The following tables summarize some typical material parameters. Most data are taken from Velten et al. (1991) and Shackelford & Alexander (1992).

Table A.1. Material parameters of some liquid metals and semiconductors near their melting point T_m

		Ga	Hg	Sn	Si	Ge
T_m	[°C]	29.78	−38.87	231.97	1410	937
ρ	[g/cm³]	5.91	13.6	7.0	2.33	5.6
$\nu \times 10^2$	[cm²/s]	0.335	0.11	0.264	0.397	0.11
κ	[cm²/s]	0.105	0.044	0.171	0.128	0.18
σ_0	[dyn/cm]	718	450	550	865	600
γ	[dyn/cm°C]	0.1	0.17	0.1	0.2	0.6
$\beta \times 10^4$	[°C⁻¹]	1.01	1.8	0.88	2.0	0.94
Pr		0.02	0.025	0.017	0.03	0.006

Table A.2. Material parameters of several transparent fluids. The values refer to the working temperature T_0

		Silicone oil	KCl	NaNO₃	C₂₄H₅₀	C₂H₅OH
T_m	[°C]		772	307	51	
ρ	[g/cm³]	0.758	1.52	1.88	0.753	0.789
$\nu \times 10^2$	[cm²/s]	0.65	0.74	1.21	3.24	1.52
$\kappa \times 10^4$	[cm²/s]	7.7	72.5	17.3	6.59	8.9
σ_0	[dyn/cm]	15.9	975	117	26	27.5
γ	[dyn/cm°C]	0.08	0.073	0.055	0.067	0.09
$\beta \times 10^4$	[°C⁻¹]	13.4	3.85	3.74	8.76	10.6
T_0	[°C]	25	790	360	90	20
Pr		8.4	1.02	7.0	49	17

A.2 Stokes Flow Near the Contact Line

Here the solution of (3.1) with boundary conditions (3.4) and (3.5) by means of the Mellin transformation is outlined. More details can be found in Kuhlmann et al. (1998b). The method is based on the work of Tranter (1948) and follows closely that of Moffatt & Duffy (1980). The Mellin transform of ψ is given by

$$\bar{\psi}(p,\theta) \;=\; \int_0^\infty r^{p-1}\psi(r,\theta)\,\mathrm{d}r\,, \tag{A.1}$$

with the back-transformation

$$\psi(r,\theta) \;=\; \frac{1}{2\pi\mathrm{i}} \int_{c-\mathrm{i}\infty}^{c+\mathrm{i}\infty} r^{-p}\bar{\psi}(p,\theta)\,\mathrm{d}p\,, \tag{A.2}$$

and where $p \in \mathbb{C}$ and $c \in \mathbb{R}$. As we shall see later on, we will have to use $\Re(p) = -1$. Therefore, the transform $\bar{\psi}$ exists if ψ satisfies the asymptotic relations

$$\psi(r,\theta) \;=\; \begin{cases} \mathcal{O}(r^{1+\epsilon}), & r \to 0, \\ \mathcal{O}(r^{1-\epsilon}), & r \to \infty, \end{cases} \tag{A.3}$$

which must be checked a posteriori.

Multiplying (3.1) by r^{p+3} and integrating from 0 to ∞, we obtain the transformed biharmonic equation

$$\left\{ \partial_\theta^4 + \left[(p+2)^2 + p^2\right]\partial_\theta^2 + p^2(p+2)^2 \right\}\bar{\psi} \;=\; 0\,. \tag{A.4}$$

Similarly, we obtain the transformed boundary conditions

$$\theta = 0: \qquad \bar{\psi} = 0, \qquad \partial_\theta^2\bar{\psi} = \omega_0 \int_0^a r^{p-1}r^2\,\mathrm{d}r = \omega_0\frac{a^{p+2}}{p+2}\,,$$

$$\theta = \alpha: \qquad \bar{\psi} = 0, \qquad \partial_\theta\bar{\psi} = 0\,. \tag{A.5}$$

The unique solution of (A.4) is

$$\bar{\psi}(p,\theta) \;=\; \frac{\omega_0 a^{p+2}}{4}\,\frac{F(\theta,p)}{W(p)(p+1)(p+2)}\,, \tag{A.6}$$

where

$$\begin{aligned} F(\theta,p) &= (p+1)\sin[(p+2)\theta - 2\alpha] - \sin[(p+2)\theta - 2(p+1)\alpha] \\ &\quad -p\sin[(p+2)\theta] - (p+2)\sin(p\theta) + (p+1)\sin(p\theta + 2\alpha) \\ &\quad + \sin[p\theta - 2(p+1)\alpha] \end{aligned} \tag{A.7}$$

and

$$W(p) \;=\; (p+1)\sin(2\alpha) - \sin[2(p+1)\alpha]\,. \tag{A.8}$$

Since we shall have to integrate $\bar{\psi}$ along $\Re(p) = -1$, it is important to note that $\bar{\psi}$ has a removable singularity at $p = -1$. The functions F and W satisfy the symmetry conditions

$$F(\theta, -p) = F(\theta, p-2) \,, \tag{A.9}$$

$$W(-p) = -W(p-2) \,. \tag{A.10}$$

From these properties we obtain

$$\bar{\psi}(-p, \theta) = -\frac{p}{p-2} a^{-2(p-1)} \bar{\psi}(p-2, \theta) \,. \tag{A.11}$$

Using (A.11) and selecting $c = -1$, one can show that

$$\partial_r \left(\frac{1}{r^2} \psi(r, \theta) \right) = -\frac{1}{r^2} \partial_r \left[\left(\frac{r}{a} \right)^2 \psi \left(\frac{a^2}{r}, \theta \right) \right] \,. \tag{A.12}$$

This formula enables us to establish the behavior for small $r' = a^2/r \to 0$, once the asymptotic behavior for $r \to \infty$ is known.

The solution (A.2) with $c = -1$ reads

$$\psi(r, \theta) = \frac{1}{2\pi i} \int_{-1-i\infty}^{-1+i\infty} r^{-p} \bar{\psi}(p, \theta) \, dp$$

$$= \frac{\omega_0 a^2}{4} \frac{1}{2\pi i} \int_{-1-i\infty}^{-1+i\infty} \left(\frac{a}{r} \right)^p \frac{F(\theta, p)}{W(p)(p+1)(p+2)} \, dp \,. \tag{A.13}$$

For $r > a$ the integrand vanishes at infinity and the integral can be closed over the right-half plane. The residue theorem yields

$$\psi(r, \theta) = -\frac{\omega_0 a^2}{4} \sum_{\substack{n \\ \Re(p_n) > -1}} \operatorname{Res} \left\{ \left(\frac{a}{r} \right)^{p_n} \frac{F(\theta, p_n)}{W(p_n)(p_n + 1)(p_n + 2)} \right\} \,, \tag{A.14}$$

where p_n denotes the poles of the integrand in (A.13). Since we can use the residue theorem only for $r > a$, the symmetry relation (A.12) is necessary to obtain the behavior for $r < a$.

Up to the values of p_n for which $F(\theta, p)$ is zero, the poles of the integrand are determined by the zeros of $W(p)$ for $\Re(p) > -1$. The zeros have been analyzed by Moffatt & Duffy (1980), who showed that the zeros are simple except for special intersection points. For simple poles we get

$$\psi(r, \theta) = -\frac{\omega_0 a^2}{4} \sum_{\substack{n \\ \Re(p_n) > -1}} \left(\frac{a}{r} \right)^{p_n} \frac{F(\theta, p_n)}{W'(p_n)(p_n + 1)(p_n + 2)} \,, \tag{A.15}$$

where

$$W'(p_n) = \sin 2\alpha - 2\alpha \cos[2(p_n + 1)\alpha] \neq 0 \,. \tag{A.16}$$

The leading-order term for $r \to \infty$ is due to the residue of the pole that has the smallest real part of p. For $\alpha < \alpha_c$, the corresponding pole of $W(p)$ is $p_0 = 0$ (Moffatt & Duffy (1980)). It is real and simple for $\alpha \neq \alpha_c$. For the present problem, $F(\theta, p_0) = 0$ and $F'(\theta, p_0) \neq 0$ with

$$\partial_p F(\theta, p \to 0) = \sin(2\theta - 2\alpha) + 2\alpha \cos(2\theta - 2\alpha) - \sin(2\theta) - 2\theta$$
$$+ \sin(2\alpha) + \theta \cos(2\alpha) + (\theta - 2\alpha) \cos(2\alpha) , \qquad \text{(A.17)}$$

and the integrand $\sim F(\Theta, p)/W(p)$ has no pole at $p_0 = 0$, in contrast to the problem considered by Moffatt & Duffy (1980). Thus for $\alpha < \alpha_c$ all poles have $\Re(p_n) > 0$.

We are now in a position to calculate the solution for small distances from the vertex. Integrating the symmetry relation (A.12), we get, with $r' = a^2/r < a$ and dropping the prime,

$$\psi(r, \theta) = \omega_0 r^2 \left(g(\theta) + \frac{1}{4} \sum_n \left(\frac{r}{a}\right)^{p_n} \frac{F(\theta, p_n)}{W'(p_n)(p_n + 1)p_n} \right). \qquad \text{(A.18)}$$

Since $\Re(p_n) > 1$ for $\alpha < \alpha_c$, the leading-order term for $r \to 0$ is produced by $g(\theta)$. The function $g(\theta)$ satisfying the boundary conditions on the walls for $r \to 0$ is just the similarity solution (3.7), i.e. $g(\theta) = f(\theta)$.

In the case $\alpha > \alpha_c$ the pole with the smallest real part, p_1, is again simple and real and satisfies (Moffatt & Duffy (1980))

$$0 > p_1 \geq -\frac{1}{2}, \qquad \text{for} \quad \alpha_c < \alpha \leq \pi . \qquad \text{(A.19)}$$

Thus the leading-order term in (A.18) is due to the pole p_1, yielding the asymptotic form of the stream function

$$\psi(r, \theta) \sim \frac{\omega_0}{4} r^2 \left(\frac{r}{a}\right)^{p_1} \frac{F(\theta, p_1)}{W'(p_1)(p_1 + 1)p_1}, \qquad r \to 0. \qquad \text{(A.20)}$$

We now consider the critical angle $\alpha = \alpha_c$. Here the pole is simple and the corresponding residue with $p_1 = 0$ is given by

$$\text{Res} \left\{ \left(\frac{a}{r}\right)^{p_1} \frac{F(\theta, p_1)}{W(p_1)(p_1 + 1)(p_1 + 2)} \right\} = \frac{\partial_p F(\theta, 0)}{W''(0)} . \qquad \text{(A.21)}$$

Using (A.21) and the symmetry relation (A.12), the stream function for small $r \to 0$ takes the asymptotic form

$$\psi(r, \theta) \sim \frac{\omega_0}{2} \frac{\partial_p F(\theta, 0)}{W''(0)} r^2 \ln \frac{r}{a}, \qquad r \to 0 . \qquad \text{(A.22)}$$

As an example, the angular dependence of the supercritical solution for $\alpha = 7\pi/8 > \alpha_c$ ($p_1(7\pi/8) = -0.3403$) is shown in Fig. A.1. It is very similar to that of the subcritical solution for $\alpha = \pi/2 < \alpha_c$.

A.3 Biorthogonal Functions for Oseen-Type Flows

We consider the Oseen-type equation

$$\tilde{L} \left(\tilde{L} - 4\beta \partial_z \right) \psi = 0 \qquad \text{(A.23)}$$

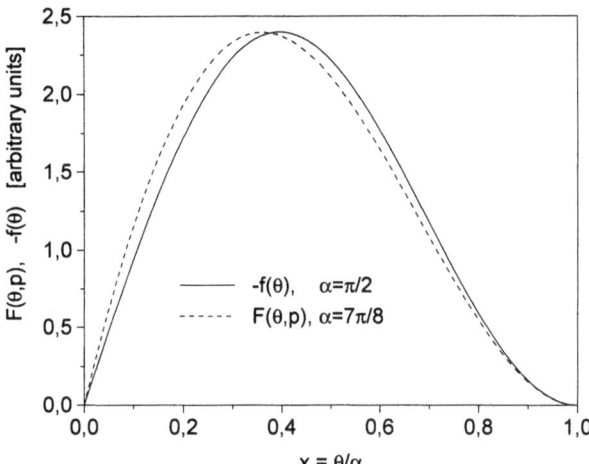

Fig. A.1. Angular dependence of the asymptotic solution for the stream function (normalized to the same maximum value). (——) $\alpha = \pi/2$; (- - -) $\alpha = 7\pi/8$

for $0 \le r \le 1/\Gamma$ and $1/2 \le z \le 1/2$. Here

$$\tilde{L} = DD_* + \partial_z^2 \tag{A.24}$$

is a linear operator, with $D_* = D + 1/r = \partial_r + 1/r$ and $\beta \in \mathbb{R}$ is a real constant. The function ψ must satisfy the boundary conditions

$$\psi = \partial_z \psi = 0 \quad \text{on} \quad z = \pm \frac{1}{2}. \tag{A.25}$$

The values of β and Γ result from the specific scaling of the physical problem (see Sect. 4.3). The separable solution of (A.23) which is regular at $r=0$ has the form

$$\psi = I_1(2\lambda r)\phi_1(z, \lambda), \tag{A.26}$$

where I_1 is the modified Bessel function of first order and ϕ_1 is

$$\phi_1(z, \lambda) = A_1 \cos(2\lambda z) + A_2 \sin(2\lambda z)$$
$$+ \frac{A_3}{2\beta} \left[e^{2\beta z} \cos(2\lambda_1 z) - \cos(2\lambda z) \right]$$
$$+ \frac{A_4}{2\beta} \left[e^{2\beta z} \sin(2\lambda_1 z) - \sin(2\lambda z) \right] . \tag{A.27}$$

Here, the abbreviation $\lambda_1^2 = \lambda^2 - \beta^2$ has been introduced. The solution of a similar problem for which $\tilde{L}$ is defined as $\tilde{L} = \nabla^2 = D_* D + \partial_z^2$ can be obtained simply by multiplication of the solution (A.26) with r. Solution (A.27) is written in a form that directly allows to take the limit $\beta \to 0$, in which the solutions given by Joseph & Sturges (1975) are recovered (see Chap. 4). Inserting the boundary conditions (A.25), one obtains the linear system for the coefficients A_1, A_2, A_3, and A_4

$$F_{ij} A_j = 0,\qquad\qquad\qquad\qquad\qquad\text{(A.28)}$$

with

$$
F_{ij} =
\begin{bmatrix}
c & s & \frac{1}{2\beta}\left(c_1 e^{\beta} - c\right) & \frac{1}{\beta}\left(s_1 e^{\beta} - s\right) \\[4pt]
c & -s & \frac{1}{2\beta}\left(c_1 e^{-\beta} - c\right) & \frac{1}{\beta}\left(-s_1 e^{-\beta} + s\right) \\[4pt]
-2\lambda s & 2\lambda c & \frac{\left(\beta c_1 e^{\beta} - \lambda_1 s_1 e^{\beta} + \lambda s\right)}{\beta} & \frac{\left(\beta s_1 e^{\beta} + \lambda_1 c_1 e^{\beta} - \lambda c\right)}{\beta} \\[4pt]
2\lambda s & 2\lambda c & \frac{\left(\beta c_1 e^{-\beta} + \lambda_1 s_1 e^{-\beta} - \lambda s\right)}{\beta} & \frac{\left(-\beta s_1 e^{-\beta} + \lambda_1 c_1 e^{-\beta} - \lambda c\right)}{\beta}
\end{bmatrix},
$$

$$\text{(A.29)}$$

where the abbreviations $s = \sin\lambda$, $s_1 = \sin\lambda_1$, $c = \cos\lambda$, and $c_1 = \cos\lambda_1$ have been used. The roots of the solvability condition

$$\det(F_{ij}) = 0 \qquad\qquad\qquad\qquad\qquad\text{(A.30)}$$

yield an infinite, countable set of eigenvalues λ_n, $n \in \mathbb{N}$. They lie symmetrically in the complex λ plane and must be calculated numerically. Here we use the eigenvalues with positive real and imaginary parts, i.e. the roots located in the first quadrant, ordered according to the magnitude of their real parts. The eigenvalues with negative imaginary parts and the associated eigenfunctions are defined as

$$\lambda_{-n} = \lambda_n^* \qquad \text{and} \qquad \phi_1^{(-n)} = \phi_1^{(n)*} = \phi_1(z, \lambda_{-n}),\qquad\text{(A.31)}$$

where the asterisk $(*)$ denotes the complex conjugate. The first ten characteristic roots for $\beta = 0.0$, 4.0, and 8.0 are listed in Table A.3. Up to a common factor the coefficients $A_i^{(n)}$, $i \in [1, 2, 3, 4]$, in (A.27) read

Table A.3. The first ten characteristic roots λ_n for $\beta = 0.0$, 4.0, and 8.0

	$\beta = 0.0$		$\beta = 4.0$		$\beta = 8.0$	
n	$\Re(\lambda_n)$	$\Im(\lambda_n)$	$\Re(\lambda_n)$	$\Im(\lambda_n)$	$\Re(\lambda_n)$	$\Im(\lambda_n)$
1	2.10620	1.12536	2.68699	1.15032	2.98376	0.88587
2	3.74884	1.38434	4.35903	1.85247	5.22515	1.88441
3	5.35627	1.55157	5.86930	2.21949	6.90984	2.69318
4	6.94998	1.67610	7.37972	2.44316	8.40349	3.24204
5	8.53668	1.77554	8.90303	2.59853	9.85415	3.61629
6	10.1193	1.85838	10.4372	2.71596	11.3066	3.88104
7	11.6992	1.92940	11.9795	2.80987	12.7722	4.07767
8	13.2773	1.99157	13.5276	2.88796	14.2522	4.22972
9	14.8541	2.04685	15.0800	2.95479	15.7451	4.35134
10	16.4299	2.09663	16.6356	3.01319	17.2491	4.45138

$$A_1^{(n)}(\lambda_n, \beta) = \frac{1}{4\sin^2\lambda_n} \begin{vmatrix} F_{12} & F_{13} & F_{14} \\ F_{32} & F_{33} & F_{34} \\ F_{42} & F_{43} & F_{44} \end{vmatrix}, \tag{A.32}$$

$$A_2^{(n)}(\lambda_n, \beta) = \frac{1}{4\sin^2\lambda_n} \begin{vmatrix} -F_{11} & F_{13} & F_{14} \\ -F_{31} & F_{33} & F_{34} \\ -F_{41} & F_{43} & F_{44} \end{vmatrix}, \tag{A.33}$$

$$A_3^{(n)}(\lambda_n, \beta) = \frac{1}{4\sin^2\lambda_n} \begin{vmatrix} F_{12} & -F_{11} & F_{14} \\ F_{32} & -F_{31} & F_{34} \\ F_{42} & -F_{41} & F_{44} \end{vmatrix}, \tag{A.34}$$

$$A_4^{(n)}(\lambda_n, \beta) = \frac{1}{4\sin^2\lambda_n} \begin{vmatrix} F_{12} & F_{13} & -F_{11} \\ F_{32} & F_{33} & -F_{31} \\ F_{42} & F_{43} & -F_{41} \end{vmatrix}. \tag{A.35}$$

The $A_i^{(n)}$ have been selected such that, in the limit $\beta \to 0$, the corresponding eigenfunctions[1] are obtained, for which $2\lambda_n \pm \sin(2\lambda_n) = 0$. For $\beta = 0$, (A.23) is symmetric with respect to the transformation $z \to -z$ and the functions are either even or odd in z. This symmetry is removed when $\beta \neq 0$ and the eigenfunctions $\phi_1^{(n)}$ no longer separate into even and odd functions. However, (A.30) is invariant under $\beta \to -\beta$. Hence the eigenvalues λ_n do not depend on the sign of β, unlike from the amplitudes $A_i^{(n)}$. From (A.23) and (A.27) we obtain the differential equation for $\phi_1^{(n)}$,

$$\left(\frac{d^2}{dz^2} + 4\lambda_n^2 \right) \left(\frac{d^2}{dz^2} - 4\beta \frac{d}{dz} + 4\lambda_n^2 \right) \phi_1^{(n)} = 0. \tag{A.36}$$

Using the definition of $\phi_2^{(n)}$

$$\phi_2^{(n)}(z) = \frac{1}{4\lambda_n^2} \frac{d^2\phi_1^{(n)}}{dz^2}, \tag{A.37}$$

(A.36) can be written as a second-order equation

$$\frac{d^2\phi_2^{(n)}}{dz^2} - 4\beta \frac{d}{dz} \left(\phi_1^{(n)} + \phi_2^{(n)} \right) + 4\lambda_n^2 \left(2\phi_2^{(n)} + \phi_1^{(n)} \right) = 0. \tag{A.38}$$

The last two equations, (A.37) and (A.38), read in vector form

$$L\boldsymbol{\Phi}^{(n)} = 0, \tag{A.39}$$

with

---

[1] The even eigenfunctions are given in (4.27).

$$\boldsymbol{\Phi}^{(n)} = \begin{pmatrix} \phi_1^{(n)} \\ \phi_2^{(n)} \end{pmatrix} \tag{A.40}$$

and

$$L = \frac{d^2}{dz^2} - 4\beta \frac{d}{dz} \begin{pmatrix} 0 & 0 \\ 1 & 1 \end{pmatrix} + 4\lambda_n^2 \begin{pmatrix} 0 & -1 \\ 1 & 2 \end{pmatrix} . \tag{A.41}$$

To obtain the eigenvalue problem adjoint to (A.39) we define the generalized Wronskian

$$W = \boldsymbol{\Psi} \cdot \partial_z \boldsymbol{\Phi} - \boldsymbol{\Phi} \cdot \partial_z \boldsymbol{\Psi} = \psi_1 \frac{\partial \phi_1}{\partial z} + \psi_2 \frac{\partial \phi_2}{\partial z} - \frac{\partial \psi_1}{\partial z} \phi_1 - \frac{\partial \psi_2}{\partial z} \phi_2 . \tag{A.42}$$

The adjoint eigenfunction $\boldsymbol{\Psi} = (\psi_1, \psi_2)$ is the function which, for given λ and $\boldsymbol{\Phi}$, makes the Wronski determinant vanish identically ($W \equiv 0$). It can be seen that $W(z = \pm \frac{1}{2})$ vanishes at the boundaries if

$$\psi_2 = \partial_z \psi_2 = 0 \qquad \text{on} \quad z = \pm \frac{1}{2} . \tag{A.43}$$

The adjoint eigenvalue problem is

$$L^\dagger \boldsymbol{\Psi} = 0 , \tag{A.44}$$

with

$$L^\dagger = \frac{d^2}{dz^2} + 4\beta \frac{d}{dz} \begin{pmatrix} 0 & 1 \\ 0 & 1 \end{pmatrix} + 4\lambda_n^2 \begin{pmatrix} 0 & 1 \\ -1 & 2 \end{pmatrix} . \tag{A.45}$$

This equation may be written in components:

$$\frac{d^2 \psi_1^{(n)}}{dz^2} + 4\beta \frac{d\psi_2^{(n)}}{dz} + 4\lambda_n^2 \psi_2^{(n)} = 0 , \tag{A.46}$$

$$\frac{d^2 \psi_2^{(n)}}{dz^2} + 4\beta \frac{d\psi_2^{(n)}}{dz} + 4\lambda_n^2 \left(2\psi_2^{(n)} - \psi_1^{(n)} \right) = 0 , \tag{A.47}$$

where ψ_2 must satisfy the boundary conditions (A.43) and also the differential equation (A.36) in which, however, β is replaced by $-\beta$. Therefore, the eigenvalues of the adjoint problem are also given by (A.30). The eigenfunctions are readily calculated. From (A.43), (A.46), and (A.47) one gets

$$\psi_2^{(n)} = B_1^{(n)} \cos(2\lambda_n z) + B_2^{(n)} \sin(2\lambda_n z)$$
$$+ \frac{B_3^{(n)}}{2\beta} \left[\cos(2\lambda_n z) - e^{-2\beta z} \cos(2\lambda_{1n} z) \right]$$
$$+ \frac{B_4^{(n)}}{2\beta} \left[\sin(2\lambda_n z) - e^{-2\beta z} \sin(2\lambda_{1n} z) \right] , \tag{A.48}$$

with the coefficients

$$B_i^{(n)}(\lambda_n, \beta) = A_i^{(n)}(\lambda_n, -\beta) , \qquad i \in [1, 2, 3, 4] . \tag{A.49}$$

For completeness all functions and adjoint functions are specified here:

$$\phi_1^{(n)} = A_{11}^{(n)} \cos(2\lambda_n z) + A_{21}^{(n)} \sin(2\lambda_n z)$$
$$+ A_{31}^{(n)} e^{2\beta z} \cos(2\lambda_{1n} z) + A_{41}^{(n)} e^{2\beta z} \sin(2\lambda_{1n} z) \,, \tag{A.50}$$

$$\phi_2^{(n)} = A_{12}^{(n)} \cos(2\lambda_n z) + A_{22}^{(n)} \sin(2\lambda_n z)$$
$$+ A_{32}^{(n)} e^{2\beta z} \cos(2\lambda_{1n} z) + A_{42}^{(n)} e^{2\beta z} \sin(2\lambda_{1n} z) \,, \tag{A.51}$$

$$\psi_1^{(n)} = B_{11}^{(n)} \cos(2\lambda_n z) + B_{21}^{(n)} \sin(2\lambda_n z)$$
$$+ B_{31}^{(n)} e^{-2\beta z} \cos(2\lambda_{1n} z) + B_{41}^{(n)} e^{-2\beta z} \sin(2\lambda_{1n} z) \,, \tag{A.52}$$

$$\psi_2^{(n)} = B_{12}^{(n)} \cos(2\lambda_n z) + B_{22}^{(n)} \sin(2\lambda_n z)$$
$$+ B_{32}^{(n)} e^{-2\beta z} \cos(2\lambda_{1n} z) + B_{42}^{(n)} e^{-2\beta z} \sin(2\lambda_{1n} z) \,, \tag{A.53}$$

where the constants that appear are given by

$$A_{11}^{(n)} = A_1^{(n)} - A_3^{(n)}/2\beta \,,$$
$$A_{21}^{(n)} = A_2^{(n)} - A_4^{(n)}/2\beta \,,$$
$$A_{31}^{(n)} = A_3^{(n)}/2\beta \,,$$
$$A_{41}^{(n)} = A_4^{(n)}/2\beta \,, \tag{A.54}$$
$$A_{12}^{(n)} = -A_{11}^{(n)} \,, \quad A_{22}^{(n)} = -A_{21}^{(n)} \,,$$
$$A_{32}^{(n)} = \left[(2\beta^2 - \lambda_n^2) A_{31}^{(n)} + 2\beta\lambda_{1n} A_{41}^{(n)} \right]/\lambda_n^2 \,,$$
$$A_{42}^{(n)} = \left[(2\beta^2 - \lambda_n^2) A_{41}^{(n)} - 2\beta\lambda_{1n} A_{31}^{(n)} \right]/\lambda_n^2 \,, \tag{A.55}$$

$$B_{11}^{(n)} = B_{12}^{(n)} + 2\beta B_{22}^{(n)}/\lambda_n \,,$$
$$B_{21}^{(n)} = B_{22}^{(n)} - 2\beta B_{12}^{(n)}/\lambda_n \,,$$
$$B_{31}^{(n)} = B_{32}^{(n)} \,, \quad B_{41}^{(n)} = B_{42}^{(n)} \,, \tag{A.56}$$
$$B_{12}^{(n)} = B_1^{(n)} + B_3^{(n)}/2\beta \,,$$
$$B_{22}^{(n)} = B_2^{(n)} + B_4^{(n)}/2\beta \,,$$
$$B_{32}^{(n)} = -B_3^{(n)}/2\beta \,,$$
$$B_{42}^{(n)} = -B_4^{(n)}/2\beta \,. \tag{A.57}$$

To obtain the orthogonality relation we consider, as in Sect. 4.1, the integral (4.28), with the result, formally identical to (4.30),

$$\int_{-1/2}^{1/2} \left[\psi_1^{(m)}, \psi_2^{(m)} \right] \cdot \mathbf{A} \cdot \begin{pmatrix} \phi_1^{(n)} \\ \phi_2^{(n)} \end{pmatrix} \, \mathrm{d}z = K_n \delta_{nm} \,. \tag{A.58}$$

Here, the normalizing constant is

$$K_n = \int_{-1/2}^{1/2} \left[\psi_1^{(n)}, \psi_2^{(n)} \right] \cdot \mathbf{A} \cdot \begin{pmatrix} \phi_1^{(n)} \\ \phi_2^{(n)} \end{pmatrix} \, \mathrm{d}z$$

$$= \frac{2\beta}{\lambda_n} \left(A_{11}^{(n)} B_{22}^{(n)} g_1 - A_{21}^{(n)} B_{12}^{(n)} g_2 \right)$$

$$+ \left(A_{31}^{(n)} + A_{32}^{(n)} \right) B_{32}^{(n)} g_3 + \left(A_{41}^{(n)} + A_{42}^{(n)} \right) B_{41}^{(n)} g_4$$

$$+ \left[A_{31}^{(n)} B_{12}^{(n)} + A_{32}^{(n)} \left(\frac{-2\beta}{\lambda_n} B_{22}^{(n)} + B_{12}^{(n)} \right) \right] g_5$$

$$+ \left[A_{41}^{(n)} B_{22}^{(n)} + A_{42}^{(n)} \left(\frac{2\beta}{\lambda_n} B_{12}^{(n)} + B_{22}^{(n)} \right) \right] g_6$$

$$+ \left[A_{41}^{(n)} B_{12}^{(n)} + A_{42}^{(n)} \left(\frac{-2\beta}{\lambda_n} B_{22}^{(n)} + B_{12}^{(n)} \right) \right] g_7$$

$$+ \left[A_{31}^{(n)} B_{22}^{(n)} + A_{32}^{(n)} \left(\frac{2\beta}{\lambda_n} B_{12}^{(n)} + B_{22}^{(n)} \right) \right] g_8 . \tag{A.59}$$

In the expression for K_n the following abbreviations have been used:

$$g_1 = \frac{2\lambda_n + \sin 2\lambda_n}{4\lambda_n} , \qquad g_2 = \frac{2\lambda_n - \sin 2\lambda_n}{4\lambda_n} ,$$

$$g_3 = \frac{2\lambda_{1n} + \sin 2\lambda_{1n}}{4\lambda_{1n}} , \qquad g_4 = \frac{2\lambda_{1n} - \sin 2\lambda_{1n}}{4\lambda_{1n}} ,$$

$$g_5 = \frac{1}{2\lambda_n} \left[\frac{\sinh \beta}{\beta} \left(\lambda_n \cos \lambda_n \cos \lambda_{1n} + \lambda_{1n} \sin \lambda_n \sin \lambda_{1n} \right) \right.$$
$$\left. + \cosh \beta \sin \lambda_n \cos \lambda_{1n} \right] ,$$

$$g_6 = \frac{1}{2\lambda_n} \left[\frac{\sinh \beta}{\beta} \left(\lambda_n \sin \lambda_n \sin \lambda_{1n} + \lambda_{1n} \cos \lambda_n \cos \lambda_{1n} \right) \right.$$
$$\left. - \cosh \beta \cos \lambda_n \sin \lambda_{1n} \right] ,$$

$$g_7 = \frac{1}{4\lambda_n} \left[\beta \cosh \beta \left(\frac{\sin(\lambda_n + \lambda_{1n})}{\lambda_n + \lambda_{1n}} - \frac{\sin(\lambda_n - \lambda_{1n})}{\lambda_n - \lambda_{1n}} \right) \right.$$
$$\left. + 2 \sinh \beta \sin \lambda_n \sin \lambda_{1n} \right] ,$$

$$g_8 = \frac{1}{4\lambda_n} \left[\beta \cosh \beta \left(\frac{\sin(\lambda_n + \lambda_{1n})}{\lambda_n + \lambda_{1n}} - \frac{\sin(\lambda_n - \lambda_{1n})}{\lambda_n - \lambda_{1n}} \right) \right.$$
$$\left. - 2 \sinh \beta \cos \lambda_n \cos \lambda_{1n} \right] . \tag{A.60}$$

Finally, the integral J_n appearing in (4.83) is given by

$$J_n = B_{12}^{(n)} \frac{\sin \lambda_n}{\lambda_n} + \frac{B_{32}^{(n)}}{\lambda_n^2} \left(\beta \sinh \beta \cos \lambda_{1n} + \lambda_{1n} \cosh \beta \sin \lambda_{1n} \right)$$

$$+ \frac{B_{42}^{(n)}}{\lambda_n^2} \left(\lambda_{1n} \sinh \beta \cos \lambda_{1n} - \beta \cosh \beta \sin \lambda_{1n} \right) , \tag{A.61}$$

and the constants entering the temperature field (4.85) are

$$A_*^{(n)} = \frac{1}{8\beta\lambda_n^2 \cos\lambda_n} \Big[A_{11}^{(n)} \beta\lambda_n \sin\lambda_n$$
$$+ \left(\beta A_{31}^{(n)} - \lambda_{1n} A_{41}^{(n)} \right) \cosh\beta \cos\lambda_{1n}$$
$$+ \left(\beta A_{41}^{(n)} + \lambda_{1n} A_{31}^{(n)} \right) \sinh\beta \sin\lambda_{1n} \Big], \tag{A.62}$$

$$B_*^{(n)} = \frac{1}{8\beta\lambda_n^2 \sin\lambda_n} \Big[-A_{21}^{(n)} \beta\lambda_n \cos\lambda_n$$
$$+ \left(\beta A_{31}^{(n)} - \lambda_{1n} A_{41}^{(n)} \right) \sinh\beta \cos\lambda_{1n}$$
$$+ \left(\beta A_{41}^{(n)} + \lambda_{1n} A_{31}^{(n)} \right) \cosh\beta \sin\lambda_{1n} \Big], \tag{A.63}$$

$$B_{nk} = -\frac{2\left[2\lambda_n I_1(2\lambda_n/\Gamma) + \mathrm{Bi}\, I_0(2\lambda_n/\Gamma) \right]}{2\mu_k I_1(2\mu_k/\Gamma) + \mathrm{Bi}\, I_0(2\mu_k/\Gamma)}$$
$$\times \int_{-1/2}^{1/2} \left(-\bar{\phi}_1^{(n)}(z) + A_*^{(n)} \cos(2\lambda_n z) + B_*^{(n)} \sin(2\lambda_n z) \right) \cos(2\mu_k z)\, \mathrm{d}z, \tag{A.64}$$

$$C_{nk} = -\frac{2\left(2\lambda_n I_1(2\lambda_n/\Gamma) + \mathrm{Bi}\, I_0(2\lambda_n/\Gamma) \right)}{2\nu_k I_1(2\nu_k/\Gamma) + \mathrm{Bi}\, I_0(2\nu_k/\Gamma)}$$
$$\times \int_{-1/2}^{1/2} \left(-\bar{\phi}_1^{(n)}(z) + A_*^{(n)} \cos(2\lambda_n z) + B_*^{(n)} \sin(2\lambda_n z) \right) \sin(2\nu_k z)\, \mathrm{d}z, \tag{A.65}$$

with μ_k and ν_k as in (4.85). The auxiliary functions $\bar{\phi}_1^{(n)}$ are

$$\bar{\phi}_1^{(n)}(z) = \frac{A_{11}^{(n)}}{4\lambda_n} z \sin(2\lambda_n z) - \frac{A_{21}^{(n)}}{4\lambda_n} z \cos(2\lambda_n z)$$
$$+ \frac{\left(\beta A_{31}^{(n)} - \lambda_{1n} A_{41}^{(n)} \right)}{8\beta\lambda_n^2} e^{2\beta z} \sin(2\lambda_{1n} z)$$
$$+ \frac{\left(\beta A_{41}^{(n)} + \lambda_{1n} A_{31}^{(n)} \right)}{8\beta\lambda_n^2} e^{2\beta z} \cos(2\lambda_{1n} z). \tag{A.66}$$

A.4 Chebyshev–Galerkin Method for Plane Layers

In order to solve the linear stability problem (6.18–6.20) the coordinate is transformed according to $z \to 2z$ and all unknowns are expanded into Chebyshev polynomials. The coefficients are determined by projecting the equations onto the basis functions $\langle T_m(z)|$ by means of the scalar product

$$\langle T_m \,|\, T_n \rangle = \int_{-1}^{1} \frac{T_m(z)\,T_n(z)}{\sqrt{1-z^2}}\, dz = \frac{\pi}{2} c_m \delta_{nm},$$

$$\text{with} \quad \begin{cases} c_m = 2,\, n = 0, \\ c_m = 1,\, n \neq 0. \end{cases} \tag{A.67}$$

Using the summation convention one obtains for $n \in [0, N]$

$$\left[(\sigma - i\omega + k^2)\left(4\langle T_m T_n'' \rangle - \frac{\pi}{2} c_m k^2 \delta_{nm} \right) - 16\langle T_m T_n'''' \rangle \right.$$
$$+ 4k^2 \langle T_m T_n'' \rangle + 4ikc\mathrm{Re}\Big(-4\langle T_m\, g''T_n'' \rangle + k^2 \langle T_m\, g''T_n \rangle$$
$$\left. + 4\langle T_m\, g''''T_n \rangle \Big) \right]\hat{\psi}_n = 0, \qquad \text{for} \quad m \in [0, N-4], \tag{A.68}$$

$$\left[(\sigma - i\omega + k^2)\frac{\pi}{2} c_m \delta_{nm} - 4\langle T_m T_n'' \rangle - 4ikc\mathrm{Re}\langle T_m\, g''T_n \rangle \right]\hat{v}_n$$
$$+ 8iks\mathrm{Re}\langle T_m\, g'''T_n \rangle \hat{\psi}_n = 0, \qquad \text{for} \quad m \in [0, N-2], \tag{A.69}$$

$$\left[\left(\sigma - i\omega + \frac{k^2}{\mathrm{Pr}} \right)\frac{\pi}{2} c_m \delta_{nm} - \frac{4}{\mathrm{Pr}}\langle T_m T_n'' \rangle - 4ikc\mathrm{Re}\langle T_m\, g''T_n \rangle \right]\hat{\theta}_n$$
$$- s\mathrm{Re}\frac{\pi}{2} c_m \delta_{nm}\hat{v}_n - 2\mathrm{Re}\Big(c\langle T_m T_n' \rangle + ik\mathrm{RePr}\langle T_m\, g'T_n \rangle \Big)\hat{\psi}_n = 0,$$
$$\text{for} \quad m \in [0, N-2]. \tag{A.70}$$

The boundary conditions (6.15) and (6.16) for ψ, v, and θ yield $8N$ additional equations

$$\hat{\psi}_n T_n(-1) = 0, \qquad\qquad\qquad \hat{\psi}_n T_n(1) = 0, \tag{A.71}$$
$$\hat{\psi}_n T_n'(-1) = 0, \qquad 4\hat{\psi}_n T_n''(1) + ik\hat{\theta}_n T_n(1) = 0, \tag{A.72}$$
$$\hat{v}_n T_n(-1) = 0, \qquad\qquad\qquad \hat{v}_n T_n'(1) = 0, \tag{A.73}$$
$$\hat{\theta}_n T_n'(-1) = 0, \qquad\qquad\qquad \hat{\theta}_n T_n'(1) = 0. \tag{A.74}$$

Equations (A.68–A.74) constitute $3(N+1)$ equations for the $3(N+1)$ complex unknowns $\hat{\psi}_n$, $\hat{v}_n$, and $\hat{\theta}_n$. If real and imaginary parts are considered separately one obtains the $6(N+1) \times 6(N+1)$ real system (6.23).

References

Abramowitz, M. & Stegun, I. A. 1972 *Handbook of Mathematical Functions*. Dover, New York.

Acheson, D. J. 1990 *Elementary Fluid Dynamics*. Oxford University Press, Oxford.

Afrid, M. & Zebib A. 1990 Oscillatory three-dimensional convection in rectangular cavities and enclosures. *Phys. Fluids* A **2**, 1318.

Alexander, J. I. D., Ouazzani, J. & Rosenberger, F. 1989 Analysis of the low gravity tolerance of Bridgman–Stockbarger crystal growth. *J. Crystal Growth* **97**, 285.

Anderson, D. M. & Davis, S. H. 1993 Two-fluid viscous flow in a corner. *J. Fluid Mech.* **257**, 1.

Anderson, D. M. & Davis, S. H. 1994 Fluid flow, heat transfer, and solidification near tri-junctions. *J. Crystal Growth* **142**, 245.

Anilkumar, A. V., Grugel, R. N., Shen, X. F., Lee, C. P. & Wang, T. G. 1993 Control of thermocapillary convection in a liquid bridge by vibration. *J. Appl. Phys.* **73**, 4165.

Ashgriz, N. & Mashayek, F. 1995 Temporal analysis of capillary jet breakup. *J. Fluid Mech.* **291**, 163.

Batchelor, G. K. 1956 On steady laminar flow with closed streamlines at large Reynolds numbers. *J. Fluid Mech.* **1**, 177.

Batishchev, V. A., Kuznetsov, V. V. & Pukhnachov, V. V. 1989 Marangoni boundary layers. *Prog. Aerospace Sci.* **26**, 353.

Bauer, H. F. 1982 Velocity distribution in a liquid bridge due to the thermal Marangoni effect. *Z. Flugwiss. Weltraumforsch.* **6**, 252.

Bauer, H. F. 1992 Natural damped frequencies and axial response of a rotating finite viscous liquid column. *Acta Mech.* **93**, 29.

Baumgartl, J. 1992 Numerische und experimentelle Untersuchungen zur Wirkung magnetischer Felder in Kristallzüchtungsanordnungen, Ph.D. thesis, Universität Erlangen-Nürnberg.

Baumgartl, J., Gewald, M., Rupp, R., Stierlen, J. & Müller, G. 1990 The use of magnetic fields and microgravity in melt growth of semiconductors: A comparative study. In *Proceedings of the VIIth European Symposium on Materials and Fluid Sciences in Microgravity*, ESA SP-295, p. 47.

Baumgartl, J., Hubert, A. & Müller, G. 1993 The use of magnetohydrodynamic effects to investigate fluid flow in electrically conducting melts. *Phys. Fluids* A **5**, 3280.

Baumgartl, J. & Müller, G. 1992 Calculation of the effects of magnetic field damping on fluid flow – comparison of magnetohydrodynamic models of different complexity. *Microgravity Q.* **2**, 197.

Bayly, B. J. 1986 Three-dimensional instability of elliptical flow. *Phys. Rev. Lett.* **57**, 2160.

Bénard, H. 1900 Les tourbillons cellulaires dans une nappe liquide. *Revue Gén. Sci. Pur. Appl.* **11**, 1261 & 1309.

Ben Hadid, H. & Henry, D. 1997 Numerical study of convection in the horizontal Bridgman configuration under the action of a constant magnetic field. Part 2. Three-dimensional flow. *J. Fluid Mech.* **333**, 57.

Ben Hadid, H. & Roux, B. 1990 Thermocapillary convection in long horizontal layers of low-Prandtl-number melts subject to a horizontal temperature gradient. *J. Fluid Mech.* **221**, 77.

Ben Hadid, H. & Roux, B. 1992 Buoyancy- and thermocapillary-driven flows in differentially heated cavities for low-Prandtl-number fluids. *J. Fluid Mech.* **235**, 1.

Ben Hadid, H., Henry, D. & Kaddeche, S. 1997 Numerical study of convection in the horizontal Bridgman configuration under the action of a constant magnetic field. Part 1. Two-dimensional flow. *J. Fluid Mech.* **333**, 23.

Benz, S., Hintz, P., Riley, R. J. & Neitzel, G. P. 1998 Instability of thermocapillary–buoyancy convection in shallow layers. Part 2. Suppression of hydrothermal waves. *J. Fluid Mech.* **359**, 165.

Bergman, T. L. & Keller, J. R. 1988 Combined buoyancy, surface tension flow in liquid metals. *Numer. Heat Transfer* **13**, 49.

Bergman, T. L. & Ramadhyani, S. 1986 Combined buoyancy- and thermocapillary driven convection in open square cavities. *Numer. Heat Transfer* **9**, 441.

Bergman, T. L. & Webb, B. W. 1990 Simulation of pure metal melting with buoyancy and surface tension forces in the liquid phase. *Int. J. Heat Mass Transfer* **33**, 139.

Bikerman, J. J. 1973 *Foams*. Springer, Berlin, Heidelberg.

Birikh, R. V. 1966 Thermocapillary convection in a horizontal layer of liquid. *J. Appl. Mech. Tech. Phys.* **7**, 43.

Block, M. J. 1956 Surface tension as the cause of Bénard cells and surface deformation in a liquid film. *Nature* **178**, 650.

Bödewadt, U. T. 1940 Die Drehströmung über festem Grunde. *Z. Angew. Math. Mech.* **20**, 241.

Bohm, J., Lüdge, A. & Schröder, W. 1994 Crystal growth by floating zone melting. In *Handbook of Crystal Growth*, vol. 2a, *Basic Techniques*, Hurle, D. T. J. (ed.), North-Holland, Amsterdam, p. 213.

Bois-Reymond, P. du 1858 Experimentaluntersuchung über die Erscheinungen, welche die Ausbreitung von Flüssigkeiten auf Flüssigkeiten hervorruft. *Ann. Phys. Chem.* **104**, 193.

Braunsfurth, M. G. & Homsy, G. M. 1997 Combined thermocapillary–buoyancy convection in a cavity. Part II. An experimental study. *Phys. Fluids* **9**, 1277.

Braunsfurth, M. G. & Mullin, T. 1996 An experimental study of oscillatory convection in liquid gallium. *J. Fluid Mech.* **327**, 199.

Brenner, H. & Leal, L. G. 1982 Conservation and constitutive equations for adsorbed species undergoing surface diffusion and convection at a fluid–fluid interface. *J. Colloid Interface Sci.* **88**, 136.

Bückle, U. & Perić, M. 1992 Numerical simulation of buoyant and thermocapillary convection in a square cavity. *Numer. Heat Transfer* A **21**, 121.

Buell, J. C. & Catton, I. 1983 The effect of wall conduction on the stability of a fluid in a right circular cylinder heated from below. *ASME J. Heat Transfer* **105**, 255.

Burelbach, J. P., Bankoff, S. G. & Davis, S. H. 1990 Steady thermocapillary flows of thin liquid layers. II. Experiment. *Phys. Fluids* A **2**, 322.

Burggraf, O. R. 1966 Analytical and numerical studies of the structure of steady separated flows. *J. Fluid Mech.* **24**, 113.

Busse, F. H.   1978 Non-linear properties of thermal convection. *Rep. Prog. Phys.* **41**, 1929.

Canright, D.   1994 Thermocapillary flow near a cold wall. *Phys. Fluids* **6**, 1415.

Canuto, C., Hussaini, M. Y., Quarteroni, A. & Zhang, T. A.   1988 *Spectral Methods in Fluid Dynamics*. Springer, Berlin, Heidelberg.

Cao, Z. H., Xie, J. C., Tang, Z. M. & Hu, W. R.   1992 Experimental study on oscillatory thermocapillary convection. *Sci. China* A **35**, 725.

Carotenuto, L., Albanese, C., Castagnolo, D. & Monti, R.   1996 Onset of oscillatory Marangoni convection in a liquid bridge. In *Materials and Fluids under Low Gravity*, Lecture Notes in Physics **464**, Ratke, L., Walter, H. & Feuerbacher, B. (eds.), Springer, Berlin, Heidelberg, p. 331.

Carotenuto, L., Castagnolo, D., Albanese, C. & Monti, R.   1998 Instability of thermocapillary convection in liquid bridges. *Phys. Fluids* **10**, 555.

Carpenter, B. M. & Homsy, G. M.   1989 Combined buoyant–thermocapillary flow in a cavity. *J. Fluid Mech.* **207**, 121.

Carpenter, B. M. & Homsy, G. M.   1990 High Marangoni number convection in a square cavity: Part II. *Phys. Fluids* A **2**, 137.

Chandrasekhar, S.   1961 *Hydrodynamic and Hydromagnetic Stability*. Oxford University Press, Oxford.

Chang, C. E. & Wilcox, W. R.   1975 Inhomogeneities due to thermocapillary flow in floating zone melting. *J. Crystal Growth* **28**, 8.

Chang, C. E. & Wilcox, W. R.   1976 Analysis of surface tension driven flow in floating zone melting. *Int. J. Heat Mass Transfer* **19**, 355.

Charlson, G. S. & Sani, R. L.   1970 Thermoconvective instability in a bounded cylindrical fluid layer. *Int. J. Heat Mass Transfer* **13**, 1479.

Charlson, G. S. & Sani, R. L.   1971 On thermoconvective instability in a bounded cylindrical fluid layer. *Int. J. Heat Mass Transfer* **14**, 2157.

Charlson, G. S. & Sani, R. L.   1975 Finite amplitude axisymmetric thermoconvective flows in a bounded cylindrical layer of fluid. *J. Fluid Mech.* **71**, 209.

Chen, Y. Y.   1992a Boundary conditions and linear analysis of finite-cell Rayleigh–Bénard convection. *J. Fluid Mech.* **241**, 549.

Chen, Y. Y.   1992b Finite-size effects on linear stability of pure-fluid convection. *Phys. Rev.* A **45**, 3727.

Chen, Q. S. & Hu, W. R.   1998 Influence of liquid bridge volume on instability of floating half zone convection. *Int. J. Heat Mass Transfer* **41**, 825.

Chen, J.-C. & Hwu, F.-S.   1993 Oscillatory thermocapillary flow in a rectangular cavity. *Int. J. Heat Mass Transfer* **36**, 3743.

Chen, G. & Roux, B.   1991 An analytical study of thermocapillary flow and surface deformations in floating zones. *Microgravity Q.* **1**, 73.

Chen, T.-Y. & Tsamopoulos, J.   1993 Nonlinear dynamics of capillary bridges: Theory. *J. Fluid Mech.* **255**, 373.

Chen, J. C., Sheu, J. C. & Lee, Y. T.   1990 Maximum stable length of nonisothermal liquid bridges. *Phys. Fluids* A **2**, 1118.

Chen, H., Saghir, M. Z., Quon, D. H. H. & Chehab, S.   1995a Numerical study on transient convection in float zone induced by g-jitter. *J. Crystal Growth* **142**, 362.

Chen, Q., Ramé, E. & Garoff, S.   1995b The breakdown of asymptotic models of liquid spreading at increasing capillary number. *Phys. Fluids* **7**, 2631.

Chen, G., Liźee, A. & Roux, B.   1998 Bifurcation analysis of the thermocapillary convection in cylindrical liquid bridges. *J. Crystal Growth* **180**, 638.

Chun, C.-H.   1980 Experiments on steady and oscillatory temperature distribution in a floating zone due to the Marangoni convection. *Acta Astronautica* **7**, 479.

Chun, C.-H. & Siekmann, J. 1995 Higher modes and their instabilities of the oscillating Marangoni convection in a large cylindrical liquid column. In *Scientific Results of the German Spacelab Mission D2*, Sahm, P. R., Keller, M. H. & Schiewe, B. (eds.), p. 235.

Chun, C.-H. & Wuest, W. 1979 Experiments on the transition from steady to oscillatory Marangoni convection in a floating zone under reduced gravity effect. *Acta Astronautica* **6**, 1073.

Chun, C.-H. & Wuest, W. 1982 Suppression of temperature oscillations of thermal Marangoni convection in a floating zone by superimposing of rotating flows. *Acta Astronautica* **9**, 225.

Colinet, P., Legros, J. C., Kamotani, Y., Dauby, P. C. & Lebon, G. 1995 Finite-amplitude regimes of the short-wave Marangoni–Bénard convective instability. *Phys. Rev. E* **52**, 2603.

Coriell, S. R. & McFadden, G. B. 1993 Morphological stability. In *Handbook of Crystal Growth*, vol. 1b, *Fundamentals. Transport and Stability*, Hurle, D. T. J. (ed.), North-Holland, Amsterdam, p. 785.

Cowley, S. J. & Davis, S. H. 1983 Viscous thermocapillary convection at high Marangoni numbers. *J. Fluid Mech.* **135**, 175.

Crawford, J. D. & Knobloch, E. 1991 Symmetry and symmetry-breaking bifurcations in fluid dynamics. *Ann. Rev. Fluid Mech.* **23**, 341.

Cröll, A., Müller-Sebert, W. & Nitsche, R. 1989 The critical Marangoni number for the onset of time-dependent convection in silicon. *Materials Res. Bull.* **24**, 995.

Cröll, A., Müller-Sebert, W., Benz, K. W. & Nitsche, R. 1991 Natural and thermo-capillary convection in partially confined silicon melt zones. *Microgravity Sci. Technol.* **3**, 204.

Cröll, A., Dold, P. & Benz, K. W. 1994 Segregation in Si floating-zone crystals grown under microgravity and in a magnetic field. *J. Crystal Growth* **137**, 95.

Cröll, A., Szofran, F. R., Dold, P., Benz, K. W. & Lehoczky, S. L. 1998 Floating-zone growth of silicon in magnetic fields II. Strong static axial fields. *J. Crystal Growth* **183**, 554.

Cross, M. C. 1988 Structure of non-linear traveling-wave states in finite geometries. *Phys. Rev A* **38**, 3593.

Cross, M. C. & Hohenberg, P. G. 1993 Pattern formation outside of equilibrium. *Rev. Mod. Phys.* **65**, 851.

Cuvelier, C. & Driessen, J. M. 1986 Thermocapillary free boundaries in crystal growth. *J. Fluid Mech.* **169**, 1.

Daviaud, F. & Vince, J. M. 1993 Travelling waves in a fluid layer subjected to a horizontal temperature gradient. *Phys. Rev. E* **48**, 4432.

Davies, J. T. & Rideal, E. K. 1963 *Interfacial Phenomena*. Academic Press, New York.

Davis, S. H. 1987 Thermocapillary instabilities. *Ann. Rev. Fluid Mech.* **19**, 403.

Davis, A. M. J. 1989 Thermocapillary convection in liquid bridges: Solution structure and eddy motions. *Phys. Fluids A* **1**, 475.

Davis, S. H. 1990 Hydrodynamic interactions in directional solidification. *J. Fluid Mech.* **212**, 241.

Davis, S. H. 1993 Effects of flow on morphological stability. In *Handbook of Crystal Growth*, vol. 1b, *Fundamentals. Transport and Stability*, Hurle, D. T. J. (ed.), North-Holland, Amsterdam, p. 860.

DebRoy, T. & David, S. A. 1995 Physical processes in fusion welding. *Rev. Mod. Phys.* **67**, 85.

Deville, M., Lê, T.-H. & Morchoisne, Y. (eds.) 1992 Numerical simulation of 3-D incompressible unsteady viscous laminar flows. *Notes on Numerical Fluid Mechanics* **36**, Vieweg.

Dijkstra, H. A. 1998 Pattern selection in surface tension driven flows. In *Free Surface Flows*, Kuhlmann, H. C. & Rath, H. J. (eds.), Proceedings of CISM, Springer, Vienna, New York (to appear).

Doi, T. & Koster, J. N. 1993 Thermocapillary convection in two immiscible liquid layers with free surface. *Phys. Fluids* A **5**, 1914.

Dold, P., Cröll, A. & Benz, K. W. 1998 Floating-zone growth of silicon in magnetic fields II. Weak static axial fields. *J. Crystal Growth* **183**, 545.

Drazin, P. G. 1992 *Nonlinear Systems*. Cambridge University Press, Cambridge.

Drazin, P. G. & Reid, W. H. 1981 *Hydrodynamic Stability*. Cambridge University Press, Cambridge.

Drummond, J. E. & Korpela, S. A. 1987 Natural convection in a shallow cavity. *J. Fluid Mech.* **182**, 543.

Dupret, F. & Bogaert, N. van den 1994 Modelling Bridgman and Czochralski growth. In *Handbook of Crystal Growth*, vol. 2b, *Bulk Crystal Growth*, Hurle, D. T. J. (ed.), North-Holland, Amsterdam, p. 875.

Dussan V., E. B. 1979 On the spreading of liquids on solid surfaces. *Ann. Rev. Fluid Mech.* **11**, 371.

Dussan V., E. B., Ramé, E. & Garoff, S. 1991 On identifying the appropriate boundary conditions at a moving contact line: An experimental investigation. *J. Fluid Mech.* **230**, 97.

Ehmann, M. & Siekmann, J. 1995 Numerical study of the oscillations of axially excited liquid annuli with rotational symmetry enclosed in revolving circular cylindrical containers. *J. Fluid Mech.* **297**, 215.

Erhard, P. & Davis, S. H. 1991 Non-isothermal spreading of liquid drops on horizontal plates. *J. Fluid Mech.* **229**, 365.

Ezersky, A. B., Garcimartín, A., Mancini, H. L. & Pérez-García, C. 1993 Spatiotemporal structure of hydrothermal waves in Marangoni convection. *Phys. Rev.* E **48**, 4414.

Fadle, J. 1941 Die Selbstspannungs-Eigenwertfunktionen der quadratischen Scheibe. *Ing. Archiv* **11**, 125.

Farooq, A. & Homsy, G. M. 1994 Streaming flows due due g-jitter induced natural convection. *J. Fluid Mech.* **271**, 351.

Favre, E., Blumenfeld, L. & Daviaud, F. 1997 Instabilities of a liquid layer locally heated on its free surface. *Phys. Fluids* **9**, 1473.

Finlayson, B. A. 1972 *The Method of Weighted Residuals and Variational Principles*. Academic Press, New York.

Finn, R. 1986 *Equilibrium Capillary Surfaces*. Grundlehren der mathematischen Wissenschaften **284**, Springer, Berlin, Heidelberg.

Floryan, J. M. & Chen, C. 1994 Thermocapillary convection and existence of continuous layers in the absence of gravity. *J. Fluid Mech.* **277**, 303.

Floryan, J. M. & Rasmussen, H. 1989 Numerical methods for viscous flows with moving boundaries. *Appl. Mech. Rev.* **42**, 323.

Fomenko, A. T. 1989 *The Plateau Problem*. Part I & II. Studies in the Development of Modern Mathematics **1**, Gordon and Breach, Philadelphia, New York.

Frank, S. & Schwabe, D. 1997 Temporal and spatial elements of thermocapillary convection in floating zones. *Exp. Fluids* **23**, 234.

Fu, B.-I. & Ostrach, S. 1985 Numerical solution of thermocapillary flows in floating zones. In *Transport Phenomena in Materials Processing*. PED-Vol. 10 / HTD-Vol. 29, ASME, New York, p. 1.

Garandet, J. P., Favier, J. J. & Camel, D. 1994 Segregation phenomena in crystal growth from the melt. In *Handbook of Crystal Growth*, vol. 2b, *Bulk Crystal Growth*, Hurle, D. T. J. (ed.), North-Holland, Amsterdam, p. 659.

Garcimartín, A., Mukolobwiez, N. & Daviaud, F. 1997 Origin of waves in surface-tension-driven convection. *Phys. Rev. E* **56**, 1699.

Garr-Peters, J. M. 1992a The neutral stability of surface-tension driven cavity flows subject to buoyant forces – I. Transverse and longitudinal disturbances. *Chem. Eng. Sci.* **47**, 1247.

Garr-Peters, J. M. 1992b The neutral stability of surface-tension driven cavity flows subject to buoyant forces – II. Oblique disturbances. *Chem. Eng. Sci.* **47**, 1265.

Gershuni, G. Z. & Zhukhovitsky, E. M. 1979 Free thermal convection in a vibratory field in weightlessness. *Dokl. Acad. Nauk. SSSR* **249**, 580.

Gershuni, G. Z. & Zhukhovitsky, E. M. 1986 Vibration-induced thermal convection in weightlessness. *Fluid Mech. Soviet Res.* **15**, 63.

Gershuni, G. Z., Lyubimova, T. V., Lyubimov, D. V. & Roux, B. 1992 Coupled thermovibrational and thermocapillary convection in liquid bridge (floating zone system). In *Symposium on Hydromechanics and Heat/Mass Transfer in Microgravity, Moscow–Perm 1991*, Avduevski, V. S. (ed.), Gordon and Breach, Philadelphia, p. 117.

Gieldt, W. H. 1987 Welding – An interdisciplinary science and technology. In *Interdisciplinary Issues in Materials Processing Manufacturing*, Samanta, S. K., Komanduri, R., McMeeking, R., Chen, M. M. & Tseng, A. (eds.), ASME, New York.

Gillon, P. & Homsy, G. M. 1996 Combined thermocapillary–buoyancy convection in a cavity: An experimental study. *Phys. Fluids* **8**, 2953.

Glicksman, M. E., Coriell, S. R. & McFadden, G. B. 1986 Interaction of flows with the crystal–melt interface. *Ann. Rev. Fluid Mech.* **18**, 307.

Gollub, H. G. & Loan, C. F. van 1989 *Matrix Computations*. Johns Hopkins University Press, Baltimore.

Goodrich, J. W., Gustafson, K. & Halasi, K. 1990 Hopf bifurcation in the driven cavity. *J. Comput. Phys.* **90**, 219.

Gray, D. D. & Giorgini, A. 1976 The validity of the Boussinesq approximation for liquids and gases. *Int. J. Heat Mass Transfer* **19**, 545.

Greenspan, H. P. 1969 *The Theory of Rotating Fluids*. Cambridge University Press, Cambridge.

Gröbner, H., Erk, S. & Grigull, U. 1963 *Die Grundgesetze der Wärmeübertragung*. Springer, Berlin, Heidelberg.

Grotberg, J. 1994 Pulmonary flow and transport phenomena. *Ann. Rev. Fluid Mech.* **26**, 529.

Grugel, R. N., Shen, X. F., Anilkumar, A. V. & Wang, T. G. 1994 The influence of vibration on microstructural uniformity during floating zone crystal growth. *J. Crystal Growth* **142**, 209.

Guckenheimer, J. & Holmes, P. 1983 *Nonlinear Oscillations, Dynamical Systems, and Bifurcations of Vector Fields*. Applied Mathematical Sciences **42**, Springer, Berlin, Heidelberg.

Guérin, R. Z., Billia, B. & Haldenwang, P. 1991 Onset of solutal convection during directional solidification of a binary alloy in a cylinder. *Phys. Fluids A* **3**, 1873.

Hardin, G. R. & Sani, R. L. 1993 Buoyancy-driven instability in a vertical cylinder: Binary fluids with Soret effect. Part 2: Weakly non-linear solutions. *Int. J. Numer. Meth. Fluids* **17**, 755.

Hardy, S. C. 1985 The surface tension of liquid gallium. *J. Crystal Growth* **71**, 602.

Harriott, G. M. & Brown, R. A. 1983 Flow in a differentially rotated cylindrical drop at low Reynolds number. *J. Fluid Mech.* **126**, 269.

Hart, J. E. 1983 A note on the stability of low-Prandtl-number Hadley circulations. *J. Fluid Mech.* **132**, 271.

Hasimoto, H. & Sano, O. 1980 Stokeslets and eddies in creeping flow. *Ann. Rev. Fluid Mech.* **12**, 335.

Henkes, R. A. W. M. & Le Quéré, P. 1996 Three-dimensional transition of natural-convection flows. *J. Fluid Mech.* **319**, 281.

Herrmann, F. M., Baumgartl, J., Feulner, T. & Müller, G. 1992 The use of magnetic fields for damping unsteady Marangoni convection in GaAs. In *Proceedings of the VIIIth European Symposium on Materials and Fluid Sciences in Microgravity*, ESA SP-333, p. 57.

Hillman, A. P. & Salzer, H. E. 1943 The roots of $\sin z = z$. *Phil. Mag.* **34**, 575.

Homsy, G. M. & Meiburg, E. 1984 The effect of surface contamination on thermocapillary flow in a two-dimensional slot. *J. Fluid Mech.* **139**, 443.

Hsieh, K.-C. 1992 A numerical study of three-dimensional surface tension driven convection with free surface deformations. In *Fluid Mechanics Phenomena in Microgravity*, AMD-Vol. 154 / FED-Vol. 142, ASME, New York, p. 85.

Hu, W. R. & Tang, Z. M. 1990 Excitation mechanism of thermocapillary oscillatory convection. *Sci. China* A **33**, 934.

Hu, W. R., Shu, J. Z., Zhou, R. & Tang, Z. M. 1994 Influence of liquid bridge volume on the onset of oscillation in floating zone convection. *J. Crystal Growth* **142**, 379.

Huerre, P. & Monkewitz, P. A. 1990 Local and global instabilities in spatially developing flows. *Ann. Rev. Fluid Mech.* **22**, 473.

Huerre, P. & Rossi, M. 1998 Hydrodynamic instabilities in open flows. In *Hydrodynamics and Nonlinear Instabilities*, Godréche, C. & Manneville, P. (eds.), Cambridge University Press, Cambridge, p. 81.

Hunter, R. J. 1989 *Foundations of Colloid Science*, vol. II. Clarendon Press, Oxford.

Hurle, D. T. J. (ed.) 1994 *Handbook of Crystal Growth*. North-Holland, Amsterdam.

Hurle, D. T. J. & Cockayne, B. 1994 Czochralski growth. In *Handbook of Crystal Growth*, vol. 2a, *Basic Techniques*, Hurle, D. T. J. (ed.), North-Holland, Amsterdam, p. 99.

Hurle, D. T. J. & Series, R. W. 1994 Use of a magnetic field in melt growth. In *Handbook of Crystal Growth*, vol. 2a, *Basic Techniques*, Hurle, D. T. J. (ed.), North-Holland, Amsterdam, p. 259.

Hyer, J. R., Jankowski, D. F. & Neitzel, G. P. 1991 Thermocapillary convection in a model float zone. *AIAA J. Thermophys. Heat Transfer* **5**, 577.

Iooss, G. 1987 Reduction of the dynamics of a bifurcation problem using normal forms and symmetries. In *Instabilities and Nonequilibrium Structures*, Tirapegui, E. & Villarroel, D. (eds.), Reidel, London, p. 3.

Israelachvili, J. 1992 *Intermolecular and Surface Forces*. Academic Press, New York.

Janssen, J. A. & Henkes, R. A. W. M. 1995 Influence of Prandtl number on instability mechanisms and transition in a differentially heated square cavity. *J. Fluid Mech.* **290**, 319.

Jones, C. A. 1985 Numerical methods for the transition to wavy Taylor vortices. *J. Comput. Phys.* **61**, 321.

Jordan, D. W. & Smith, P. 1987 *Nonlinear Ordinary Differential Equations*. Oxford Applied Mathematics and Computing Science Series, Oxford University Press, Oxford.

Joseph, D. D. 1976 *Stability of Fluid Motions*, vols. I & II. Springer, Berlin, Heidelberg.

Joseph, D. D. 1977 The convergence of biorthogonal series for biharmonic and Stokes flow edge problems. Part I. *SIAM J. Appl. Math.* **33**, 337.

Joseph, D. D. & Sturges, L. 1975 The free surface on a liquid filled trench heated from its side. *J. Fluid Mech.* **69**, 565.

Joseph, D. D. & Sturges, L. 1978 The convergence of biorthogonal series for biharmonic and Stokes flow edge problems: Part II. *SIAM J. Appl. Math.* **34**, 7.

Kakimoto, K., Eguchi, M., Watanabe, H. & Hibiya, T. 1988 Direct observation by X-ray radiography of convection of molten silicon in the Czochralski growth method. *J. Crystal Growth* **88**, 365.

Kamotani, Y. & Kim, J. 1988 Effect of zone rotation on oscillatory thermocapillary flow in simulated floating zones. *J. Crystal Growth* **87**, 62.

Kamotani, Y. & Ostrach, S. 1998 Theoretical analysis of thermocapillary flow in cylindrical columns of high Prandtl number fluids. *ASME J. Heat Transfer* (to appear).

Kamotani, Y. & Platt, J. 1992 Effect of free surface shape on combined thermocapillary and natural convection. *AIAA J. Thermophys. Heat Transfer* **6**, 721.

Kamotani, Y., Ostrach, S. & Vargas, M. 1984 Oscillatory thermocapillary convection in a simulated floating-zone configuration. *J. Crystal Growth* **66**, 83.

Kanouff, M. & Greif, R. 1994 Oscillations in thermocapillary convection in a square cavity. *Int. J. Heat Mass Transfer* **37**, 885.

Kazarinoff, N. D. & Wilkowski, J. S. 1989 A numerical study of Marangoni flows in zone-refined silicon crystals. *Phys. Fluids* A **1**, 625.

Kazarinoff, N. D. & Wilkowski, J. S. 1990 Bifurcations of numerically simulated thermocapillary flows in axially symmetric float zones. *Phys. Fluids* A **2**, 1797.

Keller, J. R. & Bergman, T. L. 1990 Thermocapillary cavity convection in wetting and nonwetting liquids. *Numer. Heat Transfer* A, **18**, 33.

Kelvin, Lord 1880 On a disturbing infinity in Lord Rayleigh's solution for waves in a plane vortex stream. *Nature* **23**, 45.

Kenning, D. B. R. 1968 Two-phase flow with nonuniform surface tension. *Appl. Mech. Rev.* **21**, 1101.

Kerr, O. S. & Dold, J. W. 1994 Periodic steady vortices in a stagnation point flow. *J. Fluid Mech.* **276**, 307.

Koplik, J. & Banavar, J. R. 1995 Corner flow in the sliding plate problem. *Phys. Fluids* **7**, 3118.

Koschmieder, E. L. 1966 On convection on a uniformely heated plane. *Beitr. Phys. Atmos.* **39**, 1.

Koschmieder, E. L. 1993 *Bénard Cells and Taylor Vortices.* Cambridge University Press, Cambridge.

Kothe, D. B. 1998 Perspective on Eulerian finite volume methods for incompressible interfacial flows. In *Free Surface Flows*, Kuhlmann, H. C. & Rath, H. J. (eds.), Proceedings of CISM, Springer, Vienna, New York (to appear).

Kozhoukharova, Zh. & Slavchev, S. 1986 Computer simulation of the thermocapillary convection in a non-cylindrical floating zone. *J. Crystal Growth* **74**, 236.

Kozhoukharova, Zh., Kuhlmann, H. C., Wanschura, M. & Rath, H. J. 1998 Influence of variable viscosity on the onset of hydrothermal waves in thermocapillary liquid bridges. *Z. Angew. Math. Mech.* (submitted).

Kuhlmann, H. 1989 Small amplitude thermocapillary flow and surface deformations in a liquid bridge. *Phys. Fluids* A **1**, 672.

Kuhlmann, H. C. 1994 Thermocapillary flows in finite size systems. *Math. Comput. Modelling* **20**, 145.

Kuhlmann, H. C. 1995 Some scaling aspects of thermocapillary flows. *Microgravity Q.* **5**, 29.

Kuhlmann, H. C. & Adabala, R. R. 1993 Biorthogonal series method for Oseen type flows. *Int. J. Engng Sci.* **31**, 1243.

Kuhlmann, H. C. & Rath, H. J. 1993a Hydrodynamic instabilities in cylindrical thermocapillary liquid bridges. *J. Fluid Mech.* **247**, 247.

Kuhlmann, H. C. & Rath, H. J. 1993b On the interpretation of phase measurements of oscillatory thermocapillary convection in liquid bridges. *Phys. Fluids* A **5**, 2117.

Kuhlmann, H. C., Wanschura, M. & Rath, H. J. 1996 Flow in two-sided lid-driven cavities: non-uniqueness, instabilities, and cellular structures. *J. Fluid Mech.* **336**, 267.

Kuhlmann, H. C., Wanschura, M. & Rath, H. J. 1998a Elliptic instability in two-sided lid-driven cavity flow. *Eur. J. Mech.* B/*Fluids* **17**, 561.

Kuhlmann, H. C., Nienhüser, Ch. & Rath, H. J. 1998b The local flow in a wedge between a rigid wall and a surface of constant shear stress. *J. Eng. Math.* (submitted).

Kuo, H. P. & Korpela, S. A. 1988 Stability and finite amplitude natural convection in a shallow cavity with insulated top and bottom and heated from aside. *Phys. Fluids* **31**, 33.

Lagnado, R. R., Phan-Thien, N. & Leal, L. G. 1984 The stability of two-dimensional linear flows. *Phys. Fluids* **27**, 1094.

Lan, C. W. & Kou, S. 1990 Thermocapillary flow and melt/solid interfaces in floating-zone crystal growth under microgravity. *J. Crystal Growth* **102**, 1043.

Lan, C. W. & Kou, S. 1991a Heat transfer, fluid flow, and interface shapes in floating-zone crystal growth. *J. Crystal Growth* **108**, 351.

Lan, C. W. & Kou, S. 1991b Effects of rotation on heat transfer, fluid flow and interfaces in normal gravity floating-zone crystal growth. *J. Crystal Growth* **114**, 517.

Lan, C. W. & Kou, S. 1993a Radial dopant segregation in zero-gravity floating-zone crystal growth. *J. Crystal Growth* **132**, 578.

Lan, C. W. & Kou, S. 1993b Effect of rotation on radial dopant segregation in microgravity floating-zone crystal growth. *J. Crystal Growth* **133**, 309.

Lanczos, C. 1956 *Applied Analysis*. Prentice-Hall, Englewood Cliffs, New York.

Landau, L. D. & Lifshitz, E. M. 1959 *Fluid Mechanics*. Pergamon, Oxford, New York.

Landau, L. D. & Lifshitz, E. M. 1985 *Elektrodynamik der Kontinua*. Lehrbuch der theoretischen Physik, Bd. VIII. Akademie Verlag, Berlin.

Langer, J. S. 1980 Instabilities and pattern formation in crystal growth. *Rev. Mod. Phys.* **52**, 1.

Langlois, W. E. 1985 Buoyancy-driven flows in crystal-growth melts. *Ann. Rev. Fluid Mech.* **17**, 191.

Laplace, P. S. 1806 *Traité de mécanique céleste*. Suppléments au Livre X. In *Œuvres Complete* 4, Gauthier-Villars, Paris.

Laure, P. & Roux, B. 1989 Linear and non-linear analysis of the Hadley circulation. *J. Crystal Growth* **97**, 226.

Laure, P., Roux, B. & Ben Hadid, H. 1990 Nonlinear study of the flow in a long rectangular cavity subjected to thermocapillary effect. *Phys. Fluids* A **2**, 516.

Lee, C. P., Anilkumar, A. V. & Wang, T. G. 1996 Streaming generated in a liquid bridge due to nonlinear oscillations driven by the vibration of an endwall. *Phys. Fluids* **8**, 3234.

Leppinen, D. M., Renksizbulut, M. & Haywood, R. J. 1996 The effects of surfactants on droplet behaviour at intermediate Reynolds numbers – I. The numerical model and steady-state results. *Chem. Ing. Sci.* **51**, 479.

Leray, J. 1933 Etude de diverses equations integrales non lineaires et de quelques problemes que pose l'Hydrodynamique. *J. de Math. Pures et Appl.* **12**, 1.

Levenstam, M. 1994 Thermocapillary convection in floatzones. Ph.D. thesis, Royal Institute of Technology, Stockholm.

Levenstam, M. & Amberg, G. 1995 Hydrodynamic instabilities of thermocapillary flow in a half-zone. *J. Fluid Mech.* **297**, 357.

Levich, V. G. 1962 *Physicochemical Hydrodynamics*. Prentice-Hall, New York.

Levich, V. G. & Krylov, V. S. 1969 Surface tension-driven phenomena. *Ann. Rev. Fluid Mech.* **1**, 293.

Leypoldt, J., Kuhlmann, H. C. & Rath, H. J. 1998 Three-dimensional numerical simulation of thermocapillary flows in cylindrical liquid bridges. *J. Fluid Mech.* (submitted).

Li, J., Sun, J. & Saghir, Z. 1993 Buoyant and thermocapillary flow in liquid encapsulated floating zone. *J. Crystal Growth* **131**, 83.

Liakopoulos, A. & Brown, G. W. 1993 Thermocapillary and natural convection in a square cavity. In *Surface Tension Driven Flows*, Neitzel, G. P. & Smith, M. K. (eds.), AMD vol. **170**, p. 57.

Liang, S. F., Vidal, A. & Acrivos, A. 1969 Buoyancy-driven convection in cylindrical geometries. *J. Fluid Mech.* **36**, 239.

Liu, Q. S. & Roux, B. 1994 Marangoni convection in immiscible double liquid layers. *Microgravity Sci. Technol.* **7**, 103.

Liu, Q. S., Chen, G. & Roux, B. 1993 Thermogravitational and thermocapillary convection in a cavity containing two superposed immiscible liquid layers. *Int. J. Heat Mass Transfer* **36**, 101.

Longuet-Higgins, M. S. 1953 Mass transport in water waves. *Phil. Trans.* A **245**, 535.

Lüdtge, R. 1869 Ueber die Ausbreitung von Flüssigkeiten auf einander. *Ann. Phys. Chem.* **137**, 362.

Lyubimov, D., Lyubimova, T. & Roux, B. 1997 Mechanisms of vibrational control of heat transfer in a liquid bridge. *Int. J. Heat Mass Transfer* **40**, 4031.

Malkus, W. V. R. & Veronis, G. 1958 Finite amplitude cellular convection. *J. Fluid Mech.* **4**, 225.

Mansell, G., Walter, J. & Marschall, E. 1994 Liquid–liquid driven cavity flow. *J. Comput. Phys.* **110**, 274.

Marangoni, C. 1871 Ueber die Ausbreitung der Tropfen einer Flüssigkeit auf der Oberfläche einer anderen. *Ann. Phys. Chem.* **143**, 337.

Mashayek, F. & Ashgriz, N. 1995 Non-linear instability of liquid jets with thermocapillarity. *J. Fluid Mech.* **283**, 97.

Masud, J., Kamotani, Y. & Ostrach, S. 1997 Oscillatory thermocapillary flow in cylindrical columns of high Prandtl number fluids. *AIAA J. Thermophys. Heat Transfer* **11**, 105.

McGillis, W. R. & Carey, V. P. 1996 On the role of Marangoni effects on the critical heat flux for pool boiling of binary mixtures. *ASME J. Heat Transfer* **118**, 103.

Menke, Ch. 1994 Numerische Analyse der Marangoni-Konvektion beim Schmelzzonenverfahren. Ph.D. thesis, University of Ulm.

Mercier, J. F. & Normand, C. 1996 Buoyant–thermocapillary instabilities of differentially heated liquid layers. *Phys. Fluids* **8**, 1433.

Meseguer, J. & Sanz, A. 1985 Numerical and experimental study of the dynamics of axisymmetric slender liquid bridges. *J. Fluid Mech.* **153**, 83.

Metzger, J. & Schwabe, D.   1988   Coupled buoyant thermocapillary convection. *Physico-Chem. Hydrodyn.* **10**, 263.

Mittelmann, H. D.   1990   Computing stability bounds for thermocapillary convection in a crystal-growth free boundary problem. In *International Series of Numerical Mathematics* **95**, Birkhäuser, Basel, Boston.

Mittelmann, H. D.   1994   Hydrodynamic stability of thermocapillary convection in liquid bridges. *Math. Comput. Modelling* **20**, 175.

Mittelmann, H. D., Law, C. C., Jankowski, D. F. & Neitzel, G. P.   1991   Stability of thermocapillary convection in float-zone crystal growth. In *International Series of Numerical Mathematics* **99**, Birkhäuser, Basel, Boston.

Moffatt, H. K.   1964a   Viscous and resistive eddies near a sharp corner. *J. Fluid Mech.* **18**, 1.

Moffatt, H. K.   1964b   Viscous eddies near a sharp corner. *Arch. Mech. Stosowanej* **16**, 365.

Moffatt, H. K. & Duffy, B. R.   1980   Local similarity solutions and their limitations. *J. Fluid Mech.* **96**, 299. (Corrigendum, 1980, in *J. Fluid Mech.* **99**, 860)

Mollot, D. J., Tsamopoulos, J., Chen, T.-Y. & Ashgriz, N.   1993   Nonlinear dynamics of capillary bridges: Experiments. *J. Fluid Mech.* **255**, 411.

Monberg, E.   1994   Bridgman and related growth techniques. In *Handbook of Crystal Growth*, vol. 2a, *Basic Techniques*, Hurle, D. T. J. (ed.), North-Holland, Amsterdam, p. 51.

Moreau, R.   1990   *Magnetohydrodynamics.* Kluwer, Dordrecht, Boston.

Morthland, T. E. & Walker, J. S.   1996   Thermocapillary convection during floating-zone silicon growth with a uniform or non-uniform magnetic field. *J. Crystal Growth* **158**, 471.

Morthland, T. E. & Walker, J. S.   1997   Thermocapillary convection in a cylindrical liquid-metal floating zone with a strong axial magnetic field and with a non axisymmetric heat flux. *J. Fluid Mech.* **345**, 31.

Muehlner, K. A., Schatz, M. F., Petrov, V., McCormick, W. D., Swift, J. B. & Swinney, H. L.   1997   Observation of helical traveling-wave convection in a liquid bridge. *Phys. Fluids* **9**, 1850.

Mühlbauer, A., Muiznieks, A., Virbulis, J., Lüdge, A. & Riemann, H.   1995   Interface shape, heat transfer and fluid flow in the floating zone growth of large silicon crystals with the needle-eye technique. *J. Crystal Growth* **151**, 66.

Müller, G.   1986   Über die Entstehung von Inhomogenitäten in Halbleiterkristallen bei der Herstellung aus Schmelzen. Habilitationsschrift, Selisch Fachbuch-Verlag, Erlangen.

Müller, G.   1989   Convection and inhomogeneities in crystal growth from the melt. In *Crystals* **12**, Freyhardt, H. C. (ed.), Springer, Berlin, Heidelberg, p. 1.

Müller, G. & Ostrogorsky, A.   1994   Convection in melt growth. In *Handbook of Crystal Growth*, vol. 2b, *Bulk Crystal Growth*, Hurle, D. T. J. (ed.), North-Holland, Amsterdam, p. 709.

Müller, G., Neumann, G. & Weber, W.   1984   Natural convection in vertical Bridgman configurations. *J. Crystal Growth* **70**, 78.

Mullins, W. W. & Sekerka, R. F.   1964   Stability of a planar interface during solidification of a dilute binary alloy. *J. Appl. Phys.* **35**, 444.

Mundrane, M. & Zebib, A.   1993   Two- and three-dimensional buoyant thermocapillary convection. *Phys. Fluids* A **5**, 810.

Mundrane, M. & Zebib, A.   1994   Oscillatory buoyant thermocapillary flow. *Phys. Fluids* **6**, 3294.

Mundrane, M., Xu, J. & Zebib, A.   1995   Thermocapillary convection in a rectangular cavity with a deformable interface. *Adv. Space Res.* **16**(7), 41.

Myshkis, A. D., Babskii, V. G., Kopachevskii, N. D., Slobozhanin, L. A. & Tyuptsov, A. D. 1987 *Low-gravity Fluid Mechanics*. Springer, Berlin, Heidelberg.

Nakamura, S., Kakimoto, K. & Hibiya, T. 1996 Convection visualization and temperature fluctuation measurement in a molten silicon column. In *Materials and Fluids under Low Gravity*, Lecture Notes in Physics **464**, Ratke, L., Walter, H. & Feuerbacher, B. (eds.), Springer, Berlin, Heidelberg, p. 343.

Napolitano, L. G. 1979 Marangoni boundary layers. In *Proceedings of the IIIrd European Symposium on Materials Science in Space*, ESA SP-142, p. 349.

Nataraj, R. 1989 Thermocapillary flows in a rotating float zone under microgravity. *AIChE J.* **35**, 614.

Neitzel, G. P., Law, C. C., Jankowski, D. F. & Mittelmann, H. D. 1991 Energy stability of thermocapillary convection in a model of the float-zone crystal-growth process. II: Nonaxisymmetric disturbances. *Phys. Fluids* A **3**, 2841.

Neitzel, G. P., Chang, K.-T., Jankowski, D. F. & Mittelmann, H. D. 1993 Linear-stability theory of thermocapillary convection in a model of the float-zone crystal-growth process. *Phys. Fluids* A **5**, 108.

Nepomnyashchy, A. & Simanovskii, I. 1995 Oscillatory convection instabilities in systems with an interface. *Int. J. Multiphase Flow* **21** Suppl., 129.

Neumann, G. 1990 Three-dimensional numerical simulation of buoyancy driven convection in vertical cylinders heated from below. *J. Fluid Mech.* **214**, 559.

Newell, A. C. & Whitehead, J. A. 1969 Finite bandwidth, finite amplitude convection. *J. Fluid Mech.* **38**, 279.

Newell, A. C., Passot, Th. & Lega, J. 1993 Order parameter equations for patterns. *Ann. Rev. Fluid Mech.* **25**, 399.

Nicolás, J. A. & Vega, J. M. 1996 Weakly nonlinear oscillations of nearly-inviscid axisymmetric liquid bridges. *J. Fluid Mech.* **328**, 95.

Nienhüser, Chr., Kuhlmann, H. C. & Rath, H. J. 1998 Stabilität thermokapillarer Konvektion in nichtzylindrischen Flüssigkeitsbrücken. *Z. Angew. Math. Mech.* (submitted).

Ohnishi, M., Azuma, H. & Doi, T. 1992 Computer simulation of oscillatory Marangoni flow. *Acta Astronautica* **26**, 685.

Oron, A., Davis, S. H. & Bankoff, S. G. 1997 Long-scale evolution of thin liquid films. *Rev. Mod. Phys.* **69**, 931.

Orr, W. McF. 1907 The stability or instability of the steady motions of a liquid. Part II: A viscous liquid. *Proc. Roy. Irish Acad.* A **27**, 69.

Oseen, C. W. 1910 Über die Stokessche Formel und über eine verwandte Aufgabe in der Hydrodynamik. *Ark. Mat. Astr. Fys.* **6**, No. 29.

Ostrach, S. 1977 Motion induced by capillarity. In *Physico-Chemical Hydrodynamics*. V. G. Levich Festschrift, vol. 2, Spalding (ed.), Advance Publications, London.

Ostrach, S. 1982 Low-gravity fluid flows. *Ann. Rev. Fluid Mech.* **14**, 313.

Ostrach, S. 1983 Fluid mechanics in crystal growth – The 1982 Freeman Scholar Lecture. *J. Fluids Eng.* **105**, 5.

Ostrach, S., Kamotami, Y. & Lai, C. L. 1985 Oscillatory thermocapillary flows. *PhysicoChem. Hydrodyn.* **6**, 585.

Pan, F. & Acrivos, A. 1967 Steady flows in rectangular cavities. *J. Fluid Mech.* **28**, 643.

Papkovich, P. F. 1940 Über eine Form der Lösung des biharmonischen Problems für das Rechteck. *C. r. (Dokl.) Acad. Sci. USSR* **27**, 334.

Parmentier, P. M., Regnier, V. C. & Lebon, G. 1993 Buoyant–thermocapillary instabilities in medium-Prandtl-number fluid layers subject to a horizontal temperature gradient. *Int. J. Heat Mass Transfer* **36**, 2417.

Patankar, S. V. 1980 *Numerical Heat Transfer and Fluid Flow*. Hemisphere, Washington, London.

Pearson, J. R. A. 1958 On convection cells induced by surface tension. *J. Fluid Mech.* **4**, 489.

Peltier, L. J. & Biringen, S. 1993 Time-dependent thermocapillary convection in a rectangular cavity: Numerical results for a moderate Prandtl number fluid. *J. Fluid Mech.* **257**, 339.

Petrov, V., Schatz, M. F., Muehlner, K. A., VanHook, S. J., McCormick, W. D., Swift, J. B. & Swinney, H. L. 1996 Nonlinear control of remote unstable states in a liquid bridge convection experiment. *Phys. Rev. Lett.* **77**, 3779.

Pierrehumbert, R. T. 1986 Universal short-wave instability of two-dimensional eddies in an inviscid fluid. *Phys. Rev. Lett.* **57**, 2157.

Pimputkar, S. M. & Ostrach, S. 1981 Convective effects in crystals grown from the melt. *J. Crystal Growth* **55**, 614.

Plateau, J. 1873 *Statique expérimentale et théorique des liquides soumies aux seules forces molécularies*. Gauthier-Villars, Paris.

Polezhaev, V. I. & Ermakov, M. K. 1992 Thermosolutal Marangoni convection short-time regimes – proposals for drop tower experiments and real time computer simulation. *Microgravity Sci. Technol.* **5**, 172.

Prange, M. 1997 Stabilisierung thermokapillarer Konvektion in zylindrischen Flüssigkeitsbrücken mittels axialer Magnetfelder. Diploma thesis, University of Bremen.

Prange, M., Wanschura, M., Kuhlmann, H. C. & Rath, H. J. 1998 Linear stability of thermocapillary convection in cylindrical liquid bridges under axial magnetic fields. *J. Fluid Mech.* (submitted).

Preisser, F., Schwabe, D. & Scharmann, A. 1983 Steady and oscillatory thermocapillary convection in liquid columns with free cylindrical surface. *J. Fluid Mech.* **126**, 545.

Press, W. H., Flannery, B. P., Teukolsky, S. A. & Vetterling, W. T. 1989 *Numerical Recipes (FORTRAN)*. Cambridge University Press, Cambridge.

Prevost, M. & Gallez, D. 1986 Nonlinear stability of thin free liquid films: Rupture and Marangoni effects. *AIChE Symp. Ser.* **252**, 123.

Priede, J. & Gerbeth, G. 1995 Hydrothermal wave instability of thermocapillary driven convection in a plane layer subjected to a uniform magnetic field. *Adv. Space Res.* **16**(7), 55.

Priede, J. & Gerbeth, G. 1997a Influence of thermal boundary conditions on the stability of thermocapillary-driven convection at low Prandtl numbers. *Phys. Fluids* **9**, 1621.

Priede, J. & Gerbeth, G. 1997b Convective, absolute, and global instabilities of thermocapillary–buoyancy convection in extended layers. *Phys. Rev. E* **56**, 4187.

Priede, J. & Gerbeth, G. 1997c Hydrothermal wave instability of thermocapillary-driven convection in a coplanar magnetic field. *J. Fluid Mech.* **347**, 141.

Priede, J., Thess, A. & Gerbeth, G. 1994 Thermocapillary instabilities in liquid metals: Hartmann-number versus Prandtl-number. In *Proceedings of the Second International Conference on Energy Transfer in Magnetohydrodynamic Flows*, Aussois, p. 571.

Ramaswamy, B. & Jue, T. C. 1992 Analysis of thermocapillary and buoyancy-effected cavity flow using FEM. *Numer. Heat Transfer* **22**, 379.

Rao, A. R. & Biswal, P. C. 1995 Steady thermocapillary convection in two immiscible liquid layers with curved free surface. *Microgravity Sci. Technol.* **8**, 77.

Rayleigh, Lord 1879 On the instability of jets. *Proc. London Math. Soc.* **10**, 4.

Rayleigh, Lord 1916 On convection currents in a horizontal layer of fluid, when the higher temperature is on the under side. *Phil. Mag.* (6) **32**, 529.

Raynal, F. 1996 Exact relation between spatial mean enstrophy and dissipation in confined incompressible flows. *Phys. Fluids* **8**, 2242.

Reynolds, O. 1895 On the dynamical theory of incompressible viscous fluids and the determination of the criterion. *Phil. Trans. Roy. Soc.* A **186**, 123.

Riahi, N. 1986 Thermocapillary instability of an infinite Prandtl number fluid with negligible gravitational effects. *Acta Mech.* **64**, 155.

Riley, R. J. & Neitzel, G. P. 1998 Instability of thermocapillary–buoyancy convection in shallow layers. Part 1. Characterization of steady and oscillatory instabilities. *J. Fluid Mech.* **359**, 143.

Robbins, C. I. & Smith, R. C. T. 1948 A table of roots of $\sin z = -z$. *Phil. Mag.* **39**, 1004.

Roberts, P. H. 1967 *An Introduction to Magnetohydrodynamics.* Longman, London, New York.

Rosenblat, S. 1982 Thermal convection in a vertical circular cylinder. *J. Fluid Mech.* **122**, 395.

Roux, B. (ed.) 1990 Numerical simulation of oscillatory convection in low-Pr fluids. *Notes on Numerical Fluid Mechanics* **27**, Vieweg.

Rupp, R., Müller, G. & Neumann, G. 1989 Three-dimensional time dependent modelling of the Marangoni convection in a zone melting configuration for GaAs. *J. Crystal Growth* **97**, 34.

Rybicki, A. & Floryan, J. M. 1987a Thermocapillary effects in liquid bridges. I. Thermocapillary convection. *Phys. Fluids* **30**, 1956.

Rybicki, A. & Floryan, J. M. 1987b Thermocapillary effects in liquid bridges. II. Deformation of the interface and capillary instability. *Phys. Fluids* **30**, 1973.

Saedeleer, C. De, Garcimartin, A., Chavepeyer, G., Platten, J. K. & Lebon, G. 1996 The instability of a liquid layer heated from the side when the upper surface is open to the air. *Phys. Fluids* **8**, 670.

Salamon, T. R., Bornside, D. E., Armstrong, R. C & Brown, R. A. 1995 The role of surface tension in the dominant balance in the die swell singularity. *Phys. Fluids* **7**, 2328.

Salamon, T. R., Bornside, D. E., Armstrong, R. C & Brown, R. A. 1997 Local similarity solutions in the presence of a slip boundary condition. *Phys. Fluids* **9**, 1235.

Sanz, A. 1985 The influence of the outer bath in the dynamics of axisymmetric liquid bridges. *J. Fluid Mech.* **156**, 101.

Sasmal, G. P. & Hochstein, J. I. 1994 Marangoni convection with a curved and deforming free surface in a cavity. *J. Fluids Eng.* **116**, 577.

Saß, V., Kuhlmann, H. C. & Rath, H. J. 1996 Investigation of three-dimensional thermocapillary convection in a cubic container by a multi-grid method. *Int. J. Heat Mass Transfer* **39**, 603.

Savino, R. & Monti, R. 1996a Oscillatory Marangoni convection in cylindrical liquid bridges. *Phys. Fluids* **8**, 2906.

Savino, R. & Monti, R. 1996b Three-dimensional numerical simulation of thermocapillary instabilities in floating zones. *Appl. Sci. Res.* **56**, 19.

Schlichting, H. 1968 *Boundary-Layer Theory.* McGraw-Hill, New York.

Schneider, J. 1995 Strukturen thermokapillarer Konvektion in einem Ringspalt. Ph.D. thesis, University of Giessen.

Schreiber, R. & Keller, H. B. 1983 Driven cavity flows by efficient numerical techniques. *J. Comput. Phys.* **49**, 310.

Schultz, W. W. & Gervasio, C. 1990 A study of the singularity of the die swell problem. *Q. J. Appl. Math.* **43**, 407.

Schwabe, D. 1981a Marangoni effects in crystal growth melts. *PhysicoChem. Hydrodyn.* **2**, 263.

Schwabe, D. 1981b Marangoni Konvektion in Schmelzen. Habilitationsschrift, University of Giessen.

Schwabe, D. 1988 Surface-tension-driven flow in crystal growth melts. In *Crystals* **11**, Freyhardt, H. C. (ed.), Springer, Berlin, Heidelberg, p. 75.

Schwabe, D. & Metzger, J. 1989 Coupling and separation of buoyant thermocapillary convection. *J. Crystal Growth* **97**, 23.

Schwabe, D. & Scharmann, A. 1979 Some evidence for the existence and magnitude of a critical Marangoni number for the onset of oscillatory flow in crystal growth melts. *J. Crystal Growth* **46**, 125.

Schwabe, D. & Scharmann, A. 1984 Measurements of the critical Marangoni number of the laminar–oscillatory transition of thermocapillary convection in floating zones. In *Proceedings of the 5th European Symposium on Material Sciences under Microgravity*, p. 281.

Schwabe, D. & Scharmann, A. 1985 Messung der kritischen Marangonizahl für den Übergang von stationärer zu oszillatorischer thermokapillarer Konvektion unter Mikrogravitation: Ergebnisse der Experimente in den ballistischen Raketen TEXUS 5 und TEXUS 8. *Z. Flugwiss. Weltraumforsch.* **9**, 21.

Schwabe, D. & Scharmann, A. 1989 Marangoni and buoyant convection in an open cavity under reduced and under normal gravity. *Adv. Space Res.* **8** (12), 175.

Schwabe, D., Scharmann, A., Preisser, A. & Oeder, F. 1978 Experiments on surface tension driven flow in floating zone melting. *J. Crystal Growth* **43**, 305.

Schwabe, D., Preisser, F. & Scharmann, A. 1982 Verification of the oscillatory state of thermocapillary convection in a floating zone under low gravity. *Acta Astronautica* **9**, 265.

Schwabe, D., Möller, U., Schneider, J. & Scharmann, A. 1992 Instabilities of shallow dynamic thermocapillary liquid layers. *Phys. Fluids* A **4**, 2368.

Schwabe, D., Da, X. & Scharmann, A. 1996 Unstable flow and solidification speed due to the interaction of thermocapillary and solutocapillary forces in directional solidification. *J. Crystal Growth* **166**, 483.

Scriven, L. E. 1960 Dynamics of a fluid interface. *Chem. Eng. Sci.* **12**, 98.

Scriven, L. E. & Sternling, C. V. 1960 The Marangoni effects. *Nature* **187**, 186.

Segel, L. A. 1969 Distant sidewalls cause slow amplitude modulation of cellular convection. *J. Fluid Mech.* **38**, 203.

Sen, A. K. & Davis, S. H. 1982 Steady thermocapillary flows in two-dimensional slots. *J. Fluid Mech.* **121**, 163.

Serrin, J. 1959 On the stability of viscous fluid motions. *Arch. Rat. Mech. Anal.* **3**, 1.

Shackelford, J. F. & Alexander, W. (eds.) 1992 *CRC Materials Science and Engineering Handbook.* CRC Press.

Shen, Y. 1989 Energy stability of thermocapillary convection in a model of float-zone crystal growth. Ph.D. thesis, Arizona State University.

Shen, J. 1991 Hopf bifurcation of the unsteady regularized driven cavity flow. *J. Comput. Phys.* **95**, 228.

Shen, Y., Neitzel, G. P., Jankowski, D. F. & Mittelmann, H. D. 1990 Energy stability of thermocapillary convection in a model of the float-zone crystal-growth process. *J. Fluid Mech.* **217**, 639.

Shevtsova, V. M. & Legros, J. C. 1998 Oscillatory convective motion in deformed liquid bridges. *Phys. Fluids* **10**, 1621.

Shevtsova, V. M., Kuhlmann, H. C. & Rath, H. J. 1996 Thermocapillary convection in liquid bridges with a deformed free surface. In *Materials and Fluids under Low Gravity*, Lecture Notes in Physics **464**, Ratke, L., Walter, H. & Feuerbacher, B. (eds.), Springer, Berlin, Heidelberg, p. 323.

Shikhmurzaev, Y. D. 1997 Moving contact lines in liquid/liquid/solid systems. *J. Fluid Mech.* **334**, 211.

Shinoda, K. & Friberg, S. 1986 *Emulsions and Solubilization*. Wiley, Chichester, New York.

Shtern, V. & Hussain, F. 1993 Azimuthal instability of divergent flows. *J. Fluid Mech.* **256**, 535.

Simanovskii, I. B. & Nepomnyashchy, A. A. 1993 *Convective instabilities in systems with interface*. Gordon and Breach, Amsterdam.

Sirignano, W. A. & Glassman, I. 1970 Flame spreading above liquid fuels: Surface-tension-driven flows. *Combust. Sci. Tech.* **1**, 307.

Slobozhanin, L. A. & Perales, J. M. 1993 Stability of liquid bridges between equal disks in an axial gravity field. *Phys. Fluids* A **5**, 1305.

Slobozhanin, L. A., Alexander, J. I. D. & Resnick, A. H. 1997 Bifurcation of the equilibrium states of a weightless liquid bridge. *Phys. Fluids* **9**, 1893.

Smith, M. K. 1982 The instabilities of thermocapillary shear layers. Ph.D. thesis, Northwestern University.

Smith, R. C. T. 1952 The bending of a semi-infinite strip. *Austral. J. Sci. Res.* **5**, 227.

Smith, M. K. 1986a Instability mechanisms in dynamic thermocapillary liquid layers. *Phys. Fluids* **29**, 3182.

Smith, M. K. 1986b Thermocapillary and centrifugal-buoyancy-driven motion in a rapidly rotating liquid cylinder. *J. Fluid Mech.* **166**, 245.

Smith, M. K. 1988 The nonlinear stability of dynamic thermocapillary liquid layers. *J. Fluid Mech.* **194**, 391.

Smith, M. K. 1995 Thermocapillary migration of a two-dimensional liquid droplet on a solid surface. *J. Fluid Mech.* **294**, 209.

Smith, M. K. & Davis, S. H. 1983a Instabilities of dynamic thermocapillary liquid layers. Part 1. Convective instabilities. *J. Fluid Mech.* **132**, 119.

Smith, M. K. & Davis, S. H. 1983b Instabilities of thermocapillary liquid layers. Part 2. Surface-wave instabilities. *J. Fluid Mech.* **132**, 145.

Sternling, C. V. & Scriven, L. E. 1959 Interfacial turbulence: Hydrodynamic instability and Marangoni effect. *AIChE J.* **5**, 514.

Stone, H. A. 1990 A simple derivation of the time-dependent convective-diffusion equation for surfactant transport along a deforming interface. *Phys. Fluids* A **2**, 111.

Stork, K. & Müller, U. 1975 Convection in boxes: An experimental investigation in vertical cylinders and annuli. *J. Fluid Mech.* **71**, 231.

Strani, M., Piva, R. & Graziani, G. 1983 Thermocapillary convection in a rectangular cavity: Asymptotic theory and numerical simulation. *J. Fluid Mech.* **130**, 347.

Straughan, B. 1992 *The Energy Method, Stability, and Nonlinear Convection*. Applied Mathematical Sciences **91**, Springer, Berlin, Heidelberg.

Subramanian, R. S. 1992 The motion of bubbles and drops in reduced gravity. In *Transport Processes in Bubbles, Drops and Particles*, Chhabra, R. P. & De Kee, D. (eds.), Hemisphere, New York.

Tan, M. J., Bankoff, S. G. & Davis, S. H. 1990 Steady thermocapillary flows of thin liquid layers. I. Theory. *Phys. Fluids* A **2**, 313.

Tang, H., Lu, F. & Hu, W. R. 1995 g-jitter effects on half floating zone convection in intermediate frequency range. *Microgravity Sci. Technol.* **8**, 10.

Tao, Y., Sakidja, R. & Kou, S. 1995 Computer simulation and flow visualization of thermocapillary flow in a silicone oil floating zone. *Int. J. Heat Mass Transfer* **38**, 503.

Thess, A. & Orszag, S. A. 1995 Surface-tension-driven Bénard convection at infinite Prandtl number. *J. Fluid Mech.* **283**, 201.

Thomson, J. 1855 On certain curious motions observable at the surfaces of wine and other alcoholic liquors. *London, Edinburgh, and Dublin Phil. Mag. J. Sci.* **10**, 330.

Tison, P., Camel, D., Tosello, I. & Favier, J. J. 1992 Experimental and theoretical study of Marangoni flows in liquid metallic layers. In *Symposium on Hydromechanics and Heat/Mass Transfer in Microgravity, Moscow–Perm 1991*, Avduevski, V. S. (ed.), Gordon and Breach, Philadelphia, New York, p. 121.

Tranter, C. J. 1948 The use of the Mellin transform in finding the stress distribution in an infinite wedge. *Q. J. Mech. Appl. Math.*, **1**, 125.

Tsai, W. & Yue, D. K. P. 1996 Computation of nonlinear free-surface flows. *Ann. Rev. Fluid Mech.* **28**, 249.

VanHook, S. J., Schatz, M. F., McCormick, W. D., Swift, J. B. & Swinney, H. L. 1995 Long-wavelength instability in surface-tension-driven Bénard convection. *Phys. Rev. Lett.* **75**, 4397.

Velten, R., Schwabe, D. & Scharmann, A. 1989 Gravity-dependence of the instability of surface-tension-driven flow in floating zones. In *Proceedings of the VIIth Symposium on Materials and Fluid Sciences in Microgravity*, ESA-SP 295, p. 271.

Velten, R., Schwabe, D. & Scharmann, A. 1991 The periodic instability of thermocapillary convection in cylindrical liquid bridges. *Phys. Fluids* A **3**, 267.

Villers, D. & Platten, J. K. 1988 Thermal convection in superposed immiscible layers. *Appl. Sci. Res.* **45**, 145.

Villers, D. & Platten, J. K. 1990 Influence of interfacial tension gradients on thermal convection in two superposed immiscible liquid layers. *Appl. Sci. Res.* **47**, 177.

Villers, D. & Platten, J. K. 1992 Coupled buoyancy and Marangoni convection in acetone: Experiments and comparison with numerical simulations. *J. Fluid Mech.* **234**, 487.

Visual Numerics 1994 *IMSL – FORTRAN Subroutines for Mathematical Applications*, Houston.

Viviani, A. & Cioffi, M. 1992 Analytical solution of thermocapillary flow and interface shape in non-isothermal coaxial liquid columns. AMD-Vol. 154, *Fluid Mechanics Phenomena in Microgravity*, Siginer, D. A. & Weislogel, M. M. (eds.), ASME, New York, p. 61.

Völkl, J. 1994 Stress in the cooling crystal. In *Handbook of Crystal Growth*, vol. 2b, *Bulk Crystal Growth*, Hurle, D. T. J. (ed.), North-Holland, Amsterdam, p. 821.

Wagner, C., Friedrich, R. & Narayanan, R. 1994 Comments on the numerical investigation of Rayleigh and Marangoni convection in a vertical circular cylinder. *Phys. Fluids* A **6**, 1425.

Waleffe, F. 1990 On the three-dimensional instability of strained vortices. *Phys. Fluids* A **2**, 76.

Wang, P. & Kahawita, R. 1998 Oscillatory behaviour in buoyant thermocapillary convection of fluid layers with a free surface. *Int. J. Heat Mass Transfer* **41**, 399.

Wang, P., Kahawita, R. & Nguyen, D. L. 1994 Numerical simulation of buoyancy-Marangoni convection in two superposed immiscible liquid layers with a free surface. *Int. J. Heat Mass Transfer* **37**, 1111.

220   References

Wanschura, M. 1996 Lineare Instabilitäten kapillarer und natürlicher Konvektion in zylindrischen Flüssigkeitsbrücken. Ph.D. thesis, University of Bremen.

Wanschura, M., Shevtsova, V. M., Kuhlmann, H. C. & Rath, H. J. 1995a Convective instability mechanisms in thermocapillary liquid bridges. *Phys. Fluids* **7**, 912.

Wanschura, M., Kuhlmann, H. C., Shevtsova, V. & Rath, H. J. 1995b Thermo- and solutocapillary convection in a cylindrical liquid bridge: Stability of axisymmetric flow. *Adv. Space Res.* **16**(7), 75.

Wanschura, M., Kuhlmann, H. C. & Rath, H. J. 1996 Three-dimensional instability of axisymmetric buoyant convection in cylinders heated from below. *J. Fluid Mech.* **326**, 399.

Wanschura, M., Kuhlmann, H. C. & Rath, H. J. 1997a Linear stability of two-dimensional combined buoyant-thermocapillary flow in cylindrical liquid bridges. *Phys. Rev.* E **55**, 7036.

Wanschura, M., Kuhlmann, H. C. & Rath, H. J. 1997b Instability of thermocapillary flow in symmetrically heated full liquid zones. In *Proceedings of the Joint Xth European and VIth Russian Symposium on Physical Sciences in Microgravity*, Avduyevsky, V. S. & Polezhaev, V. I. (eds.), vol. 1, p. 172.

Weber, E. H. 1855 Mikroskopische Beobachtungen sehr gesetzmäfsiger Bewegungen, welche die Bildung von Niederschlägen harziger Körper aus Weingeist begleiten. *Ann. Phys. Chem.* **94**, 447.

Widnall, S. E. & Tsai, Ch.-Y. 1977 The instability of the thin vortex ring of constant vorticity. *Proc. Roy. Soc. Lond.* **287**, 273.

Wong, H., Rumschitzki, D. & Maldarelli, Ch. 1996 On the surfactant mass balance at a deforming fluid interface. *Phys. Fluids* **8**, 3203.

Xu, J.-J. & Davis, S. H. 1984 Convective thermocapillary instabilities in liquid bridges. *Phys. Fluids* **27**, 1102.

Xu, J.-J. & Davis, S. H. 1985 Instability of capillary jets with thermocapillarity. *J. Fluid Mech.* **161**, 1.

Xu, J. & Zebib, A. 1998 Oscillatory two- and three-dimensional thermocapillary convection. *J. Fluid Mech.* **364**, 187.

Yamaguchi, Y., Chang, C. J. & Brown, R. 1984 Multiple buoyancy-driven flows in a vertical cylinder heated from below. *Phil. Trans. Roy. Soc. London* A **312**, 519.

Yeung, R. W. 1982 Numerical methods in free surface flows. *Ann. Rev. Fluid Mech.* **14**, 395.

Yoo, J. Y. & Joseph, D. D. 1978 Stokes flow in a trench between concentric cylinders. *SIAM J. Appl. Math.* **34**, 247.

Young, T. 1805 An essay on the cohesion of fluids. *Phil. Trans. Roy. Soc. London* **95**, 65.

Zebib, A., Homsy, G. M. & Meiburg, E. 1985 High Marangoni number convection in a square cavity. *Phys. Fluids* **28**, 3467.

Zhang, Y. & Alexander, J. I. D. 1992 Surface tension and buoyancy driven flow in a non-isothermal liquid bridge. *Int. J. Numer. Meth. Fluids* **14**, 197.

Zierep, J. 1963 Zur Theorie der Zellularkonvektion V. Zellularkonvektionsströmungen in Gefässen endlicher horizontaler Ausdehnung. *Beitr. Phys. Atmos.* **36**, 70.

Index

Springer Tracts in Modern Physics

Springer
and the
environment

At Springer we firmly believe that an international science publisher has a special obligation to the environment, and our corporate policies consistently reflect this conviction.

We also expect our business partners – paper mills, printers, packaging manufacturers, etc. – to commit themselves to using materials and production processes that do not harm the environment. The paper in this book is made from low- or no-chlorine pulp and is acid free, in conformance with international standards for paper permanency.